Information Systems Engineering and Management

Volume 36

The book series "Information Systems Engineering and Management" (ISEM) publishes innovative and original works in the various areas of planning, development, implementation, and management of information systems and technologies by enterprises, citizens, and society for the improvement of the socio-economic environment.

The series is multidisciplinary, focusing on technological, organizational, and social domains of information systems engineering and management. Manuscripts published in this book series focus on relevant problems and research in the planning, analysis, design, implementation, exploration, and management of all types of information systems and technologies. The series contains monographs, lecture notes, edited volumes, pedagogical and technical books as well as proceedings volumes.

Some topics/keywords to be considered in the ISEM book series are, but not limited to: Information Systems Planning; Information Systems Development; Exploration of Information Systems; Management of Information Systems; Blockchain Technology; Cloud Computing; Artificial Intelligence (AI) and Machine Learning; Big Data Analytics; Multimedia Systems; Computer Networks, Mobility and Pervasive Systems; IT Security, Ethics and Privacy; Cybersecurity; Digital Platforms and Services; Requirements Engineering; Software Engineering; Process and Knowledge Engineering; Security and Privacy Engineering, Autonomous Robotics; Human-Computer Interaction; Marketing and Information; Tourism and Information; Finance and Value; Decisions and Risk; Innovation and Projects; Strategy and People.

Indexed by Google Scholar. All books published in the series are submitted for consideration in the Web of Science.

For book or proceedings proposals please contact Alvaro Rocha (amrrocha@gmail.com).

Tatiana Antipova
Editor

Digital Technology Platforms and Deployment

Springer

Editor
Tatiana Antipova
Agia Napa, Cyprus

ISSN 3004-958X ISSN 3004-9598 (electronic)
Information Systems Engineering and Management
ISBN 978-3-031-86546-6 ISBN 978-3-031-86547-3 (eBook)
https://doi.org/10.1007/978-3-031-86547-3

This Springer imprint is published by the registered company Springer Nature Switzerland AG
The registered company address is: Gewerbestrasse 11, 6330 Cham, Switzerland

Preface

Digital technologies have rapidly and irreversibly changed our lives. The impact of digital technologies on our civilisation can only be compared to the emergence of fire in humanity, when from the first spark it became clear that the world would never be the same again. But how will this digital fire burn: what benefits will it bring and what dangers does it hide? What dark sides of our lives can it illuminate, and where can it burn us? The answers to these questions are not clear-cut and must be the result of many scientific studies in different fields since digitalisation has stimulated the convergence and integration of all spheres of science and life.

The digital world is constantly evolving, and it is important to stay curious and informed about emerging trends and technologies, so this book has been created to keep you learning and exploring, with the aim of equipping readers with the knowledge, tools and insights they need to stay ahead of the curve in this rapidly changing environment.

The book *Digital Technology Platforms and Deployment* is a collection of chapters written by experts in their field from different countries. This book is divided into three parts: *Digital Technology Platforms for Health and Population* (nine chapters); *Artificial Intelligence Deployment* (seven chapters); *Security, Safety, and Sustainability in the Digital Era* (seven chapters). All parts of the book are linked by the digital technologies used in the research.

In line with the UN priorities of "ensuring healthy lives and promoting well-being for all at all ages", we open our book with the first part *Digital Technology Platforms for Health and Population*, as health and population are central to the Sustainable Development Goals. Several chapters in this part are dedicated to the analysis of digital data with the aim of improving treatment and reducing mortality in the human population.

In the current digital age, artificial intelligence has penetrated almost all areas of science and life. The second part of this book—*Artificial Intelligence Deployment*—shows a small part of the possible deployment of artificial intelligence tools.

The security of personal and social data is particularly important in the digital age, and many scientific studies are devoted to the development of security, safety and sustainability, some of which are presented in the third part of this book—*Security, Safety and Sustainability in Digital Era.*

The book *Digital Technology Platforms and Deployment* can be useful for a wide range of readers, including: students, researchers and academics in fields such as Health and Population, computer science, business, education, engineering; experts and scientists from various scientific fields, to stimulate brain activity in the field of application and dissemination of digital technologies. Attentive readers will find this book a valuable resource for learning and research.

Agia Napa, Cyprus Tatiana Antipova

Contents

Digital Technology Platforms for Health and Population

Application of Evolutionary-Simulative Models in Evidence-Based Medicine

Gennady Ross and Valery Konyavsky

Abstract The chapter examines the current problem associated with the application of mathematical methods, in particular the evolutionary-simulation model (ESM), in simulating the processes of the patient's treatment plan in order to prove the correctness of the decision made. From the perspective of the system approach the technology of ESM description is considered, the main stages of the development of simulation models for determining medical parameters and the decision-making process in the implementation of the treatment plan are outlined. There are examples of sufficiently detailed description of four decision-making tasks in biophysical and biomedical research using ESM: mutually compensating biophysical processes and biophysical processes, planning of experiments and processing of their results, medical and biological norms. A new approach to the implementation of the multi-agent principle based on the theory of equilibrium random processes in the field of immunology is proposed. The analysis of the model of immune process proposed by G. I. Marchuk revealed that this model does not allow to take into account the high uncertainty of running processes, their random nature, external factors and discreteness. These problems can be eliminated if the immune process is represented as the ESM. An ESM of the immune process, based on which an intelligent service for simulating and prognosticating the organism's immune response can be developed, is proposed. Prospects of the ESM development for solving other applied problems in medicine are formulated.

Keywords Multi-agent systems · Simulation in medicine · Evidence-based medicine · Evolutionary-simulation models · Simulation models · Simulation of biophysical processes

---

G. Ross (✉)
Plekhanov Russian University of Economics, Moscow, Russia
e-mail: ross-49@mail.ru

V. Konyavsky
Moscow Institute of Physics and Technology (National Research University), Moscow, Russia

T. Antipova (ed.), *Digital Technology Platforms and Deployment*, Information Systems Engineering and Management 36, https://doi.org/10.1007/978-3-031-86547-3_1

1 Introduction

Digital technology and artificial intelligence in medicine is an interdisciplinary field, where achievements of mathematics, computer science, information and communication technologies and the latest medical treatment technologies converge [1, 2]. To date, mathematical models have gained great importance in medical research as intelligent components for decision-making. Usually it is a system of equations describing the relationship between variables characterizing a real physiological process or system. Depending on the specifics of the problem, both deterministic and probabilistic models can be used. Deterministic models usually represent a system of algebraic or differential equations, and probabilistic models are developed based on the results of experimental determination of static or dynamic characteristics of objects on the basis of methods of mathematical statistics. The necessity of application in medicine of mathematical methods of simulation with the use of new information technologies is caused by the fact that with their help it is possible to generalize adequately and in short time the complicated essence of phenomena and processes, to describe and understand facts, to reveal interrelations, to find a rational decision in conditions of uncertainty with much more completeness and reliability, than it is done on the basis of verbal characteristics; elementary routine calculations [3].

The construction of modern medical information technologies is based on multi-agent principles based on a heterogeneous group of intelligent agents, the interaction between which is described using the theory of equilibrium random processes (ERP). An intelligent agent is a certain entity that exhibits activity, autonomous behavior, is able to make decisions pursuant to a particular set of rules, and interacts with the environment and other agents [4]. The study of the functioning of intelligent agents capable of self-organization and development makes it possible to improve the reliability of the treatment plan, with consideration of the individual characteristics of the patient. A patient's treatment plan is a document that contains all the information about the treatment, the list of medical services and their sequence, the names of drugs and the way they are used, and the consumables used. It allows you to monitor the course of care and its results at each stage.

It should be noted that when the multi-agent simulation principle is applied, one can expect significant results in the field of immunology. The definition of mathematical immunology as a field of applied mathematics and the basic model of immune processes were proposed by G. I. Marchuk, RAS Academician; the model is described by a system of differential equations [G. I. Marchuk. Mathematical models in immunology. Computational methods and experiments. - Moscow: Science, 1991, Nauka]. However, under the conditions of uncertainty, this model prevents from taking into account the random nature of external factors and their discreteness, as well as self-regulation of the ERPs. These mathematical problems entail methodological and technical problems. The solution to these problems can be sought in the way of representing the immune process in the form of an evolutionary-simulation model (ESM), which represents a universal methodology for mathematical simulation of the ERPs. By the designation ESM we will understand the terms “evolutionary-simulation method” and “evolutionary-simulation model” as synonyms.

Based on the ESM, intelligent services can be developed that operate within the framework of a medical information system, allowing dialogue-based research of the disease course of a particular patient, prediction of immune system reactions to various treatment modalities, and development of treatment tactics and strategies. The ESM can be adapted to the specifics of a particular disease and build a set of virtual intelligent agents of the MAS, which will be able to comprehensively simulate the immune response of the organism.

Some features of the ESM make its application especially convenient, and in some cases without the use of the ESM the corresponding problems cannot be mathematically adequately formulated. The principal of these features is that the ESM, on the one hand, assumes a statistical description of the behavior of the object under investigation and, on the other hand, features effective and simple computational procedures for finding the best solution, taking into account the consequences of possible errors inherent therein. The same features of the ESM make it a fairly universal tool for describing mutually compensating influences in a living organism.

This simulation approach for evidence-based medicine [Trisha Greenhalgh. The Basics of Evidence-Based Medicine/Translated from English under the editorship of I. N. Denisov, RAS Academician; K. I. Saitkulov; V. P. Leonov; 4th edition, revised and supplemented. Moscow: GEOTAR-Media Publishing Group, 2019 Doc_Med_Book.indb (catahub.com)] will allow in the mode of interaction between the doctor and the computer to study the course of the disease of a particular patient, to predict trends, to develop tactics and treatment strategy [3]. Under these conditions, it will be necessary to find ways to measure adequate information exchanged between organs or even cells of a living organism, most of which constitute random processes.

Therefore, the ESM [5] appears to be promising for solving a number of theoretical and applied tasks related to decision-making in medicine: dosage of therapeutic and other influences; development of the best estimates (or standards) of the parameters of human environment; diagnostics; development of standards for preventive examinations; development of acceptable norms of harmful influences; conducting experiments and processing their results.

The purpose of this chapter is to present a number of promising areas of system research in the field of biophysical and medical-biological processes to provide evidence-based treatment of various diseases based on the application of mathematical simulation methods, namely, simulation models containing the rules of statistical observations over input and output medical parameters, and evolutionary-simulation method. An adaptation of the basic model (G. I. Marchuk) of immune processes, which is described by a system of differential equations in the form of the ESM, is proposed.

2 Methodological Aspects of the Construction of the ESM and Simulation Model

2.1 Basic Terms and Definitions

At the core of the ESM, there is a systematic (constructive) approach to a comprehensive solution of fairly universal and straightforward simulation principles and universal computational procedures [6]. The ESM formulation relies on the following key terms.

An algorithmic model. An algorithm is defined as a prescription that specifies a computational process that leads from the initial data (which may vary) to the desired result. Let us assume that an algorithmic model of the treatment process is an algorithm that transforms the values of input indicators into the values of output indicators. It is assumed that the input indicators are independent quantities and that the algorithmic model of the joint realization of the input indicators puts the joint realization of the output indicators in line.

A simulation model may contain statistical observation rules over input or output indicators. Hence, a simulation model (SM) can operate not only with the indicators characterizing the treatment process, but also directly with their relationships. Moreover, the SM exhibits the same features as the algorithmic model.

An imitator is an algorithm that reproduces the realization of the SM input indicator corresponding to the expected probability distribution law of its values during the planning period.

A tuning is the selection of SM parameter values to ensure its adequacy or to solve special tasks preceding the planning of the treatment process.

A penalty function is any way a given correspondence between the plan (or standard), the comparable reporting indicator (calculated) and the expected therapeutic consequences of their mismatch.

Aprioristic level of reality is the probability of meeting or exceeding the plan or standard.

An evolutionary-simulation model is a way to systematically link the imitators, simulation model, penalty functions and aprioristic levels of reality, which provides simulation of production planning, standardization and decision-making under probabilistic conditions.

The best plan (standard). The evolutionary-simulative model represents a system of conditions. The plan (or standard) which satisfies these conditions represents the best solution to the evolutionary-simulation model.

The evolutionary-simulation model algorithms establish a sequence of references to the imitators, simulation model, penalty functions, and aprioristic levels of reality, thereby defining a certain procedure that yields an exact or approximate solution to the evolutionary-simulation model. Thus, the ESM algorithms are algorithms for finding the best plan (or standard).

We refer to the *evolutionary-simulative method* as a set of principles and techniques for simulating the treatment plan, standardization, and decision-making based on the use of the evolutionary-simulative model and the ESM algorithms as a modeling apparatus (i.e., as simulating tools).

2.2 The Main Phases of Building a Simulation Model

First of all, in order to determine a certain treatment plan for a patient using the ESM, it is necessary to develop a simulator, i.e. a set of simulation algorithms of the patient under study, to build a simulative model of his/her disease. When building a simulator, in particular, all medical indicators (MI) of the patient are to be determined, the name and dimension of each MI are to be specified. Additionally, those MIs that are treated as independent random variables in this model are to be specified. After that, all MIs that are approved or planned are to be specified.

Which MI is to be considered in this model as an independent random variable, which is a planned one, and which is a predicted one, is determined by specific features of the simulated patient's treatment plan and the properties of the given MI. If we consider a MI as an independent random variable, it is assumed that this MI takes its values regardless of what values all other MIs in this model take. If this MI is well manageable, i.e. if we are able to give this parameter any desired value (in the admissible range of values), then we should assign to this MI an optimal value, dependent on realizations of values of other MI and, thus, should treat *this* MI as a dependent random variable.

The particular complexity of simulating a disease plan is determined by the intertwining of external processes with biological processes. Dependence on external factors makes the simulating of the treatment plan stochastic. It should also be noted the multifactorial nature of the results of the treatment plan, associated with its intensification, the use of different quality of treatment agents, with the differences of the patient's organism, the treatment technology used.

Let us formulate the task of developing a treatment plan model designed to determine the planned MIs for a specific period of time. Upon that, the initial information includes the data on the diseases of the same type selected in a given territorial area over a number of years (5)–(10).

Example. The problem is based on the use of only four MIs: X_1, X_2, X_3, X_4. All four indicators are random values. Random variables include MIs that can take one or another numerical value with different probability. Determination of the random variable types is related to the specifics of the treatment of this or that disease.

Let us formulate the following hypotheses of random MIs:

(1) X_1 is an independent random variable;
(2) X_2 is a random function f_1 that depends on a random variable X_1, i.e. $X_2 = f_1(X_1)$;
(3) X_3 is a random function f_2 that depends on a random variable X_2, i.e. $X_3 = f_2(X_2)$;

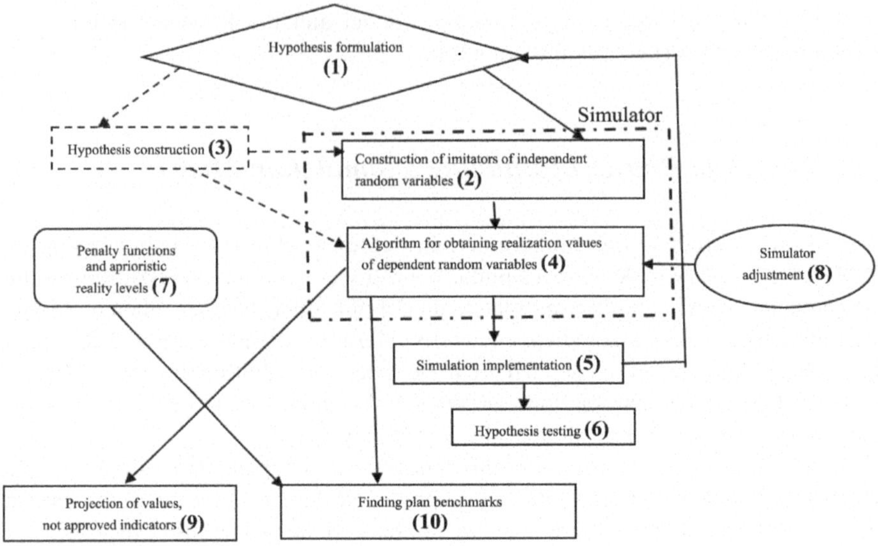

Fig. 1 Logical scheme of building a simulation model

(4) X_4 is a random function f_3 that depends on the random variables X_2 and X_3, i.e. $X_4 = f_3(X_2, X_3)$.

Thus, the formulation of the treatment plan problem involves identifying the resultant MI of the patient's health level, which satisfies the relevant requirements of optimality and simultaneously reflects the probabilistic nature of the underlying MIs. The following levels of health are distinguished: recovery (complete and incomplete); transition to a chronic form.

The logical scheme of SM construction. After determining the most important MIs (X_1, X_2, X_3, X_4) and gathering the necessary statistical information thereon, we formulate hypotheses (see Fig. 1, 1st block of the logical scheme).

The Hypotheses determine which MIs are treated as:

(a) independent random variables;
(b) random functions of other MIs (with an indication which MIs take the role of arguments);
(c) deterministic functions of other MIs (with an indication which MIs take the role of arguments).

Then, for those MIs, which are considered as independent random variables according to hypotheses, we build imitators (block 2), i.e., random number sensors, which implement the distribution law of a given MI.

The imitators construction is based on histograms (block 3) and standard sensors of pseudorandom numbers evenly distributed between zero and one [7, 8]. Then, based on the formulated hypotheses, algorithms for obtaining implementations of the remaining MIs are constructed (block 4). For this purpose, auxiliary histograms are also constructed (block 3) and pseudorandom number sensors are used.

After that, a certain number of simulations necessary to test the hypotheses are performed (blocks 5 and 6). The method of hypothesis testing depends on the criteria adopted for the adequacy of the algorithmic model (simulator) to the disease plan being simulated.

Hypotheses are considered satisfactory if, according to the accepted criteria, the simulator is adequate to the object. If the hypotheses turn out to be unsatisfactory, new hypotheses are formulated and all the above-mentioned phases of simulation construction are performed again. For those MIs, which are approved planned MIs, penalty functions are constructed or values of aprioristic reality levels P^0 are established (block 7). After that, it is possible to proceed directly to the implementation of calculations for specific households (blocks 9 and 10).

At the same time, if the statistical information characterizing a particular MI varies from the initial statistical information, on the basis of which the simulator was built, the simulator is adjusted according to the standard scheme (block 8). After that, using the ESM algorithms, the best values of the approved MIs are identified (control figures of the treatment plan – block 9) and the prognosis for the remaining MIs is developed (block 10).

The ESM is based on the developed simulation model, penalty functions or aprioristic reality levels that take into account the specific features of a particular treatment plan. Their labor intensity as well as the structure and complexity of the simulation model and penalty function can be very diverse. In some cases, implementation can be done by a single physician; in other cases, it requires the involvement of a team of physicians.

Once the ESM is constructed, it must be integrated into the patient's treatment process. When applying the ESM, the physician's behavior is composed of his/her decision-making and subsequent implementation by the patient. The results are then analyzed and taken into account when making new decisions. Under conditions of uncertainty, decision-making is associated with double-edged risks: on the one hand, there is the risk of making an over-optimistic decision (overestimation risk, i.e., the organism overreacts and wastes vital resources in an irrational manner); on the other hand, there is also the risk of making an over-pessimistic decision (underestimation risk, i.e., the organism is not able to handle the disease and its condition is getting worse) [7]. The confrontation in making the right decision ultimately determines the behavior of the physician (decision-maker) and, at the same time, the treatment process. Thus, the risk of overestimation and the risk of underestimation are the real incentives that determine the treatment process.

Since the clinical data are imperfect and the outcome of treatment is not predetermined, the physician needs to form a tentative intuitive treatment plan to be adopted before the planning period in which the plan is expected to be implemented. During the simulation process, the plan may be adjusted. The use of the Equilibrium module of the Decision system, which implements ESM, helps to reveal the intuitive content of the physician's proposed patient treatment plan and to reinforce this feeling with quantitative calculations.

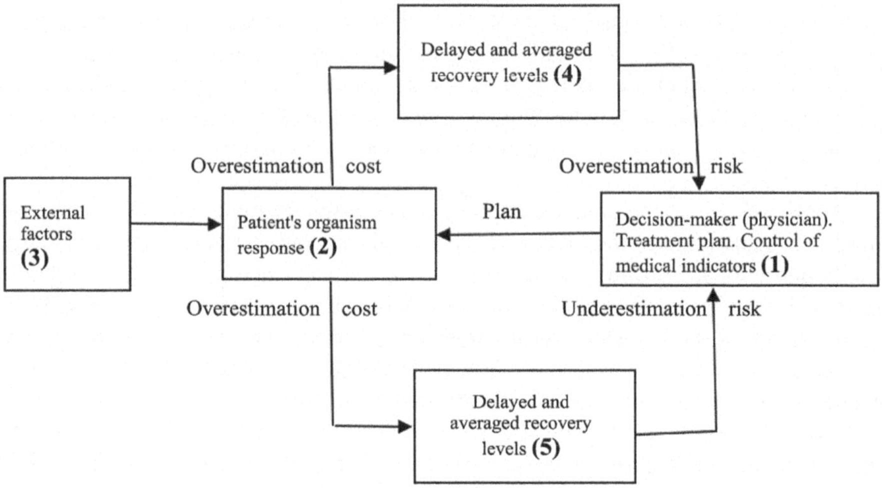

Fig. 2 Scheme for managing a treatment plan under conditions of uncertainty

An enlarged diagram of the management of a patient's treatment plan under uncertainty is shown in Fig. 2.

The physician (block 1) draws up a treatment plan, the result of which is usually expressed in numerical form. The implementation of the plan occurs under the influence of external random factors (block 3). Therefore, the actual treatment plan tends to differ from the previously established plan (block 2), which is determined by the size of the costs arising from the mismatch between the fact and the plan. In particular, if the plan turned out to be exceeded, the result is the cost of overestimation, otherwise underestimation. The decision-maker becomes aware of the amount and content of the costs not immediately, not when he/she makes the decision, but later, when the decision is not only made, but also executed, that is, at the end of the planning period. Besides, there is not only a delay in time, but also an averaging of over- and underestimation costs with those costs that were obtained from previously implemented plans (blocks 4 and 5). The average cost is the mathematical expectation, which corresponds to the risks of overestimation and underestimation.

Under conditions of uncertainty, several strategies for physician behavior can be employed. The peculiarity of planned treatment is that the plan is applied only once. For a new planning period, the plan must be adjusted or another plan must be developed. It follows that after the plan is adopted and executed, only one of the risks is realized: either the risk of overestimation or the risk of underestimation, rather than both together. In these conditions, it is necessary to apply the so-called minimax strategy, in other words, to act in such a way as to minimize the greater of the risks [8].

Let us briefly consider possible approaches to solving some of the above problems on the basis of the ESM.

3 Application of the ESM in Biophysical and Biomedical Research

3.1 The ESM of Mutually Compensating Biophysical Processes

By mutually compensating processes we mean processes that lead to opposite directional changes in some parameters of a cell, organism, or population. Mutually compensating processes constitute the physicochemical basis for regulation of organism functions. For example, the amount of oxygen in a person's blood depends on several factors, some of which contribute to its increase and some of which contribute to its decrease [A. V. Golov. Mathematical Model of Respiratory Gas Dynamics and Muscle Metabolism in the Human Organism: Abstract of the Dissertation of Candidate of Physical and Mathematical Sciences. - Moscow: NRU (MIPT), 2020].

The possibilities offered by the ESM as a means of describing mutually compensating processes are best illustrated by considering the basic structural elements of the ESM formation of some particular parameter, i.e., the amount of oxygen in human blood. In this case, we do not aim to develop any detailed model, but we will try to show the content of the main phases of research aimed at the development of the model and its application.

Under normal conditions at the current moment of time some actual amount of oxygen in the blood is established within the average norm [9]. It is possible to reproduce the implementation by constructing an appropriate simulation model ρ. Basically, ρ can be represented either as a pseudorandom number sensor, whose algorithms reproduce a known sequence of chemical reactions and physical influences that determine the oxygen content in blood, or as algorithms that have no specific chemical or physical meaning, but cumulatively reproduce statistical properties of a given MI in the norm. The well-known techniques for measuring the amount of oxygen in human blood can also be considered as a simulation model ρ.

In all cases $X^e = p(e)$ is the realization of the actual amount of oxygen in the blood, obtained using the simulation model ρ in the test (simulation) number e.

Let C_1 be a parameter quantifying the value of some known factor contributing to the increase of oxygen in the blood. Whatever the specific physicochemical mechanism of this factor, in all cases it can be said that its exposure leads to a disturbance of the equilibrium that has been achieved in the organism. This is what ultimately causes the amount of oxygen to increase, the saturation level of which rises in the blood until a new equilibrium state is reached.

A conditional measure of disequilibrium can be represented as a function of three components: X^e, the oxygen content that was prior to the factor effect; the degree of factor C_1 effect; and X, the blood oxygen content that will be established due to factor C_1 effect and all other conditions being equal.

We denote this conditional measure of disequilibrium by $f_1(C_1, X^e, X)$. By definition, we define this function at points where $X^e \leq X$, and endow the property of monotonicity, i.e. $f_1(C_1, X^e, X+\Delta) \geq f_1(C_1, X^e, X)$. This condition is quite natural and does not narrow the field of consideration, because it means that the greater the deviation of X^e from X, the greater the measure of disequilibrium.

Similarly, if C_2 is a factor contributing to the decrease of oxygen in the blood, we denote by $f_2(C_2, X^e, X)$, $X^e > X$ a measure of disequilibrium comparable to f_1 arising under the influence of factor C_2. By definition f_2 will be considered as the value region of the pair (X^e, X), where $X^e > X$, and, additionally, we will construct f_2 as a monotone function: $f_2(C_2, X^e, X+\Delta) \leq f_2(C_2, X^e, X)$.

The criterion used is the mathematical expectation of cost (MF) or, similarly, the risks associated with the fact that the reference value of blood oxygen X will be over- or underestimated.

$MF(C_1, C_2, X^e, X) = \max\{M\{f_1(C_1, X^e, X)\}; M\{f_2(C_2, X^e, X)\}\}$.

Overall, we obtain an evolutionary-simulative model (1):

$$X^e = p(e), \tag{1}$$

$$F(C_1, C_2, X^e, X) = \begin{cases} f_1(C_1, X^e, X), X^e \leq X \\ f_2(C_2, X^e, X), X^e > X \end{cases} \tag{2}$$

$$MF(C_1, C_2, X^e, X) \to min, \tag{3}$$

that describes the interaction of factors, their influence on the expected value of the amount of oxygen in the blood against the background of random environmental influences. Detailed development of the model will allow the study and calculation of these phenomena.

In particular, taking as a simulation model ρ the technique of measuring the amount of oxygen in blood and putting (in the first approximation) the functions f_1 and f_2 linear in X^e and X, we can represent the ESM in the form of

$$X^e = p(e),$$

$$F(X^e, X) = \begin{cases} C_1(X^e - X), X^e \leq X \\ C_2(X^e - X), X^e > X \end{cases}$$

$$MF \to min,$$

In this case we can relate the coefficients C_1 and C_2 by the normalizing condition $C_1 + C_2 = 1$, considering them non-negative and not exceeding one.

These assumptions about the properties of the function F are not necessary in terms of the computational procedures of the ESM, but they are justified by the fact that the measure of disequilibrium in our formulation is a conditional value.

By presenting the ESM in this form and having sufficient statistical data on the joint realizations of the X^e value, one can obtain statistical estimates of the C_t and C_2 parameters using the algorithms of the evolutionary-simulation method. With such estimates, it is possible to measure the degree of influence of factors on the amount of oxygen in the blood, i.e., to obtain commensurate estimates of one factor in UoM of another. The main thing is that within the limits of system stability we get an opportunity to make its theoretical calculation in different, in particular, extreme and unacceptable for a real organism conditions.

Along with this it is possible to calculate the probability of $X^e > X$. In other words, it is possible to establish a value characterizing the statistical stability of the system (P^0).

Graphs (or nomograms) linking the values of C_1, C_2, P^0, X,MF(X^e, X) provide complete information about the behavior of the system under different combinations of factors, about the limits of its stability, about the normal, limiting and prohibitive modes of operation.

The two mutually compensating factors and the parameter they affect together can be regarded as a control subsystem of the organism. Since the choice of factors in the first phases of research can be largely arbitrary or subjective, the allocation of the subsystem itself is in this case subjective. However, the construction of the ESM can serve as a rather universal means of distinguishing relatively isolated functional control subsystems of the organism.

Indeed, by constructing similar models for different mutually compensating factors affecting the same parameter, we can compare the relative stability of functional subsystems. Basically, since we should expect the existence of several correlated factors, it is possible, by constructing preliminary ones for different sets of significant mutually compensating ESM factors, to subsequently link these models using computational techniques and algorithms developed by large systems theory in order to coordinate solutions in multilevel hierarchical systems.

As a result, we obtain a multidimensional model, which, basically, can become a detailed and complete description of the functioning of the organism's control subsystems, suitable both for research purposes and for calculation when planning experimental or therapeutic influences.

3.2 *The ESM of Planning Experiments and Processing of Their Results*

The situation when it is possible to put a single decisive experiment in which the truth or falsity of the initial hypothesis is unambiguously established is rare. As a rule, an experiment allows us to draw conclusions with some limited reliability. Therefore, the acceptance or rejection of a hypothesis that the experiment is designed to test or on which it is based is a decision that must be made under conditions of incomplete and inaccurate source information. The situation that arises is not always reducible to the well-known formulations of the statistical theory of decision-making [10] or the theory of experiment [11].

Suppose that the purpose of the experiment is to measure the value of some indicator and the consequences of the error from its underestimation or overestimation are unequal. In this case, by constructing a penalty function that measures in conditional units the consequences of overestimation and underestimation of the value of the indicator, we can apply the ESM algorithms to process the results of the experiments and use them to get the best estimate of this indicator. Upon that, the ESM algorithms make it possible to set the necessary minimum number of experiments that provide the specified accuracy of the calculations. Suppose, for example, that C_1 measures the consequences of the error upward, and C_2 measures the consequences of the error downward of the true value of the measured indicator X, i.e.

$$F(X^e, X) = \begin{cases} C_1(X^e - X), X^e \leq X \\ C_2(X^e - X), X^e > X \end{cases}$$

where X^e is the realization obtained in the experiment "e", and X *is* the desired statistical estimate of the true value of the indicator.

In this case

$$P^0 = C_1/(C_1+C_2) \tag{4}$$

the aprioristic level of reality of the best estimate, i.e., the probability of exceeding the norm of the actual value of the indicator obtained in the experiment.

The evolutionary or convergent ESM algorithms can be applied to find the best estimate, which determine the number of experiments needed and the rules for finding the best estimate among the results of the experiments.

If values of measured quantity X during experiment depend on parameters $\alpha_1, \ldots, \alpha_n$, characterizing conditions of the experiment, then it is possible, combining ESM algorithms and tuning algorithms, to choose experimental conditions not only taking into account nature of their influences on value of measured quantity, but also taking into account consequences of deviations of received realization of quantity X from its actual value, and consequently, deviations of estimation of this variable.

The ways of combining the ESM algorithms and tuning algorithms here are the same as when applying the ESM to solve planning or standardization problems. These methods should either be chosen or specifically developed based on the specific features of the task at hand.

Suppose "e" is the number of the experiment. The set of algorithms defining the experimental procedure is denoted by $p_\alpha(e)$, where $\alpha = (\alpha_1, \ldots, \alpha_n)$ is a vector of parameters characterizing the experimental conditions. In each experiment we obtain the realization of X^e value of the measured variable, i.e.

$$X^e = p_\alpha(e). \tag{5}$$

This equality together with the penalty function F constitute a variant of the ESM formulation, which allows applying either generalized algorithms or algorithms for finding the best value X^n (depending on the substantial sense of the problem).

In particular, if it is necessary to obtain an estimate of the measured parameter that will be used only once (or if it is possible and assumed to calculate it again each time we use this estimate), we shall apply the generalized ESM algorithms. If, however, the estimate is expected to be used multiple times in different situations, then algorithms should be used to find X^n.

The meaning of these recommendations is the same as in the application of the ESM algorithms for planning or standardization [5]. In the first case, we proceed from the desire to minimize the maximum of the expected costs associated with overestimation or underestimation of this indicator, and in the second case, we seek to minimize the average total expected consequences of overestimation or underestimation of this indicator.

For each fixed value of parameters α, it is possible to find the best estimate of parameter X using ESM algorithms. We denote this estimate by X^n_α. Thus, X^n_α turns out to be an algorithmically specified function of the parameters α and conducting a sufficient number of experiments makes it possible to study this dependence and to approximate it.

Another class of problems related to the processing of experimental results is the problem of investigating statistical relationships. Existing statistical methods allow us to obtain unbiased estimates of indicators that characterize the closeness of relationships and the form of relationships, in particular correlation and regression coefficients. Along with this, it is sometimes necessary to take into account the possibility of error and its consequences in assessing the closeness and nature of the statistical relationship when estimating the form of the relationship. In this case the ESM regression can be applied.

3.3 The ESM Based on Biophysical Process Models

Mathematical models of biophysical processes are very diverse. They include both a model of cell membrane adsorption capacity and a model of an organ (e.g., kidney model) or a system (e.g., nervous system model).

A model either expresses experimentally (or otherwise) established dependencies between parameters of processes under study or reproduces with a certain accuracy the course of a process, i.e. represents a sequence of biophysical processes in the form of a sequence of transformations of parameters characterizing these processes. In both cases the model has some set of input parameters X_1, ..., X_n, output parameters X_{n+1}, ..., X_m and tuning parameters (or state parameters) α_1, ..., α_n.

Without reducing the generality of the consideration, we can assume that the input parameters characterize either the environment or an external influence (e.g., a physical impact, a nerve impulse). The output parameters characterize the state of the system, which it receives under the influence of the environment, or the amount and composition of the released substances. The tuning parameters express the closeness and nature of the relationship between the input and output parameters of the system.

For convenience and concreteness of presentation, let us turn to a simplified conditional example. The movement of blood in sufficiently wide vessels with a known accuracy can be described by a system of ordinary differential equations.

$$\frac{dXi}{dXj} = f_1(X_1, ..., X_n, X_{n+1}, ..., X_m), \; i = n + 1,...,m, \tag{6}$$

where f_i can be considered as analytical functions.

Suppose α is a function coefficient vector f_i. For each α it is possible in principle to find solutions of (6), where X_j are input and X_i are output parameters.

The first phase of ESM construction should be the construction of imitators of values of the model input parameters. The imitators can reproduce either empirical distributions or distributions that we consider possible or necessary under given experimental conditions.

In particular, in our example, based on direct measurements of the blood flow value, we can obtain statistical information about parameter X_j and then reproduce realizations of X_j^e values of this parameter with the imitator μ_j(e), which in this case appears in the form of an algorithm that converts realizations of the pseudorandom value λ^e, evenly distributed between zero and one, into corresponding realizations of X_j.

Having obtained values of input parameters by means of realization imitators, it is possible to calculate corresponding realizations of output parameters by means of the model at given values of tuning parameters. The combination of these calculations reproduces the realization of the process and allows you to accumulate statistical information about the process of any required volume.

This phase of ESM construction from the technical side usually should not cause difficulties, but can be of independent importance. In particular, the study of statistical properties of the environment parameters and the imitator construction can serve as a necessary prerequisite for checking the adequacy of the initial model and for its "tuning", if the adequacy is unsatisfactory. By comparing the statistical characteristics of the model output parameters obtained in the model experiments with the statistical characteristics obtained from the actually observed data, we can directly judge the adequacy of the model.

If the model turns out to be inadequate, its adequacy can be improved by selecting the values of the tuning parameters using the methods detailed in work [5].

The tuning conditions, i.e. the conditions according to which we want to select the values of the parameters $\alpha_1, ..., \alpha_k$, can, basically, express a wide range of requirements.

The simplest example of tuning conditions can be the requirement that the average value calculated on the array of realizations of the output parameter values obtained with the model coincides (with a given accuracy) with the average actual value. This condition can be regarded as one of the conditions for the model adequacy.

The next phase of ESM construction is the expression in the form of a penalty function of some form of intervention in the process that interests us and the assessment of its consequences. Since, basically, the ESM algorithms do not impose any restrictions on the type of the penalty function, except for monotonicity requirements, which can easily be ensured by construction, the development of the penalty function can take into account the specific features of the forms of interference. There may also be cases in which the penalty function turns out to be quite universal, namely, when the function reflects the result of the intervention rather than its form.

For example, using the piecewise linear penalty function, which is the simplest, and which has the form (7):

$$F(X_i^e, X_i) = \begin{cases} C_1(X_i^e - X_i), X_i^e \geq X_i \\ C_2(X_i - X_i^e), X_i^e < X \end{cases} \tag{7}$$

it is possible to take into account in commensurate conditional units subjective or objective estimates of the consequences of interference in the values of coefficients C_1 and C_2.

Using the ESM, it is possible to measure the effects of factors, some of which are tuning parameters, and some of which are penalty function parameters. In other words, the model makes it possible to set, for example, the value by which C_1 should be increased so that a given change in the parameter α does not lead to a change in the best value of the output parameter X_i. Thus, the parameters, characterizing the behavior of the system, become commensurable with the parameters, characterizing the consequences of the dosed influences on the system. Basically, this makes it possible to comprehensively describe the behavior of the system and the influences on it.

To construct a penalty function, it is sometimes advisable to use expert estimates. For instance, high quality assessments from a large number of experts can be obtained when our experimental influences have analogues in the field of practical medicine or veterinary medicine. The peculiarity and advantage of ESM in this case is that it comprehensively describes the behavior of the system under study, taking into account environmental influences and external interference.

3.4 The ESM of Biomedical Norms

In all fields of medicine, veterinary medicine, pharmaceutics, various and diverse norms and standards are used everywhere, which govern preventive and therapeutic effects, workloads, nutrition, human interaction with the environment, the parameters of artificial environments, etc.

The main purpose of ESM is planning and standardization. Consequently, the development of biomedical norms is directly related to the field of possible applications of the ESM. Here the application of the ESM in methodological terms practically does not differ from its applications in the development of technological, economic or financial standards and norms.

The information that can be obtained for the development of standards is statistical data on the standardized influences and their consequences. The construction of the ESM and then the application of the ESM algorithms, basically, make it possible to calculate the best (i.e., optimized under the given statistical conditions) standards.

The construction of the ESM in this case relies either on the application of special techniques, or on the development and subsequent application of techniques for the implementation of experiments in order to obtain information to calculate the best standard. Upon that, the development of the ESM and the application of the ESM algorithms here can constitute the basis for a variety of experimental and theoretical studies and processing of experimental results.

Let us turn to a conditional example. Suppose it is necessary to set the permissible share of some impurity contained in a water body. At the same time, at least the following main factors should be taken into account in mutual connection: sources of pollution, their number and intensity; permissible measures to reduce the number of such sources or their intensity; susceptibility of the environment to pollution, depending, in particular, on its ability to self-cleaning through the vital activity of certain types of microorganisms, solubility of impurities and other factors; direct and the most important indirect consequences of pollution.

There are no primary or secondary factors among those listed. Basically, the ESM provides a mathematical "description language" and computational apparatus that allows you to set and solve this problem comprehensively, taking into account the influence of all factors.

The first stage of the ESM development consists in the development of the SM describing the statistical behavior of a given indicator (the pollutant admixture fraction). The development of SM, in turn, is based on establishing the dependence of the concentration of the pollutant (Y) on its intake (X_1), the degree of solubility (X_2), the intensity of natural decontamination (X_3), etc.

Dependence

$$Y = f(X_1, X_2, \ldots, X_n) \tag{8}$$

could be presented in any form: in the form of analytical expression, in the form of systems of algorithms (including algorithms of reference to tables or nomograms), in the form of systems of equations, etc.

The next stage of ESM construction should be the study of statistical behavior of individual parameters $X_1, X_2, \ldots, X_n$, as well as the construction of imitators that reproduce the values given by these parameters.

In particular, the pollutant intake intensity (X_1) is the result of the combined action of a number of specific sources (e.g., industrial waste). By means of direct measurements, calculation or other methods, it is possible to establish the actual quantities of an incoming pollutant and then to construct an imitator of this indicator by conventional statistical methods.

If there are several parameters with which the pollutant intake can be controlled to some extent, we can account for the presence of these parameters and how they affect them as the control parameters of the random variable reproduced by this imitator when designing the imitator. For example, it is possible to invest money in the construction of sewage treatment plants (α_1), limit the intensity of certain technological operations (α_2), etc.

One of the simplest forms of an imitator might be, for example, the following:

$$X_i^e = \begin{cases} \alpha_1, 0 \le \lambda^e < P_1, \\ \dots\dots\dots\dots \\ \alpha_1, P_{k-1} \le \lambda^e \le 1, \end{cases} \tag{9}$$

where α_1, ..., α_k– control parameters.

By constructing imitators for each of the factors X_1, X_2,..., X_n, we obtain a simulation model that gives realizations of the substance concentration for any values of the control parameters (tuning parameters).

The final phase of the ESM construction should be the development of a penalty function, which measures the consequences of increasing or decreasing the permissible norm of the substance concentration. An increase in the permissible norm is not unambiguously associated with an increase in the actual concentration, but it does increase the probability of an increase in concentration. Thus, the consequence of the fact that the actual concentration of a substance is less than the norm would be the lack of measures to stabilize the concentration, and therefore the inevitable increase, which would lead to undesirable consequences. On the contrary, if the actual concentration of the substance is higher than the norm, measures will be taken, the implementation of which will require material and financial costs, and which may lead to a reduction in production output. Overall, therefore, if Y is the concentration norm of a substance, and Y^e is its actual concentration, then the function F can measure the consequences of over- or underestimating the norm.

With the ESM, using its algorithms, it is possible to calculate the best concentration norm (i.e. the norm minimizing the total expected costs of its possible under- or overestimation) for any given values of the control parameters α_1, ..., α_k.

Thus, the ESM provides an opportunity to investigate the dependence of the best norm on the control parameters. This dependence represents the necessary information for the development, along with the best norm, also the corresponding control parameters, which determine the investment in the construction of sewage treatment plants and other measures to ensure this norm.

4 Mathematical Immunology

Mathematical simulation is widely used in the tasks of modern immunology. The use of mathematical models allows a better understanding of the complex relationships between the components of the systems, numerical experiments, formulation of hypotheses about the causes of certain phenomena, identifying possible targets for drug development, and predicting the effect of different modes of therapy. However, when developing a mathematical model of the immunological process, researchers face a high uncertainty of initial parameters and a rather large dimensionality of the problem. Under these conditions, the problem of selecting the optimal model arises, so that, on the one hand, it would describe the experimental data well and, on the other hand, would be as simple as possible.

4.1 Model for Predicting Immune Processes (Model of G. I. Marchuk)

The model of analysis and prediction of the dynamics of immune processes proposed by the academician G. I. Marchuk represents a system of differential equations, which does not allow taking into account the random nature of external factors and their discreteness. These mathematical problems entail methodological and technical problems, which can be eliminated if the immune process is represented in the form of the ESM [4]. On this basis, an information system (IS) can be created, which will make it possible to study the course of a particular patient's disease, predict trends, and develop treatment tactics and strategies in a dialogue with a computer.

Let's take a closer look at Marchuk's basic model. The following most significant characteristics of the disease are identified as phase variables of the model.

1. Concentration of antigens in the affected part of the organ V(t), [pts. / ml].
2. Concentration of plasma cells C(t), [cls./ml]. This is a population of antibody producers and carriers (immunocompetent cells).
3. Concentration of antibodies in blood F(t), [pts./ml]. These are substrates of the immune system that neutralize antigens (immunoglobulins, immunocompetent cell receptors).
4. The relative characteristic of the affected organ m(t), which can be interpreted as the proportion of target organ cells destroyed by the antigen. This is a generalized characteristic of the damage the antigen causes to the target organ, $m(t) = 1 - M(t)/M^*$, where M(t) is the number of cells in the target organ at time t, and M^* is the number of cells in the norm.

The model represents a system of ordinary differential equations with a delayed argument

$$\frac{dV(t)}{dt} = (\beta - \gamma F(t))V(t) \tag{10}$$

$$\frac{dC(t)}{dt} = \xi(m)\alpha V(t-\tau)F(t-\tau) - \mu_C\big(C(t) - C'\big) \tag{11}$$

$$\frac{dF(t)}{dt} = \rho C(t) - \big(\mu_f + \eta\gamma V(t)\big)F(t) \tag{12}$$

$$\frac{dm(t)}{dt} = \sigma V(t) - \mu_m m(t). \tag{13}$$

Initial conditions:

$$V(t_0) = V_0, F(t_0) = F_0, C(t_0) = C_0, m(t_0) = m_0,$$

$$V(t) = 0, F(t) = F_0 \text{ at } t \in [t_0 - \tau, t_0),$$

where t is the time; *C'* is the constant level of plasma cells in a healthy organism; m(t) is the relative characteristic of the affected organ; β is the antigen replication factor; γ is the neutralization factor of the antigen by the antibody when they meet; ρ is the rate of antibody production by one plasma cell; η is the coefficient determining the decrease in the number of antibodies due to their reaction with antigens; μc is the coefficient determining the decrease in the number of plasma cells due to aging; μ_{m} is the proportionality factor characterizing the inverse value of organ recovery in a given number of times; μf is the coefficient inversely proportional to the decay time of antibodies; ξ(m) is the organism recovery coefficient; α is the coefficient taking into account the probability of antibodies meeting with antigens and determining the rate of new cells formation; τ is the time during which the plasma cell cascade is formed; σ is the certain constant that is unique for each disease.

The first equation describes the change in the amount of antigens in the organism. The summands in the right part of the equation describe, respectively, the increase in the number of antigens due to replication and the neutralization of antigens by antibodies. *The second equation* characterizes the change in the number of plasma cells. The first term in the right part describes the production of plasma cells, and the second term describes the decrease in their number due to aging. *The third equation* describes the increase in the number of antibodies. The terms in the right part describe the production of antibodies by plasma cells, a decrease in the number of antibodies due to aging and binding to antigens, and the intake of antibodies from the external environment due to treatment. *The fourth equation* characterizes the organ damage. The terms in the right part describe, respectively, the lesion of the organ by antigens and the restorative activity of the organism.

Let us consider the application of MAS in immunology, namely the construction of an ESM of the immune process, on the basis of which information systems (IS) for the simulation and prediction of the organism's immune response can be constructed. We will construct the ESM using the basic model (Eqs. (10)–(13)).

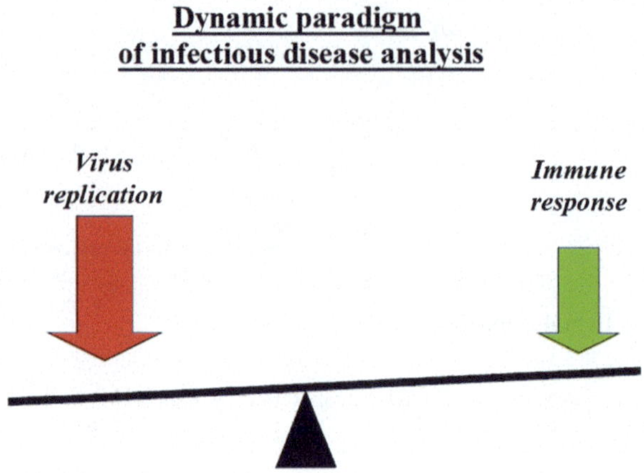

Fig. 3 Relationship between viral replication parameters (red arrow) and host immune defense (green arrow) in determining the outcome of infection

4.2 Evolutionary-Simulative Model of the Immune Response

The essential meaning of the system of differential Eqs. (10)–(13) is explained in Fig. 3, adapted from [12]. The figure illustrates the balance between virus replication, which is described by Eq. (10), and the immune response, which is characterized by antibody concentration and is described by Eq. (12). The balance between viruses and antibodies changes over time, which is described by Eq. (13).

The use of differential equations for modeling imposes a number of very significant limitations. In particular, no randomness or discreteness of variables is allowed, which in turn is necessary to express the logical interdependence of variables. These mathematical problems are eliminated if we represent the immune process in the form of the ESM.

In accordance with (10), the virus concentration V(t) depends, in particular, on the replication of viruses β and the neutralization of the antigen by the antibody when they meet γ. It is quite obvious that the values of β and γ are random and, in turn, depend on many factors, some of which are also random: body temperature, blood pressure, medications taken, medical procedures. Since V(t) depends on random variables, it is a random variable itself at any time t. By reasoning similarly, we can conclude that F(t), C(t) and m(t) are also random.

The combination of random influences and equilibrium suggests that the immune process belongs to the category of equilibrium random processes. To prove the above, it is sufficient to construct the ESM of the immune process. Let's refer to the ESM construction. Let us consider the variables $\beta, \gamma, \rho, \eta, \mu_C, \mu_m, \mu_f, \xi(m), \alpha, \tau, \sigma$ as random rather than conditionally constant. Representing a value as random, we get closer to reality by being able to select the type of probability distribution law and its parameters, reflecting the specifics of the process and the peculiarities of its course.

Once we have chosen a probability distribution law for a random quantity, we can always build a pseudorandom number sensor that allows us to obtain realizations of this quantity in statistical tests (imitator). The realization will be marked with the upper index *e*. For example, y^e is a realization of y in a statistical test. After obtaining realizations of values $\beta^e, \gamma^e, \rho^e, \eta^e, \mu_C^e, \mu_m^e, \mu_f^e, \xi^e(\mathrm{m}), \alpha^e, \tau^e, \sigma^e$ and substituting them into Eqs. (10)–(13), in the next step we can solve this system of equations and obtain realizations of functions: $\mathrm{V^e(t), F^e(t), C^e(t), m^e(t)}$. Overall, we thus have an imitational model (let us denote it by Φ) that, for any time *t*, allows us to obtain realizations of the values of the functions: *V(t), F(t), C(t), m(t)*. Suppose the notation:

$$(\mathrm{V^e(t), F^e(t), C^e(t), m^e(t)}) = \Phi(\mathrm{e}) \quad (14)$$

mean that using the imitation model Φ in the statistical test *e* in the way described above, the realizations of values of functions V(t), F(t), C(t), m(t) for the time t are obtained.

The value

$$Fa = \left.\frac{V^e(t)}{F^e(t)}\right|_{t=const} \quad (15)$$

expresses the instantaneous actual ratio at time *t* between virus replication and the immune response. Let us denote by *PL* such an unknown ratio of these quantities, which should be, in the ideal situation for the organism.

If *Fa* > *PL*, then the organism cannot handle the virus and its condition worsens, compared to some initial moment t_O, (for example, 24 h ago, or when the organism was healthy), by the value $m(t) - m(t_O)$. Suppose:

$$\Psi_1(PL, Fa) = \begin{cases} m(t) - m(t_o), ec\pi u\_Fa > PL \\ 0, ec\pi u\_Fa \leq PL \end{cases} \quad (16)$$

However, if *Fa* < *PL*, then the organism is exercising an excessive immune response and wasting vital resources in an irrational manner. Suppose *m'(t)* is a relative characteristic of the organism (or organ) that accounts for the efficiency of the immune response. Suppose:

$$\Psi_2(PL, Fa) = \begin{cases} m'(t) - m'(t_o), ec\pi u\_Fa < PL \\ 0, ec\pi u\_Fa \geq PL \end{cases} \quad (17)$$

The *PL* value is determined by the condition:

$$\min_{PL}\left\{\max_{i\in\{1,2\}}\{M\{\Psi_i(PL, Fa)\}\}\right\}, \quad (18)$$

where *M* is the sign of the mathematical expectation.

Collectively, the ratios (10)–(18) comprise the elementary structural formulation of the evolutionary-simulative model of the immune process. This model offers a fundamental opportunity to solve the following problems:

(1) model allows you to consider the influence of a variety of factors, including therapeutic (medication and other direct and indirect effects), without creating any mathematical or programming problems. A great deal of statistical information has been accumulated about various therapeutic influences and their influence on objective indicators of the organism state, in particular on the values of β, γ, ρ, η, μ_C, μ_m, μ_f, $\xi(m)$, α, τ, σ. Furthermore, mutual influences can be expressed not only in analytical functions (e.g., regression relationships are most often expressed by polynomials of low powers), but also in logical ones: if some medicine is given in combination with a certain physiotherapeutic procedure, the effect of the medicine on the virus replication and immune response is expressed by certain dependencies, otherwise by others (implication); drugs and procedures can be mutually exclusive (alternative disjunction); there are versions of obligatory in terms of the approved standards of treatment measures (logical link: "at least one of … must exist");.

(2) the patents "Method for Optimal Control of an Equilibrium Random Process" and "Method for Selecting the Parameters of the External Environment Consistent with Optimal Control of an Equilibrium Random Process" obtained by the authors allow to plot the optimal trajectory of the ERP in the phase space. If the amounts of those or other medicinal and other therapeutic influences, as well as the parameters of objective organism control are considered as coordinates of phase space, the above methods allow to develop optimal treatment tactics and strategy. This strategy is the most sparing and effective transfer of the organism from the existing state (the initial point of phase space) to the desired state (point or limited area of phase space) within a given period of time;

(3) Based on the model (10)–(18), an information system (IS) can be created that will allow to study the course of the disease of a particular patient, predict trends, develop tactics and treatment strategies in a dialogue mode with the computer. Extensive databases can be used. The IS can also be used for training and professional development, as well as for consulting. It is possible to create ISs specialized in certain diseases. The IS can also be used for research purposes, namely to study in computational experiments the dynamics of immune processes under different conditions. If a patient simultaneously suffers from several diseases, it may be reasonable to consider concurrent ERPs bound by common conditions, in particular, balance type limitations (to avoid decompensation), limitations to the acceptable limits of objective organism parameters, and limitations to the acceptable parameters of immune processes dynamics.

The multi-agent approach will be useful in extending the functionality of the IS, for example, extending its ability to model and predict the course of several diseases. To model the immune response, it will be necessary to display the interaction of different directions of immune responses. In this case, each immune response model can be considered an agent, and the simulation of the complex response will be based on the interaction of the agents.

5 Conclusion

It is shown that the evolutionary-simulative method (ESM) provides a model apparatus that allows to adequately reflect the role of mutual influence of the MI plan and the standard in the process of development of the patient's treatment technology.

One of the important features and advantages of ESM is that it is easily combined with other economic-mathematical methods and can complement them, thus allowing the construction of models that reflect various aspects in the field of evidence-based medicine.

Basically, on the basis of a combination of ESM and other economic-mathematical methods, intelligent information technologies can be created that solve a set of interrelated and complementary tasks.

The ESM algorithms are algorithms for calculating the best control plans and standards and, at the same time, algorithms for processing statistical information. The ESM algorithms provide a proof of the main properties of the problem statement and decision-making algorithms, in particular, obtaining estimates of efficiency, accuracy and reliability of algorithms, proving the convergence of procedures to the ESM solution by probability, etc.

References

1. Karpov, O.E., Khramov, A.E.: Information Technology, Computing Systems and Artificial Intelligence in Medicine, p. 480. DPK Press, Moscow (2022)
2. Karpov, O.E., Akatkin, Yu.M., Konyavsky, V.A., Mikerin, D.S.: Digital Health in a Digital Society, p. 491. Business Express, Moscow (2016)
3. Omelchenko, V.P., Demidova, A.A.: Medical Informatics, p. 528. GEOTAR-Media, Moscow (2016). https://www.studentlibrary.ru/book/ISBN9785970436455.html
4. Liechtenstein, V.E., Konyavsky, V.A., Ross, G.V., Los, V.P.: Multi-Agent Systems: Self-Organization and Development, p. 264. Finance and Statistics, Moscow (2018)
5. Liechtenstein, V.E.: Evolutionary-Simulative Method and its Application to Solving Planning and Prognostic Tasks, p. 312. Nauka, Moscow [Nauka] (1976)
6. Liechtenstein, V.E.: Discreteness and randomness in economic and mathematical problems, p. 369. Science, Moscow [Nauka] (1973)
7. Liechtenstein, V.E., Ross, G.V.: Information Technology in Business. Application of the Decision System in Micro- and Macroeconomics. Practicum, p. 509. Finance and Statistics, Moscow (2008)
8. Liechtenstein, V.E., Ross, G.V.: Information Technology in Business. Application of Decision System in Solving Applied Economic Problems, p. 550. Finance and Statistics, Moscow (2009)
9. Finogeev, Yu.P., Zakharenko, S.M., Moskalev, A.V., Sboichakov, V.B., Vinakmen, Y.A.: Recovery in infectious diseases (review). Bull. Orenburg Sci. Center Ural Branch Russian Acad. Sci. **1**, 16 [Electronic source] Recovery in Infectious Diseases (Review) (cyberleninka.ru) (2018)
10. Naumenko, A.P., Kudryavtseva, I.S., Odinets, A.I.: Probabilistic-Statistical Methods of Decision-Making: Theory, Examples, Tasks. Ministry of Education and Science of Russia, Omsk State Technical University, p. 108. OmSTU Publishing House, Omsk (2018)
11. Budnikova, I.K.: Theory and Practice of Scientific Experiment, p. 132. Kazan State Power Engineering University, Kazan (2014)

12. Bocharov, G.A.: Mathematical immunology: actual problems of simulation. Actual Problems of Numerical Mathematics. Institute of Numerical Mathematics RAS, Department of the Moscow Center for Fundamental and Numerical Mathematics at INM RAS, Institute of Personalized Medicine of the I.M. Sechenov First Moscow State Medical University Online seminar 08 May 2020. Mathematical models and parameter estimation for cell proliferation studies (nsu.ru) (2020)

National Data Bases as the Basis of Clinical Guidelines in the Provision of Medical Care in the CIS Countries

Nelli Diveeva, Lidiia Soprun, Sergey Savashinsky, and Pavel Kislyy

Abstract The World Health Organization invoke all States to switch to so-called evidence-based medicine. There are no original national evidence bases in the CIS member states. The general legal acts of the CIS concerning the provision of medical care do not contain norms on clinical recommendations. At the same time, in most cases, the procedure for developing clinical recommendations in the CIS member states is reduced to searching for existing clinical recommendations in the world (developed by various independent groups of specialists), their critical evaluation, systematic search for evidence in special databases and their evaluation. This chapter will discuss the problems and solutions, the creation of a unified approach to the use and application of national databases in the CIS countries.

Keywords National evidence bases · Clinical guidelines · World health organization · CIS countries · Health organization

1 Introduction

The World Health Organization calls on all States to switch to so-called evidence-based medicine. Decisions on the use of preventive, diagnostic, therapeutic, rehabilitation measures, palliative care are made based on the available evidence of their effectiveness and safety. Clinical guidelines (recommendations) are one of the tools to increase the evidence, and hence the validity of management decisions in healthcare [1–3]. Initially, clinical guidelines reflected the expert opinion of individual

N. Diveeva · L. Soprun (✉)
Saint-Petersburg State University, St. Petersburg, Russia
e-mail: lidas7@yandex.ru

L. Soprun · P. Kislyy
Saint-Petersburg State University Hospital, Saint-Petersburg, Russia

S. Savashinsky
Eurasian Cooperation Center for Clinical Recommendations and National Health Development, Moscow, Russia

T. Antipova (ed.), *Digital Technology Platforms and Deployment*, Information Systems Engineering and Management 36, https://doi.org/10.1007/978-3-031-86547-3_2

specialists and differed both in format and in the level of evidence. The creation of unified science-based approaches and formats of these documents has become a natural process. The process of international harmonization of clinical guidelines began in the 70 years of the twentieth century and was implemented in the WHO recommendations. Clinical guidelines summarize the most accessible information on the safety, effectiveness and effectiveness of medical technologies intended for the treatment, prevention and diagnosis of various diseases, therefore, as a rule, international teams with the necessary competencies are engaged in the preparation of these guidelines. The implementation of clinical guidelines in a particular healthcare system is determined by local conditions, legislation in general and relevant regulatory documents in particular. In the CIS member, the transition to evidence-based medicine began only in recent decades. The general legal acts of the CIS concerning the provision of medical care do not contain norms on clinical recommendations.

2 Data and Methods

Our research is a review of the evolution of health care safety management. The information was retrieved for the period 2000–2021 from medical law databases [1–8]. 50 regulatory documents regulating the introduction of clinical guidelines into the healthcare organization system in the CIS countries were analyzed [1, 3, 7, 9]. The methodology of building a strategic ladder was used to determine the strategy of building safe healthcare.

3 Results

In some national legal acts defining the provision of medical care (the Republic of Belarus, the Republic of Kazakhstan, the Russian Federation, the Republic of Uzbekistan, the Kyrgyz Republic), the place of clinical recommendations is determined when regulating the provision and evaluation of the quality of medical care, the procedure for their development and approval. In other CIS member States, national legal acts regulating the provision of medical care provide only a general reference to the need to switch to evidence-based medicine, and this type of documents (in relation to medical activities) as clinical recommendations are not mentioned at all [7–9].

At the same time, in most cases, the procedure for developing clinical recommendations in the CIS member states is reduced to searching for existing clinical recommendations in the world (developed by various independent groups of specialists), their critical evaluation, systematic search for evidence in special databases and their evaluation.

As a result, the output results in rewritten clinical recommendations for the "average world" patient, not actually adapted to the conditions of a particular country

[1, 10]. The general methodological guidelines in the CIS member States indicate that it is necessary to adapt the recommendations being developed to the conditions of the country by taking into account the peculiarities of medical care, population, culture and mentality. However, the specific mechanism of such accounting, based on the principles of evidence-based medicine is not defined.

The adaptation of existing clinical recommendations is understood by developers as a modification of recommendations so that they can be used in local conditions. This is accompanied by a change in the wording and meaning of the statement, which inevitably entails a rejection of the scientific facts on which this recommendation was based. It can be stated that there are no national evidence bases in the CIS member States that, through preclinical and clinical studies, system analysis, would contribute to a high level of evidence of clinical recommendations adapted for a particular country. A logical explanation for this process is connected with the process of creating original clinical recommendations. Original clinical recommendations require considerable time and effort. Most of this work has already been done by developers of similar guidelines from other countries. There is no need to duplicate the work already done.

However, it is quite logical that the question arises about the level of evidence of such clinical recommendations adapted for use in a particular country. The connection between each statement and scientific facts is the most important property of benign clinical recommendations.

Changes in healthcare often have the priority task of optimizing (actually reducing) healthcare costs. Introduction of new technologies, including technologies of expert support of managerial decisions based on evidence-based data, new algorithms for managerial decision-making are being introduced.

The adoption of these management decisions leads to a reduction in costs. Optimization of healthcare in the context of innovative development requires complex changes in the healthcare system, including the decision-making mechanism and its legal justification.

In this regard, it seems that a regional structuring of international cooperation is required on the territory of the CIS. This will significantly expand the volume and improve the quality of evidence-based information, create national evidence bases that meet high international standards through the connection of evidence data with the requests of a specific national health system.

It is necessary to take into account the specifics of the infrastructure and financing of each country, with the peculiarity of the historical, gender, age and genetic context of the development of nations and nationalities. Such an approach seems logical, since it avoids duplication of efforts of experts, facilitating the exchange of information, education and knowledge transfer.

Regional cooperation can and should be carried out in several directions. The first direction may be the creation of a harmonized legal framework in the CIS member [10]. The long-term work of the expert community in this direction has ended in success. The draft Model Law "On clinical recommendations in the provision of medical care in the CIS member States" was submitted to the Interparliamentary

Fig. 1 The subjects of digital healthcare

Assembly of the Member States of the Commonwealth of Independent States for discussion and approval. On April 16, 2021, the specified Model Law was submitted.

The Model Law establishes principles for the development and implementation of clinical recommendations in the CIS member. In the provision of medical care stay, comply with international standards, related to ensuring a high level of their evidence, identifying and resolving conflicts of interest in the process of formation and activities of the working group on the creation and promotion of clinical recommendations (Fig. 1).

The Model Law pays special attention to the tasks of international cooperation in the development and implementation of clinical recommendations.

We have identified the following tasks:

promotion,
systematic development,
dissemination and implementation of clinical recommendations;
exchange of independent experts in the field of development and implementation of clinical recommendations;
avoiding duplication of efforts, promoting information exchange, education and knowledge transfer.

Priority areas and support for research activities in the field of developing clinical recommendations determine the improvement and harmonization of the methodology for the systematic development, dissemination, implementation and evaluation of the quality of clinical recommendations (Fig. 2).

A demonstrative example is the experience of the states, where all the benefits of the right organizational changes were leveled by the unmanageable human factor and dilapidated infrastructure.

Promoting the introduction of scientific achievements into practice; establishing links between relevant international organizations to coordinate activities related to clinical recommendations and other initiatives in the field of quality of medical care contributes to the maintenance of a database. Databases contain information about

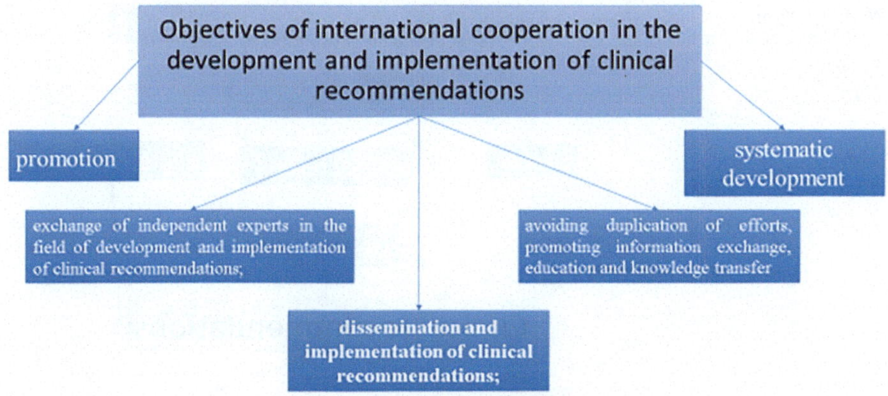

Fig. 2 The tasks of international cooperation in the development and implementation of clinical recommendations

the member organizations of the network, their work programs, methodologies and reference documentation Fig. 3.

Development of a unified methodology for systematic search of evidence for the CIS space. Creation, clinical testing and implementation of Clinical recommendations, taking into account the specifics of the sections of practical medicine and scientific research. It is necessary to recognize the importance of independent research before implementing their results in the field of healthcare and analyzing their application in real clinical practice to ensure public health and social well-being Fig. 4.

Fig. 3 Databases

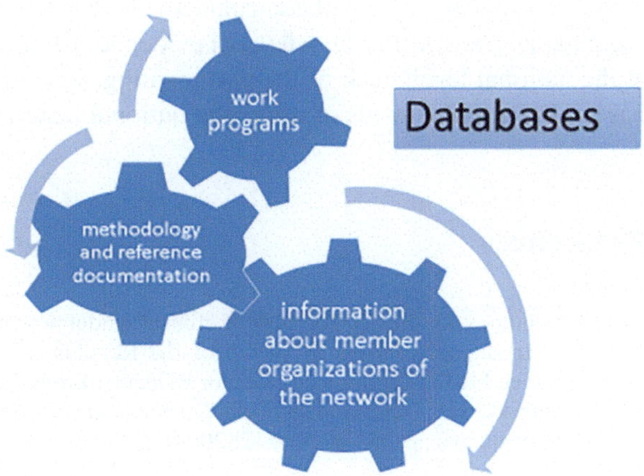

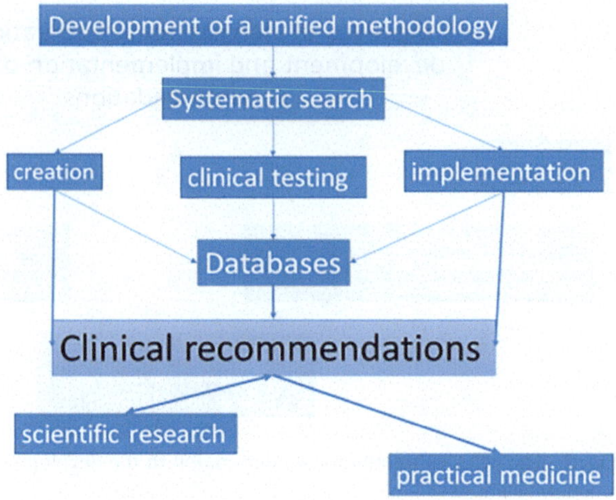

Fig. 4 Clinical recommendations for the CIS space

4 Conclusion

We have analyzed the national legal regulation of clinical recommendations in the legal systems of the CIS member states (conditions, standards, phases, implementation of state supervision/supranational supervision by regulatory authorities). A comparative description of legal regulation in the legal systems of the CIS member states according to national evidence bases has been compiled. The assessment of compliance of the requirements of the CIS member states to the DCI with the optimal ones was carried out on the basis of the data obtained at the third stage to create methodological recommendations.

Preparation, organization, promotion, holding of training courses, exhibitions, lectures, seminars, events and trainings contributes to the dissemination of scientific, popular scientific information about clinical trials. It seems that such regional international cooperation based on common historically established economic, social, cultural, linguistic roots will contribute to the development and promotion of clinical recommendations with a high level of evidence specifically for the CIS space. In turn, at the national level, each of the participating States can create a national evidence base that meets the needs of a specific national health system.

References

1. Methodology for the development of clinical guidelines based on the principles of evidence-based medicine//Ministry of Health of the Republic of Uzbekistan, Tashkent Institute of Advanced Medical Training, Center for Evidence-Based Medicine. Tashkent (2005)
2. Baschinsky, S.E.: Development of clinical practical guidelines from the standpoint of evidence-based medicine, pp. 38–39, 50–512 (2004)

3. https://iacis.ru/baza_dokumentov/modelnie_zakonodatelnie_akti_i_rekomendatcii_mpa_sng/modelnie_kodeksi_i_zakoni
4. Roitberg, G.E., Kondratova, N.V.: Medical organization according to international quality standards. In: A Practical Guide to Implementation, 152pp. (2018)
5. Voskanyan, Y., Kidalov, F., Shikina, I., Kurdyukov, S., Andreeva, O.: Model of individual human behavior in health care safety management system. In: Antipova, T. (ed.) Comprehensible Science, vol. 186, pp. 413–423. Springer International Publishing (2021). https://doi.org/10.1007/978-3-030-66093-2_40
6. Voskanyan, Y., Shikina, I., Kidalov, F., Andreeva, O., Makhovskaya, T.: Impact of macro factors on effectiveness of implementation of medical care safety management system. In: Antipova, T. (ed.) Integrated Science in Digital Age 2020, vol. 136, pp. 346–355. Springer International Publishing (2021). https://doi.org/10.1007/978-3-030-49264-9_31
7. Lukashenko, M.V., Gavrilova, N.Y., Bregovskaya, A.V., Soprun, L.A., Churilov, L.P., Petropoulos, I.N., Malik, R.A., Shoenfeld, Y.: Corneal confocal microscopy in the diagnosis of small fiber neuropathy: faster, easier, and more efficient than skin biopsy? Pathophysiology **29**, 1–8 (2022). https://doi.org/10.3390/pathophysiology29010001
8. Voskanyan, Y., Shikina, I., Kidalov, F., Davidov, D.: Medical care safety - problems and perspectives. In: Antipova, T. (ed.) Integrated Science in Digital Age, vol. 78, pp. 291–304. Springer International Publishing (2020). https://doi.org/10.1007/978-3-030-22493-6_26
9. Voskanyan, Y., Shikina, I., Kidalov, F., Davidov, D., Abrosimova, T.: Risk management in the healthcare safety management system. J. Digit. Sci. **3**(1), 41–53 (2021). https://doi.org/10.33847/2686-8296.3.1_4
10. Voskanyan, Y., Shikina, I., Kidalov, F., Sofia, S.: A strategic model of a medical organization in the context of cognitive transformation of the world. In: Antipova, T. (ed.) Comprehensible Science, vol. 315, pp. 426–437. Springer International Publishing (2022). https://doi.org/10.1007/978-3-030-85799-8_37

Detection of Tumor Pathology During Mass Health Screenings of Citizens Over 18 Years Old

Olga Zakharchenko, Dina Terenteva, Irina Shikina, Sofia Shlyafer, and Denis Altunin

Abstract This study is devoted to the identification of tumor pathology, as well as with confirmed tumor verification, during mass health screening among citizens over 18 years old. It shows a total of 312,206 tumour cases were identified between 2013 and 2020, with 183,436 being malignant. There is a marked upward trend in both benign and malignant tumor detection rates, with a CAGR of 6.6%. In 2020, due to the COVID-19 pandemic, which temporarily disrupted medical examination, the detection of tumor pathology continued to grow. This indicates a growing prevalence of cancer diagnoses, reflecting the success of extensive preventive measures, especially health screenings. In addition, the move to automated health check data collection has improved the accuracy, reliability and quality of the statistics provided.

Keywords Screening · Citizens over 18 years old · Screening · Cancer · Mass health screenings · Active detection

1 Introduction

In 2013, the World Health Organization (WHO) established a goal of reducing the mortality and morbidity rates associated with chronic non-communicable diseases (CNCDs) by 25% globally. These diseases include cardiovascular diseases, cancers, respiratory illnesses, and diabetes mellitus [1].

So, in 2013, a comprehensive preventive program was initiated among Russians to identify non-communicable diseases in the early stages among citizens over 18 years old, in the form of mass health screenings—"medical examination" [2].

This programme places particular emphasis on the identification of diseases associated with cardiovascular diseases, diseases of the respiratory system, endocrine pathology (diabetes mellitus) and tumour diseases. Our previous research has demonstrated that malignant neoplasms represent a significant cause of morbidity, disability

O. Zakharchenko · D. Terenteva · I. Shikina (✉) · S. Shlyafer · D. Altunin
Russian Institute of Health, Moscow, Russia
e-mail: shikina_irina@mail.ru

T. Antipova (ed.), *Digital Technology Platforms and Deployment*, Information Systems Engineering and Management 36, https://doi.org/10.1007/978-3-031-86547-3_3

and mortality, which has a considerable impact on occupational, productivity and economical resources at the global and Russian levels [3, 4].

Consequently, the fight against cancer is an essential component of all health systems worldwide providing health services in health organizations regardless of ownership. In our country, efforts against cancer are integrated into a specially created national Russian project "Healthcare" [5]. A principal indicator of this federal initiative is the ratio of malignant neoplasms identified during the early stages of development (I-II stages) [6]. Annual medical examinations and screenings are the main elements of the strategy for early detection of malignant tumors [7, 8]. Early diagnosis allows timely treatment and increases the chances of recovery, which can increase the level of well-being.

A distinctive feature of the Russian medical examination of the program is its two-stage approach. The first stage includes mass health screening in the early stages among citizens over 18 years old, in the form of a search for existing diseases and factors that can cause them. All this is studied for further examinations by specialists, then information is established to clarify the diagnoses. The next stage of examinations is necessary to confirm or clarify the diagnosis [9–11].

Currently, the first phase of a two-step clinical examination is focused on screening for cancer across seven of the most prevalent cancer sites:

Screening for cervical cancer (in women);
Screening for breast cancer (in women);
Screening for prostate cancer (in men);
Screening for colorectal cancer (including the rectum);

Examinations to detect visible and other forms of cancer, which include assessment of the skin, the inside of the oral cavity and lips, groping the area above the thyroid gland, palpation of superficial lymph nodes. The oncology component of the initial health check includes:

Questionnaire (survey);
Examination by a paramedic (midwife);
Cytological smear collection (for women);
Mammography (for women);
Blood test in feces (for all citizens over 18 years old);
Measurement of prostate-specific antigen (PSA) (for men);
Esophagogastroduodenoscopy (EGD) (for both men and women).

Notably, randomized trials have shown that the most effective screenings include cervical cancer screening (cytology tests and human papillomavirus testing), mammography screening for breast cancer, investigation for possible colorectal cancer (fecal blood test and sigmoid colon examination), and lung cancer screening (low-dose spiral computed tomography).

The second stage involves other types of research to confirm the diagnosed disease based on the examination and received recommendations from specialists [12].

It is worthy of note that clinical examination represents a common preventive measure in Russia, with coverage extending to more than half of the population aged

18 and above. An analysis of the evolution of oncological diseases' detection during medical examinations over an eight-year period will facilitate the determination of the optimal trajectory for the further development of this initiative [13–15].

Russian experience can be replicated in other regions of our planet, especially in countries where there is a significant level of government involvement in health care.

The aim of this study of indicators of active search for cancer diseases during the Russian clinical examination program for 8 years (2013–2020) based on data from industry statistical reporting forms.

2 Data and Methodology

Based on official statistical data from industry mandatory reporting forms (Form № 131/o) in digital format to Ministry of Health was analysed and evaluated from 2013 to 2020 [14–17].

Clinical examination in Russia is carried out in each territory of a medical organization to which citizens over 18 years old are attached. In the decree of the Russian Government this process is spelled out [12]. The present study considered only data obtained from medical organisations whose activities include the detection and mitigation of potential risk markers, as well as the implementation of necessary preventive observations to search tumour diseases [18–20].

The study used a comprehensive, systematic and functional approach, along with descriptive statistical methods. We calculated rates for dynamic series, as well as intensive and extensive rates.

3 Results

In the initial phase of our investigation, we conducted an analysis of the coverage of preventive health examinations among Russian citizens aged 18 and above over an eight-year period, spanning from 2013 to 2020. It is important to note that due to the global pandemic caused by COVID-2019 virus in 2020, clinical assessments were halted, and only data for the initial four months were included in the report.

The analysis of data from Form № 131/o for the period 2013–2020 revealed that the proportion of individuals undergoing the initial stage of preventive health examinations among those eligible for screening ranged from 77.1% in 2015 to 97.4% in 2019. This information was provided by Russian medical organisations. In 2020, this percentage dropped significantly to 44.7% of those who were supposed to participate in the program (see Fig. 1).

Figure 1 illustrates the number of newly diagnosed tumours, including malignant ones. A total of 312,206 cases of cancer were reported throughout the analysed period, of which 183,436 (58.8%) were classified as malignant. The greatest number of cancer diagnoses was recorded in 2016–2017 and 2019. This is likely attributable to

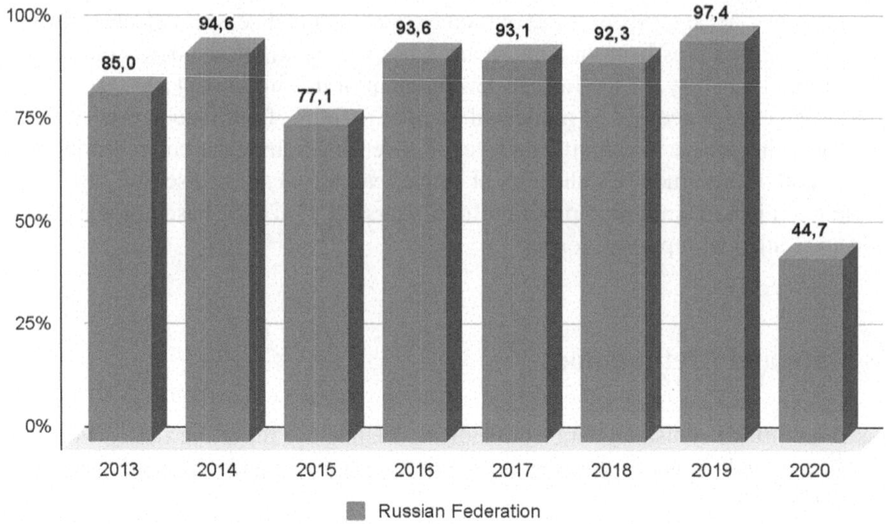

Fig. 1 The proportion of individuals who completed a Stage I clinical examination, %. *Source* Authors' compilation based on data of Forms # 131/o

alterations in the protocols governing the Russian clinical examination programme among specific demographic groups, as well as discrepancies in the volume and frequency of tests and examinations. Conversely, the lowest number of cancer cases was reported in 2013 and 2014, at the inception of the preventive healthcare project, and in 2020 amidst the ongoing pandemic. Nevertheless, the proportion of malignant neoplasms among all diagnosed cancers ranged from 50 to 69%.

The average data on the found oncological pathology, including with verification, during preventive examinations also depend on the results included in the study for 2020. Despite expectations of a decrease in the number of newly diagnosed cancers during preventive health examinations in 2020, the trend continued to be positive (see Table 1).

Table 1 Average data on tumor pathology, including those with verification, during preventive health examinations of the adult population in Russia, taking into account and excluding data from 2020

Indicators	2013–2019	2013–2020
Average value	42378.83	41901.5
Average growth rate,%	1.25	1.07
Average growth rate,%	24.6	6.6
Average absolute gain, abs	6714	1195

The percentage of persons who underwent the first stage of preventive health examinations from 2013 to 2020 in the Russian Federation showed significant fluctuations. The decline in the proportion of individuals who completed medical examination in 2015 was linked to alterations in the ongoing health study, including modifications in reporting and documentation procedures.

Considering that preventive medical examinations in 2020 were carried out from January to April, the calculation of the average coverage of Russian citizens over 18 years old, both taking into account and excluding 2020. During the entire study period, more than 21 million 481 thousand citizens took part in these exams annually, with a total increase of 1.0464 during this period. This indicates that, on average, coverage increased by 4.6% annually (by about 1.013 million individuals). However, in light of the challenging epidemiological circumstances that prevailed in 2020, the average dynamics exhibited an unanticipated decline, rather than growth. Consequently, during the period under review, there was a decline in the proportion of the adult population undergoing medical examinations, with a reduction of 0.9082, representing an average annual decrease of 9.2% (approximately 1,360 million individuals).

In the subsequent phase of our investigation, authors concentrated on the examination of the search for oncological pathology, including subsequent verification, over the period from 2013 to 2020 (see Fig. 2).

In accordance with the regulatory documents governing preventive health research, the principal parameter for the utility of the programmes is the percentage

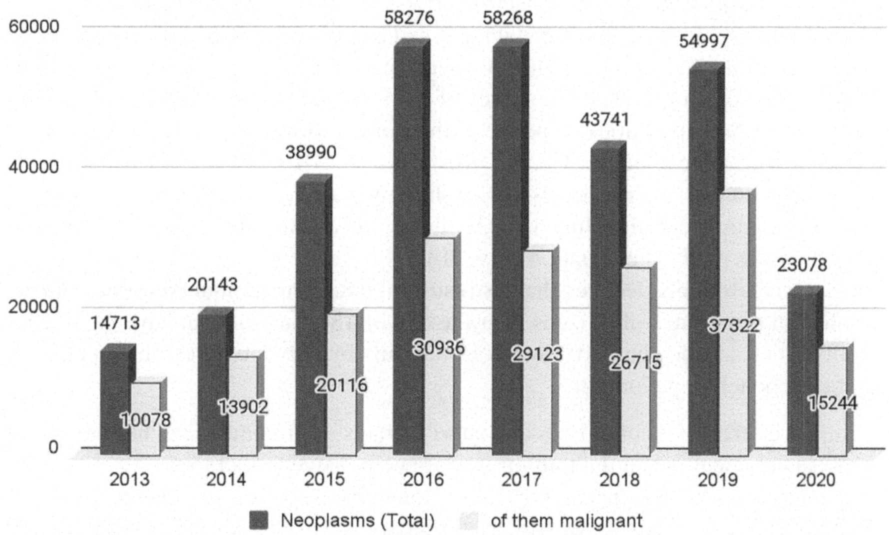

Fig. 2 The number of citizens over 18 years old with a diagnosis of oncological pathology identified during the Russian clinical examination program from 2013 to 2020 (abs. numb.). *Source* Authors' compilation

of citizens over the age of 18 undergoing medical examinations scheduled for the current year [21–24].

In our other study, it was shown that one fifth of Russian citizens over 18 years old participated in preventive health examinations annually from 2013 to 2019, while people over 60 years old are more actively involved in this process.

Analysis of Form No. 131/o showed that screening coverage is increasing every year, which positively contributes to the detection of cancer cases. A significant moderate to strong correlation was found between population coverage of preventive health surveys and the incidence of newly diagnosed malignant diseases. Nevertheless, despite the relatively high participation rate, it is imperative that public health knowledge initiatives be continued and that new methods be explored to encourage citizens to undergo medical examinations.

The examination of data from Form No. 131/o revealed that neoplasms ranked tenth among eleven disease categories identified during preventive examinations. The comparatively low incidence of oncological diseases, in comparison to other chronic non-communicable diseases, can be attributed to several factors:

Insufficient awareness about cancer among healthcare providers in primary care institutions, leading to delays in registering patients for monitoring.

Incomplete execution of the examination volume stipulated in preventive health procedures, including cancer screening components.

Inadequate organizational and methodological guidance regarding patient routing for suspected or diagnosed cancer at the regional level.

Variability in material and technical resources and staffing across medical institutions in different regions, affecting access to quality oncological care.

In order to negate these detrimental factors, it is imperative to pursue a programme of rigorous study and improvement in organisational change, with the objective of facilitating a more comprehensive survey of citizens over the age of 18. A noteworthy illustration of an expeditious response to the novel requirements for the healthcare system in the context of the Coronavirus Disease 2019 (Covid-19) pandemic is the modification of the protocols for evaluating individuals who have recovered from the infection. Starting July 1, 2021, these individuals are required to undergo comprehensive medical evaluations [19, 20].

It can be reasonably deduced that the results of detecting tumour pathology during mass health screenings of citizens over the age of 18 years old can have significant benefits. The early detection of cancerous tumours by preventive screening can have significant benefits, including:

1. improved treatment outcomes and survival rates, as treatment can be initiated at an earlier stage, when the tumour is more responsive to therapy;
2. a reduction in the need for more intensive, costly treatments, such as chemotherapy or surgery, which ultimately saves healthcare resources and reduces healthcare costs;
3. a reduction in cancer-related morbidity and mortality.
4. Furthermore, early detection can improve patient quality of life by preventing disease progression and reducing the need for more invasive treatments. It can

also increase public awareness of cancer and the importance of early detection, encouraging individuals to seek regular screenings and preventive care. Additionally, early detection can provide valuable data on cancer incidence, patterns, and trends in specific populations, which can help researchers and healthcare providers develop more effective screening, prevention, and treatment strategies. Effective screening, prevention, and treatment strategies improve patient-centred care, promoting more personalised and patient-focused healthcare. They also improve healthcare access for underserved populations, including those with limited healthcare resources, ensuring that all citizens have equal opportunities to detect and treat cancer. Furthermore, they increase the overall life expectancy of the population, promoting healthier and longer lives for individuals and society.

4 Conclusion

The objective of this work, that concentrated on examining kinetics of cancer identification for preventive adults' health examinations, was accomplished through the utilisation of descriptive statistical techniques, the analysis and statistical processing of results, the compilation of dynamic series and the calculation of means. In light of the findings of this study, a number of conclusions may be drawn:

1. There has been a positive trend in the coverage of adults with medical assessments from 2013 to 2019.
2. An increase in the detection rate of oncological diseases was recorded during this period, including in 2020 amidst challenging epidemiological conditions due to COVID-19.
3. The proportion of detected tumor pathology among all neoplasms is 58.8%.
4. Collected for the year 2020 data correlate with the results of the preliminary years of the preventive health examination programme (2013–2014) in relation to both the level of coverage achieved and cancer detection rates.
5. Our findings may be beneficial for healthcare organizers, clinicians, and epidemiologists, especially since part of this research was conducted during a complex epidemiological situation caused by COVID-19.
6. To facilitate a more comprehensive analysis of cancer detection during preventive health examinations, it is crucial to enhance continuity between outpatient healthcare providers and specialized medical organizations.
7. The automated collection of statistical data on oncological diseases detected during preventive health examinations and their risk factors requires further improvement based on cutting-edge digital technologies.

References

1. Action Plan against Non-Communicable Diseases for 2013–2020. https://apps.who.int/iris/bitstream/handle/10665/94384/9789244506233_rus.pdf?sequence=5. (Circulation Date: 10.02.2022)
2. Order of the Ministry of Health of the Russian Federation of December 3, 2012 No. 1006n "On approval of the procedure for medical examination of certain groups of the adult population". https://www.garant.ru/products/ipo/prime/doc/70255634/?ysclid=m491w7ir4y403654165. (In Russ.)
3. Kaprin A.D., Starinsky V.V., Shakhzadova A.O.: State of oncological assistance to the population of Russia in 2019. M.: MNIOI named after P.A. Herzen - a branch of the Federal State Budgetary Institution "NMITS of Radiology" of the Ministry Some methodological aspects of conducting and evaluating adult health check-up: a systematic review of regulatory public health documentation in the Russian Federation. Kuban Sci. Med. Bull. 30(6):66–80 (2023). https://doi.org/10.25207/1608-6228-2023-30-6-66-80. (In Russ.)
4. Antipova, T.V., Melnik, M.V., Nechaeva, O.B., Shikina, I.B., Vechorko, V.I., Lutseva, E.M.: Social aspects of public health **1**(47) (2016). https://doi.org/10.21045/2071-5021-2016-47-1-3 [Electronic Resource]. http://vestnik.mednet.ru/content/view/730/30/lang.ru/. (Circulation Date: 10.09.2021.) (In Russ.)
5. Passport of the national project "Healthcare" (approved Presidium of the Presidential Council for Strategic Development and National Projects, Minutes No. 16 dated December 24, 2018). https://www.garant.ru/products/ipo/prime/doc/72085920/?ysclid=m49209h296791887213. (In Russ.)
6. Efremov, S.A., Petkau, V.V., Gruzdeva, E.A.: Implementation of a system to support the work of the oncological service at the regional level. Health Manag. **2**, 25–30 (2021). https://doi.org/10.21045/1811-0185-2021-2-25-30(InRuss.)
7. Tatiana, A., Irina, S.: Informatic indicators of efficacy cancer treatment. In: 12th Iberian Conference on Information Systems and Technologies 1 (CISTI) Lisbon, Portugal, pp. 1–5, 21–24 June 2017. https://doi.org/10.23919/CISTI.2017.7976049
8. Voskanyan, Y., Shikina, I., Andreeva, O., Kidalov, F., Davidov, D.: Multifactorial model of adverse events and medical safety management system. J. Digit. Sci. **2**(1), 29–39 (2020). https://doi.org/10.33847/2686-8296.2.1_3
9. Zakharycheva, T., Makhovskaya, T., Shirokova, A., Shikina, I.: Autonomic dysregulation syndrome in covid-19 convalescents: possible causes and approaches to its correction. In: Antipova T. (ed) Comprehensible Science. ICCS 2021. Lecture Notes in Networks and Systems, vol. 315. Springer, Cham (2022). https://doi.org/10.1007/978-3-030-85799-8_34
10. Zakharycheva T., Makhovskaya T., Shirokova A., Shikina, I.: 2021 International Conference on Advances in Digital Science (ICADS 2021), AISC 1352 (Antipova, T. (ed.)), pp.191–197 (2021). https://doi.org/10.1007/978-3-030-71782-7_17
11. Chernobrovkina, A.E.: Contribution of the medical examination of the adult population to the early detection of oncological diseases of the female genital organs. Prev. Med. **25**(1), 7–13 (2022). https://doi.org/10.17116/profmed2022250117. (In Russ.)
12. Order of the Ministry of Health of the Russian Federation of April 27, 2021 No. 404n "On approval of the Procedure for conducting preventive medical examination and clinical examination of certain groups of the adult population". https://www.garant.ru/products/ipo/prime/doc/401314440/?ysclid=m4924pimxf894901519. (In Russ.)
13. Zakharchenko, O.O., Terentyeva, D.S.: Sexual age characteristics of the coverage of the adult population of the Russian Federation in 2013–2019. Soc. Aspects Popul. Health **3**(67) (2021). https://doi.org/10.21045/2071-5021-2021-67-3-1 [Electronic Resource]. http://vestnik.mednet.ru/content/view/1263/30/lang.ru. (In Russ.)
14. Bryzgalova, O.E., Armashevskaya, O.V.: Online resource as a tool for preventive measures in providing medical care to women during the Covid-19 pandemic. Obstet. Gynecol. **1**, 113–121 (2022). https://doi.org/10.18565/aig.2022.1.113-121. (In Russ.)

15. Zakharchenko, O.O., Terentyeva, D.S., Suraeva, N.A., Komarov, Y.I.: Detection of malignant neoplasms in the process of medical examination of the adult population in the North-western Federal District in 2016–2019. Soc. Aspects Public Health [online publication] **6**(67), 5 (2021). https://doi.org/10.21045/2071-5021-2021-67-6-5 [Electronic Resource]. http://vestnik.mednet.ru/content/view/1319/27/lang.ru/. (In Russ.)
16. Order of the Ministry of Health of the Russian Federation of November 10, 2020 № 1207n "On approval of the accounting form of medical documentation N 131/y" Card for recording preventive medical examination (medical examination), "the procedure for its maintenance and the form of industry statistical reporting No. 131/o" Information on conducting preventive medical examination and medical examination of certain groups of the adult population, "the procedure for filling it out and the deadlines for submission". https://base.garant.ru/400178546/. (In Russ.).
17. Kaprin, A.D., Alexandrova, L.M., Starinsky, V.V., Kalinina, A.M., Ipatov, P.V., Fighters, S.A., Drapkina, O.M.: Dispensation of certain groups of the adult population of Russia as an instrument for early detection of malignancies (results of 2015–2016). Prev. Med. **21**(4), 13–19 (2018). https://doi.org/10.17116/profmed201821413. (In Russ.)
18. Khodakova, O.V., Koshevaya, N.V.: Self-esteem of health as an element of self-preservation behavior and commitment to medical examination of adult population. Curr. Health Health Stat. Issues **4**, 309–326 (2019). https://doi.org/10.24411/2312-2935-2019-10099 [Electronic Resource]. https://healthproblem.ru/ru/magazines?text=317. (In Russian)
19. Nechaeva, O.B., Shikina, I.B., Chukhrienko, I.Y., et al.: Resource support for medical organizations providing assistance in the profile of "oncology". Curr. Health Health Stat. Issues **2**, 269–279 (2019). https://doi.org/10.24411/2312-2935-2019-10042 [Electronic Resource]. https://healthproblem.ru/ru/magazines?text=259. (In Russ.)
20. Valkova, L.E., Levit, M.L., Merabishvili, V.M., Pankratieva, A.Y., Krupina, M.V., Dubovichenko, D.M., Agaeva, A.V., Ryzhov, A.Y., Potekhina E.F., Valkov, M.Y.: Cancer mortality trends after implementation of the National Programme for Dispensarization of certain groups of the adult population: a population-based study from Arkhangelsk, North-Western Russia. Res. Pract. Med. J. (Issled. prakt. med.) **7**(4), 175–189 (2020). https://doi.org/10.17709/2409-2231-2020-7-4-14
21. Antipova T., Shikina I.: Informatic indicators of efficacy cancer treatment. In: 12th Iberian Conference on Information Systems and Technologies (CISTI) Lisbon, Portugal 21–24 June 2017, pp. 1–5. https://doi.org/10.23919/CISTI.2017.7976049; Preven. Med. **21**(4), 13–19 (2018). https://doi.org/10.17116/profmed201821413. (In Russ.)
22. Nechaeva O.B., et al. Resource support for medical organizations providing assistance in the profile of "oncology". Curr. Health Health Stat Issues **2**, 269–279 (2019). https://doi.org/10.24411/2312-2935-2019-10042 [Electronic Resource]. https://healthproblem.ru/ru/magazines?text=259. (In Russ.)
23. Voskanyan, Y., et al.: Risk management in the healthcare safety management system. J. Digit. Sci. **3**(1), 41–53 (2021). https://doi.org/10.33847/2686-8296.3.1_4
24. Zaharchenko, O.O., Shikina, I.B., Terenteva, D.S.: Some methodological aspects of conducting and evaluating adult health check-up: a systematic review of regulatory public health documentation in the Russian Federation. Kuban Sci. Med. Bull. **30**(6), 66–80 (2023). (In Russ.). https://doi.org/10.25207/1608-6228-2023-30-6-66-80

Impact of Antibiotics on the Microbiome of Aquatic Ecosystems

Cecilia Curis and Sebastian Marius Curis

Abstract The Earth's population is facing global warming and one of the dramatic repercussions of this phenomenon is the decrease in the supply of drinking water. Against the acute background of diminishing water resources, a new cause for concern has arisen, namely the pollution of drinking water sources with various types of contaminants. The present chapter aims to briefly address the issue of microbiome study in aquatic ecosystems by conducting a literature review realized through searching for articles related to this topic, published in the PubMed database between 2011 and 2024. The objective is to quantify the extent of the scientific world's concern towards this phenomenon in an indirect way, namely by analyzing the reporting of results. The study establishes the pandemic period as a reference, thus realizing a temporal distribution of the studies identified in the literature. The decision to choose the pandemic period as a benchmark was due by the objective reason of the increase in antibiotic consumption due to the existence of a large number of Covid-19 cases of illness. Even before the period mentioned there were also concerns about this issue. Today, the reasons for concern have become increasingly evident in view of the growing resistance to antibiotics.

Keywords Antibiotic · Contaminants · Aquatic ecosystems

1 Introduction

This chapter aims to be a concrete motivation to raise awareness of a problem of general interest, which for a long time has been unknown or minimized by public opinion. At present, there is not still an optimal level of awareness among the population about the risk that current generations start to face and with witch future generations will be faced, the fight against pathogens. The overuse of antibiotics/antimicrobials has led to the phenomenon of antibiotic/antimicrobial resistance so that in the future there will a potential danger; will possible we may not have capacity

C. Curis (✉) · S. M. Curis
Dunărea de Jos" University of Galați, Galați, România
e-mail: Cecilia.Curis@ugal.ro

T. Antipova (ed.), *Digital Technology Platforms and Deployment*, Information Systems Engineering and Management 36, https://doi.org/10.1007/978-3-031-86547-3_4

to treat certain diseases caused by pathogens [1]. For this reason, the limitation of antibiotic therapy has become a public health issue [2]. In this context, the pollution of aquatic ecosystems with antibiotics/antimicrobials represents an important link in the mechanism of resistance's acquisition to these classes of substances. A report published in 2019 brought to light that there were "approximately 4.95 million antimicrobial resistance-associated deaths (AMR), including 1.27 million deaths attributable to bacterial AMR" [3]. The novel elements that this chapter aims to bring are represented firstly by the awareness of the nefarious contribution of the contamination of these areas to the genesis of the phenomenon of antibiotic resistance and secondly by the correlations between the factors that contribute to the complexity of the phenomenon [4–6] with consequences on the microbiome of aquatic ecosystems. Thus, we reduced the search only to articles with direct reference to microbiome realizing a distinct delimitation for marine [7] and freshwater aquatic ecosystems, [8] antibiotic resistance, antibiotic's classification and association with other factors that potentiate the contaminating effect of antibiotics [9–11]. The structure of the chapter is based on the current state of knowledge with reference to antimicrobials and their influence on aquatic ecosystems, the antibiotic resistance determined in this way, consequences on the trophic chain, objectives and perspectives with reference to immediate and long-term implications on public health.

2 Literature Review

A first observation, as a consequence of the research approach, suggests that the relationship between contaminants and aquatic ecosystems has a fairly long history. However, the mode of action of antibiotics on aquatic ecosistems and especially the microbiome approach are relatively recent concepts considering the small number of chapters identified. Due to the importance of the subject as a part of public health and the still low degree of awareness and involvement of the population in this process of pollution with contaminants, it is necessary to mediatize and include it in general education. The problem of pharmacological contaminants can even be assimilated to a civic duty, thus creating the need to sensitize each person through bidirectional involvement both from the perspective of beneficiary of antibiotic treatments and as a target through the direct and indirect consequences of this phenomenon. We are faced with a paradox, namely, pharmacological products used in therapy are vectors that promote per se a new type of pathology determined by antibiotic resistance [12].

The biota, also referred to as the biome or microbiome, is the totality of animals and plants, i.e. the flora and fauna of a region [13] or habitat. There are similarities between the human microbiome and that of natural habitats, both having as a common denominator the fact that they represent a population of mostly beneficial microorganisms, some of which also have pathogenic potential under certain conditions. Another common feature is the fact that both types of microbiota show vulnerability to antibiotics through the same type of mechanism, i.e. the selection of antibiotic-resistant strains.

The importance of current research on the microbiome in general and the microbiome of aquatic ecosystems, in particular, is not essentially due to the concept itself. This is mainly due to the type of approach; there is a return to what was known for a long time, namely the fact that microorganisms behave differently in their living environment compared to their behavior in laboratory conditions [14, 15]. There is a need for in vivo study of the relationship between microbial and animal communities in a symbiotic relationship in aquatic ecosystems. This, in a fragile balance, is in danger of being disturbed by human activities as well as by natural phenomena (modification of salinity or climate changes). Thus, beyond the interaction with the host organisms, the microorganisms existing in the aquatic environment fulfill an essential role in the ecology of aquatic ecosystems [16, 17]. The biofilm constituted by the probiotic microbial community, resulting from the selective assembly through ecological interaction of bacterial populations, has an essential role in the prevention of bacterial infections in fish. The influence of contaminants on it contributes to the decrease of the defense mechanism of the fish population in aquatic ecosystems [18, 19].

3 Data and Methodology

We searched the PubMed database for publications on contaminants using as keywords the association of two terms, "contaminants and aquatic ecosystems". The result was a total of 15 articles in the period 1980–2024. Given the small number of chapters found on this topic, we increased the number of keywords using the triad "contaminants—aquatic ecosystems—antibiotic". The result was a total of 17,333 articles published between 1966 and 2024. Since the aim of the research carried out is to analyze the influence of antibiotics on the microbiome of aquatic systems, we also included the keyword "microbiome" among the search terms and thus identified a total of 66 articles with strict reference to the subject in the period 2011–2024. We excluded a total of three articles on contaminants in aquatic ecosystems without specific reference to antibiotics, finally including a total of 63 chapters in the analysis.

In the design of the study, we did not start from hypotheses or research questions, the purpose of the study being to demonstrate the interest in a subject belonging to public health.

Considering the small number of articles, only 63, published on the chosen topic in an interval of 14 years, i.e. 0.071 articles per year, we considered the frequency of the phenomenon studied as low. The data obtained from the search process in the mentioned database were statistically analyzed with the help of Microsoft Excell application using the Data Analysis function.

The search was conducted independently by the two investigators and was followed by data matching to limit bias.

The comparative analysis, taking as a benchmark a period of global social crisis—the pandemic, allows a paradigm shift regarding the reporting of medical systems to the needs and consequences of turning points in the evolution of mankind.

Table 1 Statistics

Average	7
Median	6
SDEV	5.29
Confid T	1.33

We have included the work in the category of systematic reviews, arguing that it corresponds to the last criteria established for this type of analysis: the academic synthesis of the evidence (63 studies on aquatic ecosystem contaminants's theme) using critical methods of identification, definition and evaluation. In other words, the systematic review extracted and interpreted data from published studies on a topic, subsequently performing an analysis, description, critical evaluation and summary of the data in an evidence-based approach.

We did not apply the PRISMA criteria because the specialized literature clearly mentions the fact that the inclusion of the PRISMA criteria can only be limited to studies related to interventional research, the present study being a non-interventional research, more precisely a non-interventional systematic evaluation.

4 Results

4.1 Statistical Analysis of the Data

The Table 1 shows the values of the mean, median, SDEV, confidence interval when $\alpha = 0.05$ (95%). The table below shows the calculated values of the mean, median, SDEV and confidence interval, related to the number of studies analyzed.

We calculated statistical significance using the one-way ANOVA test Table 2.

The normal distribution of the data is observed, they being dispersed (scattered) around the diagram line Fig. 1.

The hypothesis of a linear relationship between the independent variable time and the dependent variable, number of published articles, is respected and there is a normal distribution of the model. Homogeneity is present of the variant (homoscedascity). The criterion of independence is met, more precisely, two observations from a data set are not related to each other nor are they interconditioned in any way Fig. 2.

The first sample has the variant 16.94 and the second 28. The ratio between the larger and the smaller variant has the value 1.65. Since this ratio < 4 we can assume that the variations between the two groups are approximately equal Table 3.

With reference to the microbiome of aquatic ecosystems, which I stated is a relatively recently studied concept, the present analysis identified a number of 19 studies on the microbiome of freshwater ecosystems and 3 articles on the microbiome of saltwater ecosystems and one articles on the role of the microbiome in the food

Table 2 ANOVA—statistics

Anova: single factor						
Summary						
Groups	*Count*	*Sum*	*Average*	*Variance*		
Column 1	2	19	9.5	0.5		
Column 2	2	33	16.5	364.5		
Column 3	2	25	12.5	220.5		
Anova						
Source of variation	*SS*	*df*	*MS*	*F*	*P-value*	*F crit*
Between groups	49.33333333	2	24.6666667	0.126387703	0.885729	9.552094
Within groups	585.5	3	195.166667			
Total	634.8333333	5				

Fig. 1 Probability diagram of data distribution

chain. In total, a number of 23 articles about the microbiome were found, their percentage from the total of 63 articles being 36.5% Fig. 3.

In the previous figure, we find the breakdown by year of the studies on the importance of the microbiome at the level of freshwater ecosystems, respectively, salt water, fresh water versus salt water and its involvement in their food chain. It should be mentioned that there is only one study regarding the damage to the trophic chain by affecting the biota. In the period 2023–2024 there is also one study published annually in the form of a comparative analysis, microbiome—fresh water versus microbiome—salt water Fig. 4.

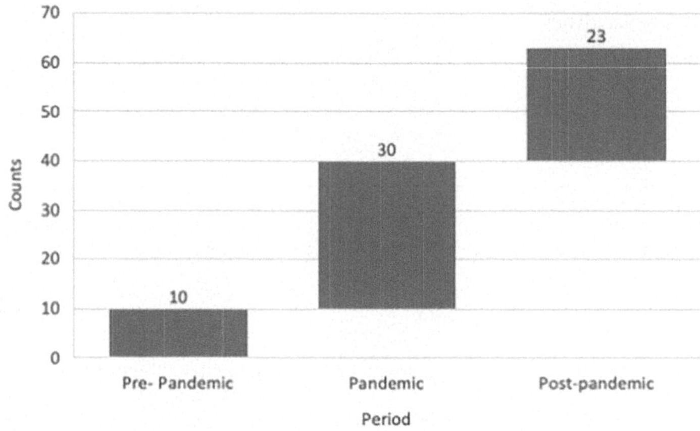

Fig. 2 Distribution of studies by period: pre-/pandemic/ post-pandemic – comparative

Table 3 t-test—statistics

t-test: two-sample assuming equal variances		
	Variable 1	Variable 2
Mean	2019.222222	7
Variance	16.94444444	28
Observations	9	9
Pooled variance	22.47222222	
Hypothesized mean difference	0	
df	16	
t Stat	900.4491394	
P (T ≤ t) one-tail	2.25744E-39	
t critical one-tail	1.745883676	
P (T ≤ t) two-tail	4.51489E-39	
t critical two-tail	2.119905299	

The distribution by years highlighted the fact that there is only one study regarding the involvement of the microbiome in the trophic chain of aquatic ecosystems. It can be noted that in the pre-pandemic period there is also, only one chapter that refers to the microbiome of aquatic ecosystems (Fig. 5).

It is notable that in 2021 the interest in the subject of the microbiome of aquatic ecosistems has increased. The largest number of articles published on this topic can be observed. In the following years, until now, the interest shown by the scientific world with reference to the contamination of aquatic ecosystems with pharmacological contaminants, especially for antibiotics, remained high and recorded a plateau, 5 articles each year. It is possible that this increased interest originating in the second year of the pandemic is due to the increase in the amount of antibiotics prescribed

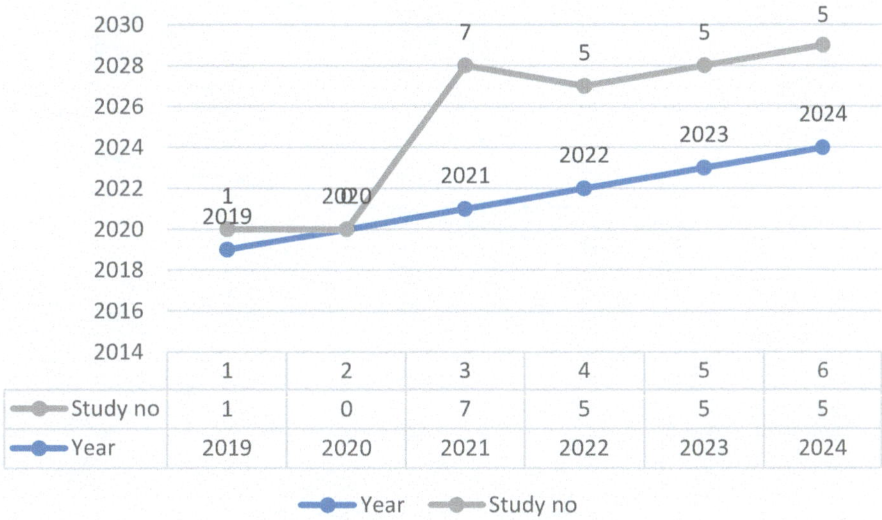

Fig. 3 Distribution by year of studies with reference to the microbiome

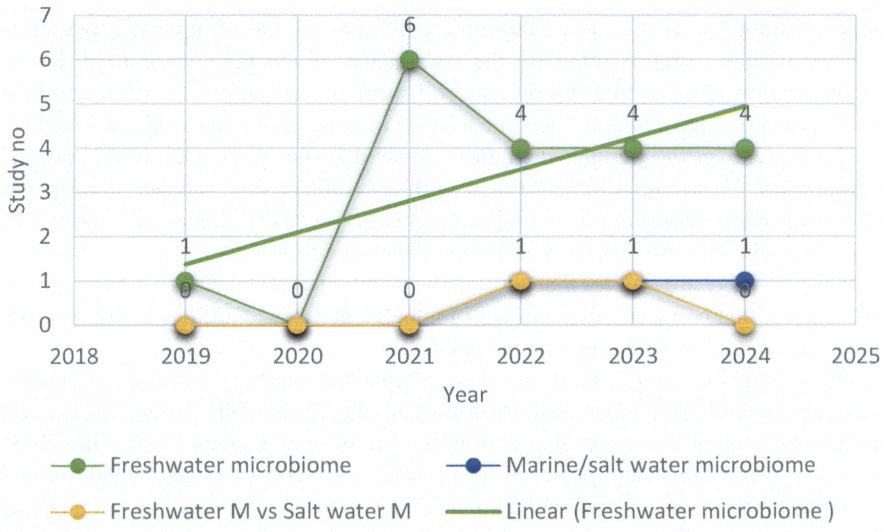

Fig. 4 Year distribution of studies microbiome—freshwater/saltwater/ freshwater versus saltwater/ trofic chain

during the pandemic, due to the large number of illnesses. Later, the research topic retained its interest due to the awareness of the immediate and long-term danger of the microbiome being affected by antibiotic contamination and its inclusion in the register of public health.

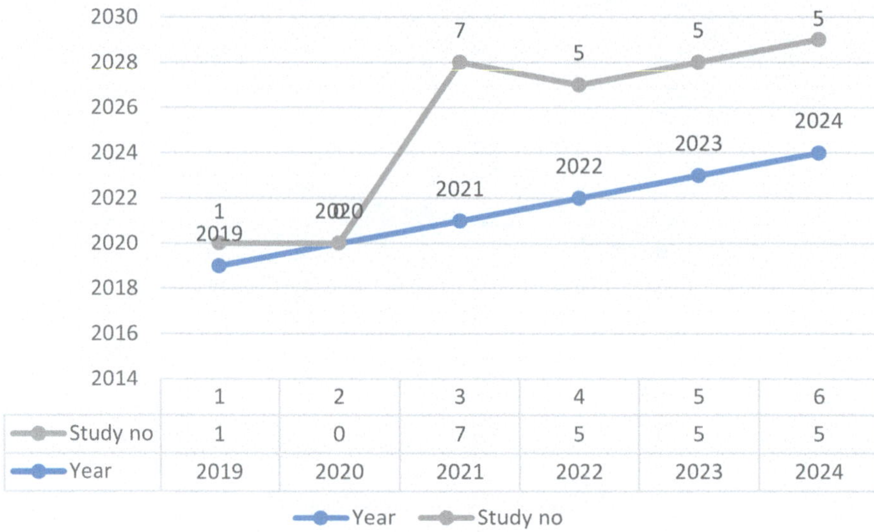

Fig. 5 Global breakdown by year of studies on the microbiome

A consequence of the contamination of the microbiome of aquatic ecosystems with antibiotics is represented by the emergence of the phenomenon of antibiotic resistance. This occurs through several mechanisms and has obvious clinical implications that have already become a cause of concern for the medical world.

The analysis of the published studies reveals that out of the total of 63 articles, a number of 21 articles (33.33%) deal with the subject of resistance determined by contaminants represented by antibiotics. More precisely, 1/3 of the number of chapters deals with the aspect of antibiotic resistance (Table 4).

Mention: there are articles that address the problem of resistance to several antibiotic classes, this being the explanation of a larger number of studies resulting from summing up the figures in the table (Figs. 6 and 7).

The analysis carried out reveals the fact that the majority of studies reported the emergence of bacterial resistance following the study of the impact of generic antibiotics on aquatic ecosystems. This aspect can be explained by the fact that there are situations in which the presence of a certain antibiotic can only be determined indirectly in the waste water of treatment plants. As an example, the category of ß-lactams can be offered, which undergo a hydrolysis process and can no longer be detected directly in wastewater, but their presence could be detected by the presence of bla genes. These are the largest class of antibiotic resistance genes (ARG) [12].

It can also be observed that a number of 4 articles (19.04%) address the topic of synergistic action of antibiotics and other contaminants (plastic/microplastics and heavy metals) on aquatic ecosystems. The high percentage of articles can be interpreted as an aspect of interest in terms of the increased degree of negative impact of these associations.

Table 4 Classification of the studies

14	3	2	3	1	1	1	1	4
Antibiotics generic	Tetracycline	Quinolone	Cephalosporins	Sulfamide	Aminoglycoside	Glycopeptide	Carbapenem	Association—microplastic or heavy metals

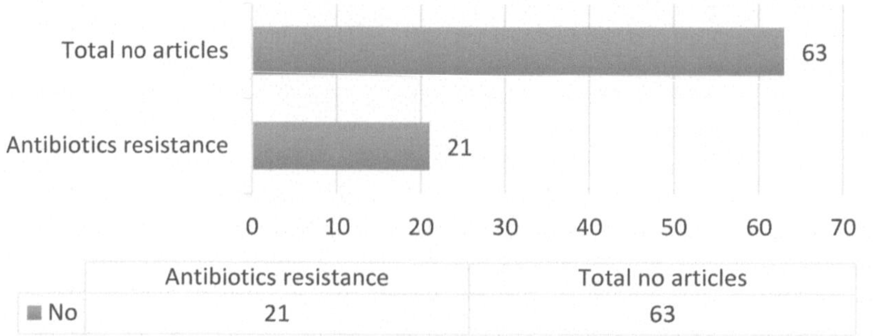

Fig. 6 Comparison between total number of articles versus number of articles about antibiotics resistance

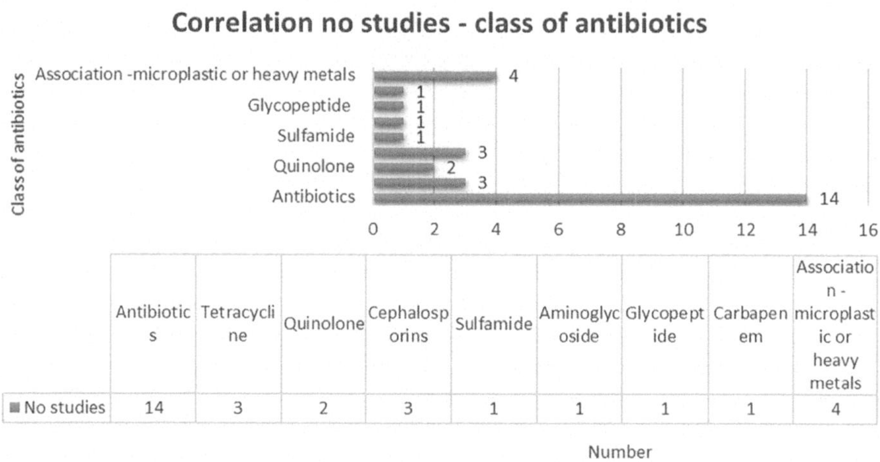

Fig. 7 Correlation between number of studies and class of antibiotics

Regarding intervention methods for neutralization of pharmacological contaminants in the form of antibiotics in wastewater, the number of chapters identified in the search is small, only 4 (6.34%). The small number of articles on wastewater treatment methods cannot be interpreted as a lack of concern of the scientific world; rather, it is about a lack of standardization of methods for determining and reporting data and technical means of chemical and biological analysis of wastewater (WWTP), associated with still limited means of intervention. Recent studies discuss the role of the microbiota in modern WWTPs and the exploitation of its ecological role by deciphering food web interactions at this level [20].

Data are not yet available for all affected areas of the globe. In Romania we have identified only two published studies on this topic [21, 22].

5 Conclusions

We carried out a systematic review by identifying some studies related to the action of antibiotics on aquatic ecosystems by searching the data in the PubMed database.

We used as a search key the addition of some terms relevant to the chosen subject, namely, "contaminants—aquatic ecosystems—antibiotics—microbiome". The time period under analysis was between 2011 and 2024. We identified and analyzed 63 studies.

The data obtained from the search process in the mentioned database were statistically analyzed with the help of Microsoft Excell application using the Data Analysis function.

The search was conducted independently by the two investigators and was followed by data matching to limit bias.

The presence of pharmacological contaminants, represented by antibiotics, in freshwater and saltwater aquatic ecosystems remains a problem of growing public health concern. A cause for concern is the additive effect of contaminants from different classes that mutually potentiate each other's effect, as well as the persistence in the form of biologically active metabolites, some of them with masked toxic effect, which intervene in the biota of aquatic ecosystems.

We quantified the extent of the scientific world's concern about this phenomenon in an indirect way, namely, by analyzing the reporting of results on it and we set as a benchmark, the pandemic period, a period with high consumption of antibiotics, thus realizing a temporal distribution of studies identified in the literature.

There is still a small number of studies carried out and published in specialized literature. There was high heterogeneity due to multiple factors, such as interest in certain variables involved in the phenomenon, methodological diversity, differences between studies, differences between outcome assessments.

The most studied aspect was the involvement of the microbiome in the phenomenon of microbial resistance, which was the subject that benefited from the largest number of studies, followed by the identification of the classes of antibiotics most commonly found in wastewater. We have seen an increase in interest since the pandemic period, and that interest is still there.

The analysis reported in this chapter can be exploited to identify and complete the missing links in the mechanism of antibiotic resistance by informing the competent authorities, while at the same time constituting an easy tool to be used to raise public awareness about the risks of unjustified antibiotic consumption.

The limits of the research process was due imited access to information (in this case by accessing a single database with free access), language fluencys, time constraints, methodological limitations, the high heterogeneity due to multiple variables, the methodological diversity, the differences between the studies, the differences between the evaluations of the results, limitations of the review process, limitation to studies in English, difficulties in the selection process.

It is important in the future to involve interdisciplinary teams (doctors, biologists, geologists, biochemists, engineers, geneticists) who, through expertise in different fields, can make correlations in accordance with the complexity of this phenomenon and carry out serial analyzes, in dynamics, regarding the evolution and ways of counteracting this anthropogenic phenomenon with consequences on its future.

Acknowledgements There was no funding for this study and the authors declare no conflict of interest.

References

1. Sosa-Hernández, J.E., Rodas-Zuluaga, L.I., López-Pacheco, I., et al.: Sources of antibiotics pollutants in the aquatic environment under SARS-CoV-2 pandemic situation. Case Stud. Chem. Environ. Eng. **4**, 100127 (2021)
2. Pinto, I., Simões, M, Gomes I.B.: An overview of the impact of pharmaceuticals on aquatic microbial communities antibiotics (Basel) (2022)
3. Murray, C., Ikuta, K.S., Sharara, F., et al.: Global burden of bacterial antimicrobial resistance in 2019: a systematic analysis. The LANCET **399**(10325), 629–655 (2022)
4. Wang, B., Zhang, S.Q., Dong, J.L., Li, Y., Jin, Y.X., Xiao, H.W., Wang, H.C., Fan, S.J., Cui, M.: Ambient temperature structures the gut microbiota of zebrafish to impact the response to radioactive pollution. Environ. Pollut.Pollut. **15**, 293 (2022)
5. Ohore, O.E., Wei, Y., Wang, Y., Nwankwegu A.S., Wang, Z.: Tracking the influence of antibiotics, antibiotic resistomes, and salinity gradient in modulating microbial community assemblage of surface water and the ecological consequences. Chemosphere **305** (2022)
6. Yu, Z., Zhang, L., Huang, Q., Dong, .S, Wang, X., Yan, C.: Combined effects of micro-/ nano-plastics and oxytetracycline on the intestinal histopathology and microbiome in zebrafish (Dan.io rerio). Sci. Total Environ. **15**, 843 (2022)
7. Wang, Z., Li, Q, Huang, H., Liu, J., Wang, J., Chen, Y., Huang, S., Luo, X., Zheng, Z.: Distribution and potential ecological risks of microplastics in Zhushan Bay, China. Chemosphere **335** (2023)
8. Liang, J., Li, C., Mo, J., Iwata, H., Rehman, F., Song, J., Guo, J.: Metatranscriptomic profiles reveal the biotransformation potential of azithromycin in river periphyton. Water Res. **1**, 251 (2024)
9. Yu, Q., Yang, J., Su, W., Li, T., Feng, T., Li, H.: Heavy metals and microbiome are negligible drivers than mobile genetic elements in determining pchapter-attached and free-living resistomes in the Yellow River. J. Hazard. Mater. **424** (2022)
10. Bhattacharyya, A., Haldar, A., Bhattacharyya, M., Ghosh, A.: Anthropogenic influence shapes the distribution of antibiotic resistant bacteria (ARB) in the sediment of Sundarban estuary in India. Sci. Total. Environ. **647**, 1626–1639 (2019)
11. Wanyan, R., Pan, M., Mai, Z., Xiong, X., Wang, S., Han, Q., Yu, Q., Wang, G., Wu, S., L,i H.: Fate of high-risk antibiotic resistance genes in large-scale aquaculture sediments: geographical differentiation and corresponding drivers. Sci. Total Environ. **905** (2023)
12. Curis, C., Rosca, R.O., Moroianu, L.A., et al.: Pharmacological contaminants in aquatic ecosystems. FARMACIA **72**(1), 44–59 (2024)
13. https://www.merriam-webster.com/dictionary/biota. Accessed 08 July 2024
14. Wang, Y., Zhai, J., Tang, B., Dong, Y., Sun, S., He, S., Zhao, W., Lancuo, Z., Jia, Q., Wang, W.: Metagenomic comparison of gut communities between wild and captive Himalayan griffons. Front. Vet. Sci. **11** (2024)
15. Huang, G., Qu, Q., Wang, M., Huang, M., Zhou, W., Wei, F.: Global landscape of gut microbiome diversity and antibiotic resistomes across vertebrates. Sci. Total Environ. **838**(2) (2022)

16. Sehnal, L., Brammer-Robbins, E., Wormington, A.M., Ludek, B., et al.: Microbiome composition and function in aquatic vertebrates: small organisms making big impacts on aquatic animal health. Front. Microbiol. **12** (2021)
17. Zhang, H., Xing, D., Wu, Y., Jin, R., et al., Editorial: Microbial ecology and function of the aquatic systems, Front Microbiol. (2022)
18. Angoshtari, R., Scribner, K.T., Marsh, L.T.: The impact of primary colonizers on the community composition of river biofilm. PLoS One **18**(11) (2023)
19. Alexander, T.J, Vonlanthen, P., Seehausen, O.: Does eutrophication-driven evolution change aquatic ecosystems? Philos. Trans. R. Soc. Lond. B Biol. Sci. **372**(1712) (2017)
20. Zhang, L., Huang, X., Zhou, J., Ju, F.: Active predation, phylogenetic diversity, and global prevalence of myxobacteria in wastewater treatment plants, pp. 671–681 (2023)
21. Voinea, C.I., Alistar, F.C., Banciu, A., et al.: Snapshot of the pollution-driven metabolic and microbiota changes in Carassius gibelio from Bucharest leisure lakes. Sci. Total. Environ. **1**, 884 (2023)
22. Alistar, C.F, Nica, I.C., Nita-Lazar M., et al.: Antioxidative defense and gut microbial changes under pollution stress in Carassius gibelio from Bucharest lakes. Int. J. Environ. Res. Public Health **19**(12) (2022)

Assessment of Patients Discharged from Hospitals

Sofia Shlyafer and Irina Shikina

Abstract A comprehensive analysis was conducted to examine the distribution of adult patients discharged from hospital, the incidence of hospitalised patients and the average length of stay of patients by disease class and nosology for the period between 2016 and 2020. This study considers and evaluates statistical data pertaining to adult hospitalised patients over a five-year period (2016–2020). In this study, statistical and analytical methods have been employed. The findings revealed a 15.4% decline in the number of patients discharged from hospital after the age of 18. Additionally, the incidence rate of hospitalised patients decreased from 205.5 to 175.8 per 1000 people, while the bed stay decreased from 11.7 to 11.0 days. There is a slight decrease in the incidence of hospitalised patients across all disease classes. In the final year of observation (2020), a shift in the structure of the incidence of hospitalised patients by class of disease was noted, resulting from the compelled re-classification of multi-disciplinary hospitals into the infectious category due to the impact of the COVID-19 pandemic.

Keywords Health patients · Hospital · Diseases classes · Morbidity hospitalized in hospitals · Treatment duration on average

1 Introduction

All health systems of the countries of the world solve the following tasks: solving national security issues, increasing the birth rate, educating children, developing social projects, economic development, financial stability, improving the quality and standard of living of their citizens or population's individual groups.

The adult population is the main group of users of health services that hospitals provide [1]. It has become possible to conduct a scientific assessment of the incidence of all adults hospitalized in hospitals or only those who work in Russia since 2010. It is possible to assess not only individual, but also in separate groups. The incidence

S. Shlyafer · I. Shikina (✉)
Russian Institute of Health, Moscow, Russia
e-mail: shikina_irina@mail.ru

T. Antipova (ed.), *Digital Technology Platforms and Deployment*, Information Systems Engineering and Management 36, https://doi.org/10.1007/978-3-031-86547-3_5

rate of hospitalized in hospitals helps to understand which pathology in patients leads to the need for hospitalization in a hospital. This indicator can demonstrate a group of diseases that lead to chronic processes, indicate the lack of timely detection and treatment of the disease on an outpatient basis [1–5].

At the beginning of 2021, 146 million 171 thousand 015 people were registered in our country, of which 115 million 787 thousand 674 people are adults over 18 years old, which is 79.2% of the total number of citizens living. From 2016 to 2021, the total population in the country older than 18 years old decreased by 1.5% (from 117 million 530 thousand 242 to 115 million 787 thousand 674 people), therefore, the number of adults as a whole decreased by 1% (from 80.2 to 79.2%) [6, 7].

2 Data and Methodology

The main source for this study is the mandatory reporting form No. 14 in digital format provided by federal medical organizations to the Ministry of Health. According to the objective of this study, the selected indicators for 5 years (from 2016 to 2020) were studied [8–10].

The incidence of hospitalised patients is defined as the frequency of all illnesses observed in those who were treated or rehabilitated in hospital during the calendar year [11–15]. The assessment of patients discharged from hospital over 18 years of age was given on the basis of calculations performed using extensive and intensive indicators.

3 Results

20 million 403 thousand 079 patients over the age of 18 were discharged from hospitals in Russia in 2020. Consequently, this is 83.7% of the total number of patients who were treated in hospitals in the current calendar year (80.2% in 2016).

18.7% of adults over 18 years old were treated and then discharged from the hospital for pathology of the circulatory system. 13.1% of women of fertile age were treated in hospitals in connection with pregnancy, childbirth and the puerperium. 9.9% of adult patients have suffered a severe COVID-19 infection, and their condition required hospitalization. A similar number of patients 9.9% were treated for neoplasms. 8.0% were hospitalized for diseases of the digestive system.

7.7% of patients were treated in hospital for urinary tract diseases. 6.9% of the condition of adult patients with diseases of the respiratory system required hospitalization. Thus, all these listed six classes of diseases brought 74.15% of cases to the total amount of discharged patients from hospitals (see Fig. 1).

Among patients discharged from hospitals with diseases of the circulatory system, the following were found, in descending order 32.8%—ischaemic heart disease, 28.1%—cerebrovascular disease, 14.2%—conditions associated with high blood

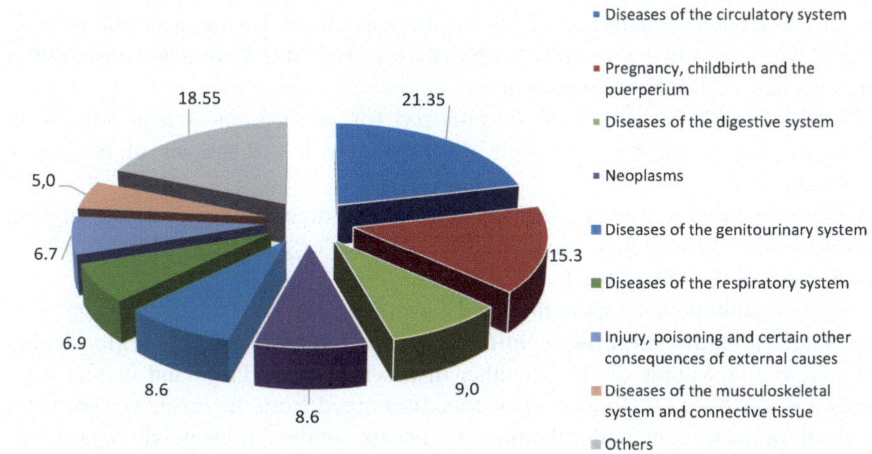

Distribution of adult patients discharged from hospitals by disease classes in Russia in 2016 (share in %)

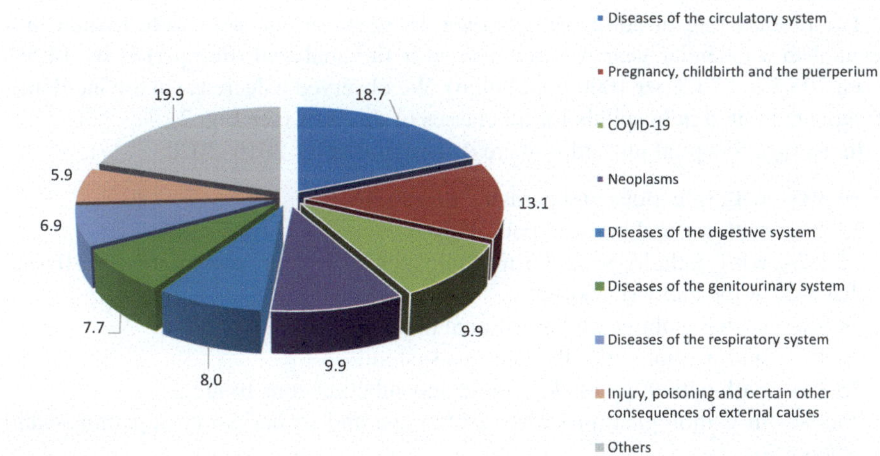

Distribution of adult patients discharged from hospitals by disease classes in Russia in 2020 (share in %).

Fig. 1 Distribution of adult patients discharged from hospitals by disease classes in Russia in 2016 and 2020 (share in %)

pressure. The latter patients had 12.5% hypertensive heart disease; a similar number (12.5%) were identified with other heart diseases. Atrial fibrillation and atrial flutter were also detected in 5.9% of patients.

79.9% of adult patients were hospitalized for verification in a hospital with a diagnosis of neoplasms, and were discharged with a diagnosis of malignant neoplasms.

Of the patients who were treated in the hospital for digestive diseases, it turned out that every fifth (17.9%) had diseases of the pancreas or gallbladder disease, 17.7% had diseases of the biliary tract.

19.9% of adult patients spent time in the hospital due to diseases of the genitourinary system, after discharge they continued to treat urolithiasis. In 12.2%, glomerular, tubulointerstitial kidney disease or other diseases of the kidneys and ureters were detected in the hospital. 11.8% of patients discharged from the hospital were there as a result of the treatment of inflammatory diseases of the female pelvic organs, etc.

Pneumonia, which belongs to the respiratory disease class, was treated in almost every second discharged adult patient in the hospital.

Throughout our study, a decrease in adult patients discharged from hospitals was recorded by a total of 15.4% (from 24 million 125 thousand 970 to 20 million 403 thousand 079).

The disease rate of adult patients over 18 years of age who were treated in a hospital in a calendar year was reduced over the analyzed time period by 14.5% (from 205.5 to 175.8 per 1000 population). We observed a decrease in the incidence of patients treated in hospitals for all classes of diseases (see Fig. 2).

In average, hospital morbidity were demonstrated for 2016–2020:

39.3%—with infectious and parasitic diseases;
33.3%—with ear and mastoid pathology;
31.1%—with pathology of the musculoskeletal system and connective tissue, diseases of the endocrine system,
28.9%—with metabolic and nutritional disorders,
26.4%—with pathology of the nervous system;
26.3%—with pathology of skin tissue and subcutaneous tissue,
26.2%—in pathological processes of the eye and its accessory apparatus—and others (Fig. 2).

According to data of Fig. 2 we can see a change in the structure of the incidence of hospitalised patients by disease class during the COVID-19 pandemic in 2020. It was the result of the re-profiling of specialist hospital departments such as therapy, cardiology, etc. into infectious disease departments.

Per 1000 adult patients treated in the hospital, the highest incidence rates of hospitalized patients were shown from decrease to increase due to injury, poisoning and certain other consequences of external causes (10.3); pathology of the respiratory system (12.2) and genitourinary system (13.5); digestive pathologies (14.0); cases of new coronavirus infection COVID-19 (17.4); neoplasms (17.5), pathology of the circulatory system (32.9), etc. (see Fig. 2).

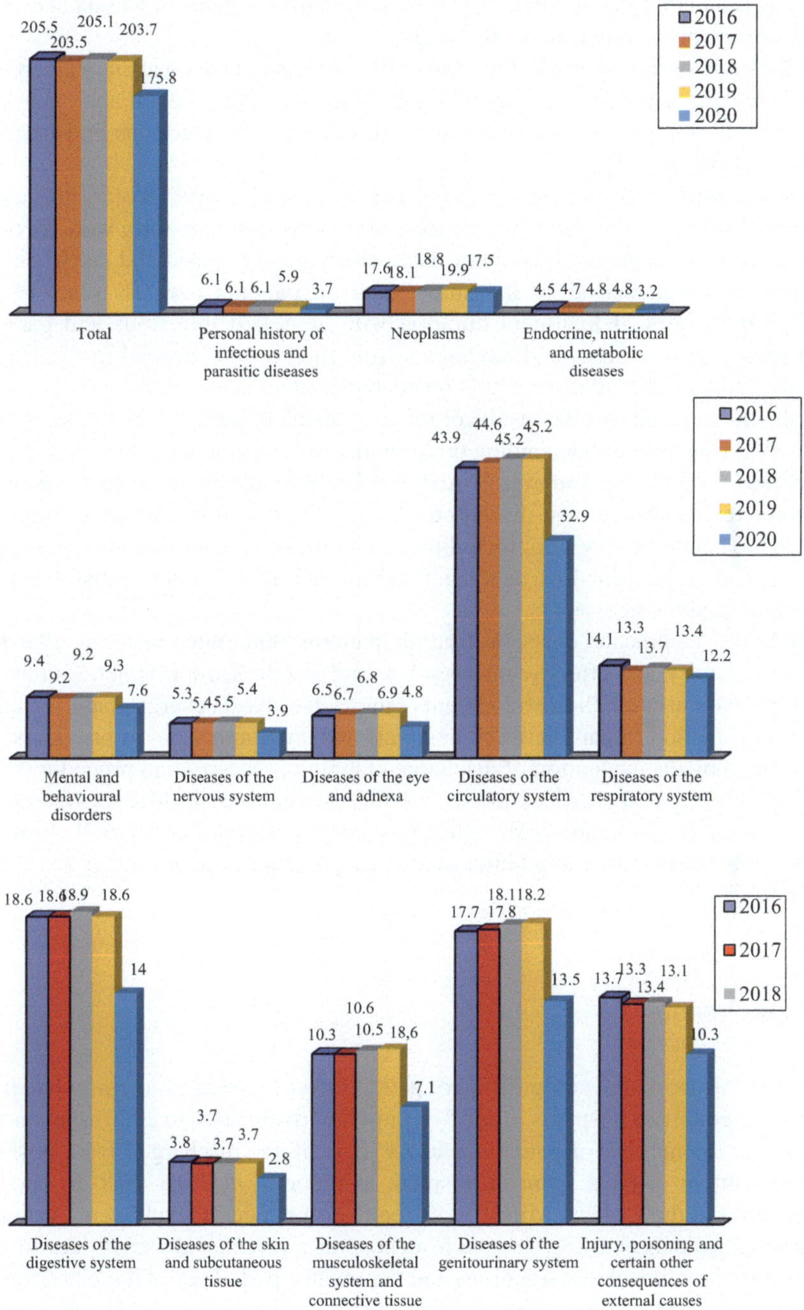

Fig. 2 Hospitalisation morbidity rate for adult patients

It was found that the duration of treatment for adult patients in Russia decreased by an average of 0.7 days during the study period.

In the study of the duration of treatment in a stationary bed, on average, increased rates were demonstrated among pathologies associated with mental and behavioral disorders (42.0 days), in cases of pathologies caused by infections and parasites (34.3 days) (see Fig. 3).

Over the entire study period, according to data on patients over 18 years of age with identified infectious and parasitic diseases, deviations from the norm were recorded in the ongoing clinical and laboratory data, which are not considered elsewhere.

Over the entire period of the study, data on patients over 18 years of age revealed high rates of treatment duration with identified infectious and parasitic diseases, symptoms, signs and deviations from the norm, identified in the clinical and laboratory studies that were not considered in other sections.

It should be noted that the results of the assessment of patients on discharge from hospital may be different depending on a number of potential outcomes. Assessment of medication adherence and overall quality of life after discharge can provide insight into the effectiveness of post-discharge care and interventions. Areas for improvement in the health care system, including communication, coordination and support services, can be identified by analysing readmission rates, patient satisfaction and patterns of health care use.

Assessing healthcare costs, utilisation patterns and outcomes can also help providers identify cost-effective interventions and practices, improving resource allocation and patient care. The development of more effective discharge plans, including better coordination of care between hospitals and community-based providers, can be informed by the evaluation of discharge planning processes and procedures.

By identifying best practices and successful interventions, health care systems can optimise their services and improve the experience of patients, which will ultimately lead to better outcomes and a better quality of life. But these are issues for further research.

4 Conclusion

As part of this work, the composition of patients over 18 years of age who had treatment in the country's hospitals over a 5-year period (from 2016 to 2020) was studied. All data on the number of patients over 18 years of age discharged from hospitals and the number of days spent in hospital, as reported in 2016–2020 by medical organization to the Russian Ministry of Health at the federal level, were examined. Discharged patients over 18 years old from hospitals were ill and were forced to go to the hospital for clinical investigations and treatments: pathology of the blood system and digestive organs, genitourinary system, respiratory organs; identified neoplasms, complications during pregnancy, childbirth and the puerperium, other.

We found out that the incidence of patients who were treated in hospitals decreased slightly in all classes of diseases. Over the period of study, there was observed to

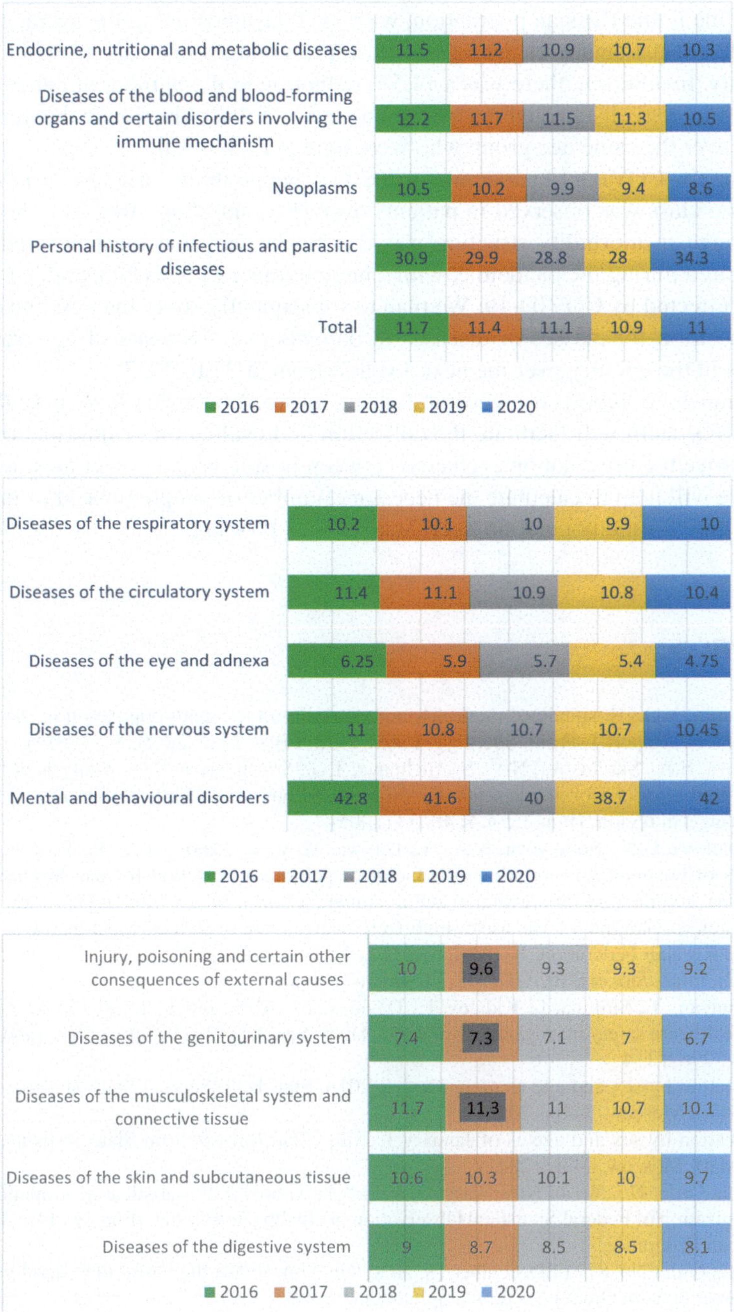

Fig. 3 Average hospital bed days for adult patients by disease class in Russia (2016–2020)

be a decline in the Russian population, with 11.7 days becoming the average length of stay for in-patients over the age of 18, representing a decrease from 11.0 days previously. In addition, there was a 14.5% reduction in the number of patients aged over 18 who were discharged from hospital, and a 15.4% reduction in the number of patients over the same age group who were treated in hospital.

During the COVID-19 pandemic in 2020, a change in the structure of morbidity by disease class was observed in patients over 18 years of age treated in hospitals. This change in morbidity structure was due to the fact that most hospitals were reconfigured during the pandemic, with a large number of beds being added to treat patients infected by COVID-19. We plan to subsequently study the possible change in composition for classes of diseases in patients over 18 years of age treated in hospitals in the country over the next 3 years (from 2021 to 2023).

The transfer of reports from hospitals using information technology in the Russian healthcare system will facilitate the collection and evaluation of information. This will improve the information exchange between hospitals and polyclinics, which in the future will help to calculate the necessary number of people in need of inpatient medical care for the population over the age of 18 [16–21].

References

1. Shchepin, V.O.: Hospitalized morbidity and mortality in the adult population of the Russian Federation. Report 1. Probl. Soc. Hyg. Public Health Hist. Med. **22**(2), 3–8 (2014)
2. Pavlova, E.A., Skrebtsova, N.V., Svetlichnaya, T.G., Sovershaeva, S.L.: Analysis of hospitalized morbidity of working-age population living in environmentally disadvantaged areas of Arkhangelsk region. Hum. Ecol. **7**, 48–53 (2005)
3. Mikhailova, U.N., Solovyova, N.A., Varlamova, M.A., Kurtanov, H.A., Pavlova, N.I.: Indicators of hospitalized morbidity as a factor of pathology selection for the development of personalized methods of prevention and treatment. Yakutsk Med. J. **4**(64), 106–108 (2018)
4. Shlyafer, S., Shikina, I.: The use of high-tech medical care in patients older than working age in the Russian Federation. In: Antipova, T. (ed.) DSIC 2021, LNNS, vol. 381, pp. 379–391 (2022). https://doi.org/10.1007/978-3-030-93677-8_33
5. Voskanyan, Y., Shikina, I., Kidalov, F., Davidov, D., Abrosimova, T.: Risk management in the healthcare safety management system. J. Digit. Sci. **3**(1) (2021). https://doi.org/10.33847/2686-8296.3.1_4
6. Population by sex and age as of January 1, 2016. Stat. Bull. Federal State Statistics Service (Rosstat), Moscow, 344p. (2016)
7. Population by sex and age as of January 1, 2021. Stat. Bull. Federal State Statistics Service (Rosstat), Moscow, 443p. (2021)
8. Rosstat Order No. 591 of November 27, 2015 "On Approval of Statistical Instruments for the Organization of Federal Statistical Observation in Health Care by the Ministry of Health of the Russian Federation"
9. Rosstat Order No. 812 of December 18, 2020 'On Approval of the Forms of Federal Statistical Observation with Guidelines for Completion to Organize Federal Statistical Observation in the Field of Health Protection by the Ministry of Health of the Russian Federation'
10. Rosstat Order No. 679 of November 19, 2018 'On Approval of the Statistical Tools for the Organization of Federal Statistical Observation in the Field of Health Protection by the Ministry of Health of the Russian Federation'

11. Kucherenko, V.Z.: Application of statistical analysis methods for the study of public health and health care: textbook for practical training. GEOTAR-Media, 256p. (2007)
12. Shlyafer, S.I., Ivanova, M.A.: Hospitalized morbidity of the population over working age of the Russian Federation. Adv. Gerontol. **26**(5), 690–694 (2016)
13. Shlyafer, S.I., Shikina, I.B.: Evaluation of indicators characterizing inpatient surgical care delivery to older patients in the Russian Federation. Soc. Aspects Popul. Health [Serial Online] **67**(5), 5 (2021). https://doi.org/10.21045/2071-5021-2021-67-5-5
14. Shlyafer, S.I.: The hospital medical care support of individuals older than able-bodied age in the Russian Federation. Probl. Soc. Hyg. Public Health Hist. Med. **29**(2):238—244 (2021). https://doi.org/10.32687/0869-866X-2021-29-2-238-244
15. Shlyafer, S., Shikina, I.: High-tech heart surgery in elderly patients in Russia. ICS. J. Digit. Art & Humanit. **4**(1) (2023). https://doi.org/10.33847/2712-8148.4.1_6, ISSN 2712-8148
16. Voskanyan, Y., Shikina, I., Kidalov, F., Davidov, D.: Medical care safety - problems and perspectives. In: Antipova, T. (eds.) Integrated Science in Digital Age. ICIS 2019. Lecture Notes in Networks and Systems, vol. 78. Springer, Cham (2020). https://doi.org/10.1007/978-3-030-22493-6_26
17. Kobyakova, O.S., Golubev, N.A., Polikarpov, A.V., Sidorov, K.V.: Data collection system automation model on health system indicators based on primary data. Russ. J. Prevent. Med. **26**(1):11–16 (2023). https://doi.org/10.17116/profmed20232601111
18. Zadvornaya, O.L., Voskanyan, Y.E., Shikina, I.B., Borisov, K.N.: Socio-economic aspects of medical errors and their consequences in medical organizations. MIR (Mod. Innov. Res.) **10**(1), 99–113 (2019). https://doi.org/10.18184/2079-4665.2019.10.1.99-113
19. Voskanyan, Y., Shikina, I., Kidalov, F., Kurdyukov, S.: Management of the flow of patients in a modern medical organization. In: Antipova, T. (ed.) DSIC 2021, LNNS, vol. 381, pp. 456–466 (2022). https://doi.org/10.1007/978-3-030-93677-8_40
20. Voskanyan, Y., Shikina, I., Kidalov, F., Shlyafer, S.: A strategic model of a medical organization in the context of cognitive transformation of the world. A strategic model of a medical organization in the context of cognitive transformation of the world. In: Antipova, T. (eds,) Comprehensible Science. ICCS 2021. Lecture Notes in Networks and Systems, vol. 315. Springer, Cham (2022). https://doi.org/10.1007/978-3-030-85799-8_37
21. Shlyafer S., Shikina I.: Analysis of morbidity in hospitals. JDAH **4**(2), 31–36 (2023). https://doi.org/10.33847/2712-8149.4.2_4

Comparative Analysis of the Urban and Rural Populations' Major Demographic Characteristics in the Russian Federation and the Republic of Uzbekistan

Marina Erugina, Galina Sazanova, Irina Krom, Elena Dolgova, and Mariya Eremina

Abstract The purpose of the study is to conduct a comparative analysis of the demographic characteristics of the urban and rural population of the Russian Federation and the Republic of Uzbekistan. Within the framework of the study we analysed the demographic characteristics of the urban and rural population of Russia and Uzbekistan for the period from 2010 to 2020. Statistical, analytical and mathematical research methods were used for the analysis. The statistical significance of the difference in relative values was calculated with Student's coefficient at $p < 0.05$ and $t > 2$. In the period from 2010 to 2021 we observed an increase in the population of the Russian Federation and the Republic of Uzbekistan. The study revealed a statistically significant decrease in the birth rate over the 10-year period, as well as an increase in the mortality rate and natural population decline in Russia. In Uzbekistan, there was a significant increase in both the birth rate and the natural population growth rate over the period studied. Until 2019 there was a statistically significant decrease in the mortality rate. In 2020, the mortality rate increased by 10.5%. The decline in fertility and mortality rates is more pronounced in Russia's rural areas, while the increase in mortality rates is more pronounced among the urban population. In Uzbekistan there was a statistically significant increase in birth rates in both urban and rural areas. The characteristics of the population of different countries determine the specifics of the state of public health of their populations, the organisation of preventive and diagnostic medical care; they require the timely development of preventive measures aimed at improving the organisation of preventive medical care in conditions of new epidemiological and other risks, especially in rural areas.

Keywords Rural and urban populations of the Russian Federation and the Republic of Uzbekistan · Demographic characteristics

---

M. Erugina · G. Sazanova (✉) · I. Krom · E. Dolgova · M. Eremina
Saratov State Medical University Named After V. I. Razumovsky, Saratov, Russia
e-mail: sazanovagu@yandex.ru

T. Antipova (ed.), *Digital Technology Platforms and Deployment*, Information Systems Engineering and Management 36, https://doi.org/10.1007/978-3-031-86547-3_6

1 Introduction

In the period from 2000 to 2018, the population of the Central Asian countries grew by an average of 40%, i.e. 60% in Tajikistan and 42% in Uzbekistan, both in urban and rural areas [1]. New trends and patterns of natural population movement in developing countries are the result of the population explosion. The proportion of young people in Asian, African and Latin American countries is significantly higher than in European or North American countries [2].

According to the estimates of Yakovets et al., the fertility rate of the urban and rural population in Russia will decrease by 2033 due to a decrease in the number of women of fertile age and a shift in the age distribution of women towards older ages. The ageing of the population will continue—the number of people of retirement age will increase. Over a fifteen-year period, the working age population will decrease by 10% and the older working age population will increase by 14% by 2030 [3].

Most countries in the world face problems in providing health care to rural populations. In Russia and Uzbekistan there are differences in the accessibility of medical care for the urban and rural population, i.e. there is a lack of quality medical care in rural areas due to insufficient funding and continuous "optimisation" of rural medical organisations. The provision of medical services to the rural population is strongly influenced by the specific social and economic conditions of rural areas and the structural characteristics of rural health care [1–3].

The health of the rural population is an important component of the development of rural areas, as these territories are characterised by a record decline in population and a worsening demographic crisis of the rural population [4]. In Russia, for example, one of the reasons for this is "natural population decline associated with the mortality rate exceeding the birth rate, negative migration processes and the age structure of the population" [5].

The purpose of the study is to carry out comparative analysis of the demographic characteristics of the urban and rural population in the Russian Federation and the Republic of Uzbekistan.

2 Materials and Methods

To obtain information regarding health status tendencies in the Russian and Uzbek rural population in the ten-year retrospective, the study has covered demographic characteristics of the urban and rural populations, reproduction coefficients, mortality rate, birth rate, natural population growth rate over the period of time from 2011 until 2020 in the populations of the Russian Federation and the Republic of Uzbekistan.

Among sources of statistical data used for the study are such official sources as Healthcare in Russia, Russia and World Countries, electronic collections of materials Russia in Figures, publications and operational data from the websites publishing official state statistics for the Russian Federation and the Republic of Uzbekistan over 10 years [6–9].

Statistical, analytical and mathematical research methods were used for analysis. The statistical significance of the difference in relative values was calculated with Student's coefficient at $p < 0.05$ and $t > 2$ and Microsoft Excel 2019 [10–13]. Data was presented in the $P \pm m$ format.

3 Results

The population of Russia as of 31.12.2020 was 146.17 million people, of which 109.25 million people belonged to the urban segment of the population (74.8%) and 36.92 million people formed the rural segment of the population (25.2%). The population of Russia in the period from 2010 to 2021 will increase by 3 286 000 people (by 2.3%), with the urban population increasing by 3.6% (3 831 000 people) and the rural population decreasing by 1.5% (525 000 people). In 2020, the rural population decreased by 0.7% compared with 2019, while the urban population decreased by 0.3%. The population of Uzbekistan on 31 December 2020 was 34.6 million people, of whom 17.5 million (50.7%) lived in urban areas and 17.1 million (49.3%) in rural areas (Table 1). During the period from 2010 to 2020, the population of the Republic will increase by 5.4 million people (15.7%), i.e. the urban population will increase by 2.6 million people (14.9%) and the rural population will increase by 2.8 million people (16.6%) (Table 1).

Females predominate in the structure of the Russian population, as of 1 January 2021 there will be 10.5 million more women than men in the Russian Federation.

The ratio of the main age groups in the total population of Russia is characteristic of the regressive type of population health (the proportion of people over working age is 35% higher than the proportion of people under working age).

Table 1 Dynamics of urban and rural population size in the Russian Federation and the Republic of Uzbekistan over the period of time from 2010 until 2020 (in thousands of people, as of end of the year)

Population size/year	2010	2015	2018	2019	2020
Russian Federation*					
Total population	142,865	146,545	146,781	146,749	146,171
Urban population	105,421	108,658	109,454	109,563	109,252
Rural population	37,444	37,887	37,327	37,186	36,919
Republic of Uzbekistan*					
Total population	29,123	31,575	33,256	33,905	34,558
Urban population	14,897	15,964	16,807	17,144	17,510
Rural population	14,226	15,611	16,449	16,761	17,048

* Russia and the world countries. 2020. Data book/Rosstat—Moscow, 2020—387 pages

When analysing the age structure of the Russian population, we paid attention to the ratio of males to females in the three oldest age groups, each of which consists of less than 1 million people (the group of 100-year-olds and over—26.7 thousand people, 95–99-year-olds—124.2 thousand people, 90–94-year-olds—612.1 thousand people). Within the structure of the above-mentioned age groups, the share of women is 76.8%, while the share of men is 23.2%.

Among the age groups of the working-age population, the smallest group is that of the 20–24 year olds (people born in 2000–2004), which consists of 6.83 million people. This group is followed by the second smallest group, that of the 15–19 year olds (people born in 2005–2010), with 7.22 million people.

The male population predominates in all 5-year age groups from 0 to 34 years, but from the age group 35–39 years, the female population predominates and grows progressively.

The most numerous groups in the population structure of the Russian Federation (more than 12 million people) are the working age groups 30–34 and 35–39, followed by the 40–44, 55–59 and 60–64 age groups, each comprising 10–12 million people. When analysing the number of children, we have paid attention to the small number of 5-year age groups considered, none of which reaches 10 million people, which is an unfavourable prognostic factor for the country's economy in general and the national health system in particular (Table 2).

Men predominate in the structure of the Uzbek population, as of 1 January 2021 there are 0.2 million more men than women in the Republic of Uzbekistan.

The ratio of the main age groups in the total population is characteristic of the progressive health type of the population of the Republic of Uzbekistan (the proportion of people under working age is 18.8% higher than the proportion of people over working age). The largest age group of working age is represented by 25–20 year olds born between 1991 and 1995 (3.22 million people), followed by the second largest age group of 30–34 year olds born between 1986 and 1990 (3.08 million people).

The male population predominates in all 5 age groups from 0 to 43 years, starting from the age groups 35–39 years and 45–49 years the number of males and females within the groups is similar, and starting from the age group 50–54 years the number of females predominates and increases progressively. In the groups of 80–84 year olds, 85 year olds and over, the number of men again predominates.

The most numerous groups in the population structure of the Republic of Uzbekistan (over 3 million people) are groups of people of working age (from 0 to 14 years), which is a favourable prognostic factor for the country's economy in general and the national health system in particular.

The comparative analysis of the population reproduction rate in Russia for the period from 2010 to 2020 showed a statistically significant decrease in the birth rate over 10 years from 12.5 ± 0.009 per 1000 people in 2010 to 9.8 ± 0.008 per 1000 people in 2020; an increase in the mortality rate from 14.2 ± 0.009 per 1000 people in 2010 to 14.6 ± 0.009 per 1000 people in 2020; and a natural decrease in the population from 1.7 ± 0.003 to 4.8 ± 0.008 per 1000 people ($t > 2$). We observed a positive statistically significant trend in infant mortality, which decreased by 36%,

Table 2 Population size in the Russian Federation and the Republic of Uzbekistan according to gender and age groups as of 01.01.2021 (million people)

Age (in years)	Russia*			Uzbekistan*		
	Total	Males	Females	Total	Males	Females
0–4	8.33	4.28	4.05	3.75	1.97	1.80
5–9	9.43	4.85	4.58	3.34	1.73	1.61
10–14	8.15	4.18	3.97	3.08	1.58	1.49
15–19	7.22	3.69	3.53	2.57	1.31	1.25
20–24	6.83	3.48	3.35	2.76	1.41	1.35
25–29	9.01	4.62	4.39	3.22	1.64	1.58
30–34	12.53	6.32	6.21	3.08	1.56	1.52
35–39	12.07	5.99	6.08	2.62	1.31	1.31
40–44	10.81	5.24	5.57	2.15	1.07	1.08
45–49	9.94	4.74	5.20	1.91	0.96	0.96
50–54	8.87	4.15	4.72	1.62	0.78	0.84
55–59	10.15	4.54	5.61	1.53	0.73	0.80
60–64	10.20	4.29	5.91	1.23	0.58	0.65
65–69	8.40	3.23	5.17	0.76	0.35	0.41
70–74	5.89	2.08	3.81	0.47	0.21	0.26
75–79	2.92	0.88	2.04	0.25	0.11	0.14
80–84	5.76	1.46	4.30	0.69	0.60	0.09
85 and over	2.15	0.51	1.64	0.27	0.26	0.05
Total population	146.46	67.99	78.47	34.56	17.38	17.18
Under working age	27.39	14.06	13.33	10.17	5.28	4.9
Of working age	81.88	42.71	39.17	20.72	9.99	10.68
Over working age	36.90	11.08	25.82	3.67	2.11	1.6

* Russia and the world countries. 2020. Data book/Rosstat—Moscow, 2020—387 pages

from 7.5 ± 0.06 per 1000 population in 2010 to 4.5 ± 0.05 per 1000 population in 2020 ($t > 2$).

In Uzbekistan, a significant increase in the birth rate from 22.0 ± 0.02 cases per 1000 people in 2010 to 24.6 ± 0.03 cases per 1000 people in 2020 and a 13.3% increase in natural population growth from 17.2 ± 0.02 to 19.5 ± 0.02 per 1000 people were observed over the study period. Until 2019, there was a statistically significant decrease in the mortality rate (in 2010—4.8 ± 0.01, in 2019—4.6 ± 0.01 cases per 1000 inhabitants). In 2020, the mortality rate increased by 10.5% and reached 5.1 ± 0.01 cases per 1000 inhabitants, $t > 2$. A high level of infant mortality is noticeable in 2020 (9.4 ± 0.02 per 1000 inhabitants), although a statistically significant trend towards its decrease by 20%, $t > 2$ was observed during the study period (Table 3).

In Russia the decrease in births is more clearly observed in the rural areas: from 14.0 ± 0.02 per 1000 people in 2010 to 9.6 ± 0.02 per 1000 people in 2020, while

Table 3 Dynamics of major demographic characteristics of the populations of the Russian Federation and the Republic of Uzbekistan over the period of time from 2010 to 2020 (per 1000 people)

Characteristic/ years	2010*	2015*	2018*	2019*	2020*
Russian Federation**					
Birth rate	12.5 ± 0.009	13.3 ± 0.009	10.9 ± 0.008	10.1 ± 0.008	9.8 ± 0.008
Mortality	14.2 ± 0.009	13.0 ± 0.009	12.5 ± 0.009	12.3 ± 0.009	14.6 ± 0.009
Infant mortality	7.5 ± 0.06	6.5 ± 0.06	5.1 ± 0.05	4.9 ± 0.05	4.5 ± 0.05
Natural population growth (decrease)	−1.7 ± 0.003	0.3 ± 0.001	−1.6 ± 0.003	−2.2 ± 0.004	−4.8 ± 0.008
The Republic of Uzbekistan***					
Birth rate	22.0 ± 0.02	23.5 ± 0.03	23.3 ± 0.03	24.3 ± 0.03	24.6 ± 0.03
Mortality	4.8 ± 0.01	4.9 ± 0.01	4.7 ± 0.01	4.6 ± 0.01	5.1 ± 0.01
Infant mortality	11.7 ± 0.02	11.4 ± 0.02	9.9 ± 0.02	9.3 ± 0.02	9.4 ± 0.02
Natural population growth (decrease)	17.2 ± 0.02	18.6 ± 0.02	18.6 ± 0.01	19.7 ± 0.02	19.5 ± 0.02

* $p < 0.5$; $t > 2$
** The Demographic Yearbook of Russia, issues 2010–2020
*** Uzbekistan Healthcare https://www.stat.uz/ru

in the urban areas we observed a decrease in birth rates from 12.0 ± 0.01 per 1000 people in the year 2010 to 9.9 ± 0.009 per 1000 people in the year 2020 ($t > 2$). The highest birth rate in the Russian Federation was recorded in 2015 (13.3 ± 0.009 per 1000 people), among urban population it was recorded in 2015 as well (13.4 ± 0.01 people per 1000), among rural population it was observed in 2010 (14.0 ± 0.02 per 1000 people of the rural population).

The mortality rate in urban population had significantly increased from 13.5 ± 0.01 per 1000 people to 14.3 ± 0.009 per 1000 people in the year 2020, at the same time we observed statistically significant decrease in the rural mortality rate over this period from 16.1 ± 0.02 per 1000 people to 15.4 ± 0.02 per 1000 people. The lowest mortality rate in the Russian Federation was recorded in 2019 (12.3 ± 0.009 per 1000 people in the Russian Federation, i.e. 11.9 ± 0.01 per 1000 people from the urban population and 13.3 ± 0.02 per 1000 people from the rural population); rural population mortality rate is 12% higher than that of urban population.

The comparative analysis of urban and rural fertility rates in the Republic of Uzbekistan revealed a high fertility rate and a statistically significant increase in the period from 2010 to 2020, i.e. by 15.8% in the urban area (2010—19.6 ± 0.03 per 1000 people, 2020—22.7 ± 0.02 per 1000 people), by 8.5% in the rural area (2010—24.5 ± 0.03 per 1000 people, 2020—26.6 ± 0.03 per 1000 people).

Table 4 Dynamics of reproduction characteristics in the urban and rural population of the Russian Federation and the Republic of Uzbekistan over the period of time from 2010 until 2020 (per 1000 people)

Year	Russian Federation**			Republic of Uzbekistan***		
	Births*	Deaths*	Natural population growth*	Births*	Deaths*	Natural population growth*
Urban population						
2010	12.0 ± 0.01	13.5 ± 0.01	−1.5 ± 0.003	19.6 ± 0.03	5.1 ± 0.02	14.5 ± 0.02
2015	13.4 ± 0.01	12.6 ± 0.06	0.8 ± 0.002	20.5 ± 0.03	5.1 ± 0.01	15.4 ± 0.02
2018	11.0 ± 0.01	12.0 ± 0.01	−1.0 ± 0.003	21.2 ± 0.02	5.0 ± 0.01	16.2 ± 0.02
2019	10.2 ± 0.01	11.9 ± 0.01	−1.7 ± 0.003	22.3 ± 0.03	4.9 ± 0.01	17.4 ± 0.02
2020	9.9 ± 0.009	14.3 ± 0.009	−4.4 ± 0.009	22.7 ± 0.02	5.6 ± 0.01	17.1 ± 0.02
Rural population						
2010	14.0 ± 0.02	16.1 ± 0.02	−2.1 ± 0.007	24.5 ± 0.03	4.5 ± 0.01	20.0 ± 0.03
2015	12.8 ± 0.02	14.4 ± 0.02	−1.6 ± 0.006	26.5 ± 0.03	4.6 ± 0.01	22.1 ± 0.03
2018	10.7 ± 0.02	13.6 ± 0.02	−2.9 ± 0.009	25.5 ± 0.03	4.4 ± 0.01	21.1 ± 0.02
2019	9.8 ± 0.02	13.3 ± 0.02	−3.5 ± 0.01	26.3 ± 0.03	4.3 ± 0.01	22.0 ± 0.03
2020	9.6 ± 0.02	15.4 ± 0.02	−5.8 ± 0.01	26.6 ± 0.03	4.7 ± 0.01	21.9 ± 0.02

* $p < 0.5$; $t > 2$
** The Demographic Yearbook of Russia, issues 2010–2020
*** Uzbekistan Healthcare https://www.stat.uz/ru

We observed a low and statistically significant decrease in the total mortality of the urban population of Uzbekistan by 3.9% between 2010 and 2019 (2010—5.1 ± 0.02, 2019—4.9 ± 0.01 per 1000 people). However, in 2020 this parameter increases by 14.3%—up to 5.6 ± 0.09 per 1000 inhabitants ($t > 2$). We also observed a statistically significant decrease of 4.4% in the rural mortality rate between 2010 and 2019 (2010—4.5 ± 0.01, 2019—4.3 ± 0.01 per 1000 inhabitants). In 2020, the mortality rate will increase by 9.3% to 4.7 ± 0.01 per 1000 inhabitants (Table 4).

4 Discussion

The comparative analysis of urban and rural population dynamics in the Russian Federation over 10 years shows the continuing urbanisation of the population (the urban population in the Russian Federation increased by 2.3%, while the rural population decreased by 1.5%). In Uzbekistan, the urban and rural population grew by 14.9 and 16.6% respectively.

In Russia, structural imbalances were observed when comparing the number of age groups: those born between 2000 and 2010 were 43 and 59% less than those born between 1985 and 1995, respectively, which determined the predicted decline in the working age population of 10–15% in the fifteen-year perspective. In Uzbekistan, on the other hand, the projected increase in the working-age population was determined by people born between 2000 and 2010.

Characteristics of the state of public health of the population are largely determined by the type of health of the population in a given country, specific aspects of life and national characteristics of the peoples of these countries. Comparative analysis of the main statistical characteristics of public health of the urban and rural population of Russia shows unfavourable trends in fertility, mortality, natural population increase with positive dynamics in infant mortality rate over the last ten years. The comparative analysis of the public health of the urban and rural population of Uzbekistan shows a significant increase in the birth rate, natural population increase and a positive trend in the total mortality rate. However, despite the overall positive dynamics, infant mortality remains relatively high. In both countries, an increase in the population mortality rate was recorded at the end of 2020, most likely related to the COVID-19 pandemic, which affected mortality rates directly and indirectly through the use of force.

5 Conclusion

In order to achieve the objective of the study, i.e. to carry out a comparative analysis of the demographic characteristics of the rural and urban population of the Russian Federation and the Republic of Uzbekistan, we have studied the size of the urban and rural population, the reproduction rates of the population of the Russian Federation and the Republic of Uzbekistan, i.e. mortality, fertility, natural population increase for the period from 2011 to 2020. The ten-year data were taken from official sources of the Federal State Statistics Service of the Russian Federation and the Republic of Uzbekistan.

The characteristics of the population of Russia and Uzbekistan determine the characteristics of the state of public health, the organisation of preventive, diagnostic and curative medical care. The study predicts a general decrease in the Russian working age population over the next 10–15 years with a simultaneous increase in the need for medical care due to the demographic characteristics of the older age groups of the working population. This poses a serious challenge to the healthcare system in terms of ensuring the availability of medical care for the population and requires the timely development of preventive measures aimed at solving organisational and human resource problems in healthcare, especially in rural areas. In Uzbekistan, the increase in the birth rate exceeds the increase in the mortality rate; the population is growing, but there is a relatively high infant mortality rate, which requires the development of preventive programmes aimed at improving medical care for women and children.

An increase in mortality during the pandemic requires measures to improve the organisation of preventive health care in the face of new epidemiological and other risks.

Population characteristics of different countries determine peculiarities of planning the organisation of preventive, therapeutic and diagnostic medical care for the population and require constant monitoring.

Acknowledgements The work was carried out within the framework of the advanced scientific research project No. INTL-SARGMU-SAMGMI-2021-05 dated August 02, 2021 (No. NIOKTR 121090200094-8 dated September 1, 2021).

References

1. United Nations World Population Prospects 2019, vol. II: demographic profiles. United Nations, Department of Economic and Social Affairs, Population Division (2019). https://population.un.org/wpp/Publications/Files/WPP2019_Volume-II-Demographic-Profiles.pdf. Accessed 25 Jan 2022
2. Nureev, R.M.: Demographic explosion and its echo in the 21st century. In: The phenomenon of market economy: from origins to present day. Synthesis of digital technologies and innovative solutions: collection of scientific papers of the IX International Scientific and Practical Conference on Economics, dedicated to the memory of the famous scientist and major organizer of economic science in the South of Russia and Central Asia, Doctor of Economic Sciences, Professor A.F. Sidorova, Sochi, March 31 – 04, 2021. Electronic Publishing Technologies LLC, Maykop, pp. 278–288. EDN NVSFSD (2021)
3. Yakovets, T.Y., Golubkov, V.V.: Forecast of demographic situation in Russia until the year 2033. Econ. Math. Methods **54**(4), 71–87 (2018). https://doi.org/10.31857/S042473880003321-2. EDN YSQHNR
4. Suslin, S.A., Kashirin, A.K., Katin, A.A.: Medical demographic tendencies in the health state of the rural population]. Printed text. J. Sci. 'Health & Educ. Millenn. **19**(10), 260–263 (2017)
5. Kalininskaya, A.A., Muftahova, A.V., Bayanova, N.A., Sulkina, F.A.: Lifestyle and health status of the rural population living in different medical care accessibility zones. Printed text. Manager Healthc. **8**, 55–61 (2019)
6. LexUZ: Presidential decree of the republic of Uzbekistan dated December 7, 2018 No. UP-5590 'On complex measures for radical enhancement of health care system of the Republic of Uzbekistan' (Electronic Resource) (2018). https://lex.uz/docs/4096199. Accessed 16–28 Jan 2022
7. Merkulov, AA: Some demographic indicators of the commonwealth of independent states and prospects for creating conditions for population growth. Int. Coop. Eurasian States: Polit. Econ. Law **3**, 40–48 (2019)
8. Studennikova, N.S.: Health of rural residents as a determining factor in rural development. Printed text. Bull. Rural Dev. Soc. Policy **3**(15), 91–93 (2017)
9. Rosstat: Heatlhcare in Russia. Data Book. Mocsow, Rosstat, 171p. Printed text (2021)
10. Rosstat: The demographic yearbook of Russia, (Electronic Resource), issues 2010–2020 (2020). https://rosstat.gov.ru/folder/210/document/13207. Accessed 16–28 Jan 2022
11. UzStat: Healthcare in Uzbekistan, (Electronic Resource) (2022). https://www.stat.uz/ru. Accessed 16–28 Jan 2022
12. Rosstat: Russia and World Counties. Data Book. Moscow, 387p. Printed text (2020)
13. Paniotto, V.I.: Quantitative methods in sociological research. Paniotto V.I., Maximenko V.S.. Kyiv, 170p. (2003)

Russia and Uzbekistan Population Health: Comparative Results of Morbidity, Mortality and Disability Rate Research

Marina Erugina, Bakhrom Tuhtarov, Galina Sazanova, Bathtier Nizomov, and Elena Dolgova

Abstract The purpose of the study was to examine health trends in the Russian Federation and the Republic of Uzbekistan. The material used was statistical data presented on the official websites of WHO, the Interstate Statistical Committee of the CIS, the Federal State Statistics Service of the Russian Federation and the Republic of Uzbekistan for the period 2010–2020. Analytical and statistical research methods were applied. Reliability of differences between 2010 and 2020 indicators was assessed using Student's t-criterion with $t > 2$ and $p < 0.05$. Russia has a regressive type of population age structure, while the Republic of Uzbekistan has a progressive type. In the period from 2010 to 2020 in the Russian Federation there was a statistically reliable decrease in the birth rate by 12.8%, as well as an increase in mortality by 34% and a corresponding increase in negative values of the natural population growth rate by three times. In Uzbekistan over the same period there was a statistically significant increase in the birth and death rates (11% each) and a 14% increase in the positive values of the natural population growth rate. In the Russian Federation, the primary morbidity rate for respiratory and circulatory diseases increased. In Uzbekistan, since 2010 there has been an increase in primary morbidity in the classes of infectious and parasitic diseases, diseases of the circulatory system, respiratory system and digestive system. In the Russian Federation, there was a statistically significant decrease in the rate of primary disability, while in Uzbekistan this indicator increased. Thus, like the demographic situation in Russia and Uzbekistan—regressive and progressive—the health characteristics of the population are clear indicators of the effectiveness of national health systems.

Keywords Demographics · Morbidity · Disability

M. Erugina · G. Sazanova (✉) · E. Dolgova
Saratov State Medical University Named After V. I. Razumovsky, Saratov, Russia
e-mail: sazanovagu@yandex.ru

B. Tuhtarov · B. Nizomov
Samarkand State Medical Institute, Samarkand, Uzbekistan

T. Antipova (ed.), *Digital Technology Platforms and Deployment*, Information Systems Engineering and Management 36, https://doi.org/10.1007/978-3-031-86547-3_7

1 Introduction

In 2015, UN member states adopted Resolution 70/1 'Transforming our world: the 2030 Agenda for Sustainable Development', which includes 17 goals aimed at ending poverty, conserving the planet's resources and ensuring people's well-being. Health is central to the Sustainable Development Goals (SDGs). The SDGs are either directly related to health or indirectly aimed at improving health.

According to the World Health Organization (WHO), of the 55.4 million deaths in the world in recent years, 55% were caused by 10 leading causes of death, including, in descending order, cardiovascular, respiratory and neonatal conditions. The three categories of causes of death are "injuries, noncommunicable diseases and communicable diseases" [1].

The new agenda, based on the development goals formulated in the Millennium Declaration, is relevant to all countries and aims to improve equity in terms of meeting the needs of women, children, the poor and the disadvantaged. One of the indicators that characterise the SDGs is mortality caused by non-infectious diseases. These diseases are dominated by diseases of the circulatory system (CSDs), which account for 31.5% of total population mortality and about 45% of non-infectious disease mortality [2].

The highest CSD mortality rates are observed in Central Europe, Eastern Europe and the CIS. In 2015, 600 people from the CIS and 400 people from Central and Eastern Europe (per 100 000 population) died from CSD [3, 4]. The burden of CSD has a serious impact on national economies, i.e. the economic losses due to CSD in the European Union (EU) amount to approximately 210 billion euros per year, which is associated with disability, mortality, high treatment costs, low effectiveness of primary and secondary prevention measures [5]. Russia (RF) has adopted national health and demographic projects aimed at reducing mortality and increasing life expectancy of the population, including the federal projects "Combating cardiovascular diseases", "Combating cancer", "Strengthening public health" [6, 7]. There are also similar programmes in Uzbekistan (Uz) [8]. The health status of the population is determined by the different types of population [9].

WHO recommends introduction of strategic trends for their prevention, which will allow the achievement of the stated UN SDGs to reduce the burden of non-infectious diseases at the national level [10].

Due to the given circumstances, epidemiological studies that include morbidity, disability and mortality analysis are one of the most important mechanisms for monitoring the health status of the population. It supports and improves the implementation of strategies aimed at improving the demographic situation.

The purpose of the study is to examine trends in the health status of the populations of the Russian Federation and the Republic of Uzbekistan.

2 Materials and Methods

Statistical data available on the official website of the WHO, the Interstate Statistical Committee of the CIS, the Federal State Statistics Service of the RF and Uz for 2010–2020 served as the material for conducting the study of demographic indicators and population health indicators for the population of the RF and Uz, which are members of the Commonwealth of Independent States (CIS) [11–14].

Analytical and statistical research methods were used for comparative analysis of demographic indicators, as well as primary morbidity and disability indicators. Statistical significance of 2010 and 2020 data was assessed using Student's t-test [15]. The statistical error for all values studied is in the range to $p < 0.05$.

3 Results

We have carried out a statistical analysis of the age structure of the population of the RF and the UZ in the period from 2010 to 2020 (Table 1).

Despite population growth between 2010 and 2020 in both the RF (3%) and the UZ (15%), the age structure did not change in either country. Both in 2010 and in 2020, the RF had a regressive age structure, while the UZ had a progressive age structure. Thus, the total population structure in Russia consists of 16% of people under working age and 24% of people over working age; in the UAE these values are 30 and 4% respectively.

The comparative analysis of the demographic characteristics of the population in the countries under study shows that the situation in the Russian Federation is similar with regard to the birth and death rates, i.e. the birth rate is about 13% and the death rate is more than 30% ($p < 0.05$). Thus, it can be said that there is a negative trend in

Table 1 Comparative characteristic of the population's age groups in the Russian Federation and Uzbekistan in 2010 and 2020 (in thousands of people and in %)

Age structure of the population	Russian Federation				Uzbekistan			
	2010		2020		2010		2020	
	Absolute value	%	Absolute value	%	Absolute value	%	Absolute value	%
Under working age	22,856	16	27,442	16	8561	30	10,117	30
Working age	85,710	60	82,678	60	18,477	66	22,258	66
Above working age	34,284	24	36,629	24	1159	4	1349	4
Total population	142,850	100	146,749	100	28,562	100	33,724	100

population growth in RF. The situation in UZ is different: both the birth rate and the mortality rate are lower than the birth rate (see Table 2).

Causes of death in RF and UZ were studied in detail in 2020. In RF the analysis gave the following results: the first place is CSD, the second—malignant tumours and the third—external causes. In UZ, the first rank is CSD, the second—malignant tumours, and the third—diseases of the respiratory system. A significant decrease in the mortality rate has been defined in RF from 2011 to 2018. From 2018 to 2020, the mortality rate will increase in both RF and UZ (Table 3).

The analysis of the primary disease was carried out in both countries during the study period. The results show the following trends CSD and respiratory diseases are increasing in RF. At the same time, the situation in UZ is different: the increase of infectious and parasitic diseases, CSD and diseases of the digestive system (Table 4).

The analysis of disability rates for the period 2010–2020 was studied in RF and UZ. The results showed the following situation: positive rates in RF and negative rates in UZ (Table 5).

Table 6 shows that there is a tendency for musculo-skeletal diseases and malignant tumours to increase in the UZ. Whereas in the RF there is a tendency to decrease all causes of death (Table 6).

Table 2 Demographic indicators in the populations the RF and Uz from 2010 until 2020: comparative characteristic (per 1000 people)

Country	Birth rate*		Mortality rate*		Natural population growth*	
	2010	2020	2010	2020	2010	2020
Russian Federation	12.5	9.8	14.2	14.6	−1.7	−4.8
Uzbekistan	21.9	24.6	4.6	5.1	17.3	19.5

* $p < 0.05$; $t > 2$

Table 3 Comparative characteristic of mortality rates by causes in the Russian Federation and Uzbekistan over the period form 2011 until 2020 (per 1000 people)

Year	Causes of death*									
	Russian Federation					Uzbekistan				
	Total mortality	Circulatory system diseases	Malignant tumors	Respiratory system diseases	External causes	Total mortality	Circulatory system disease	Malignant tumors	Respiratory system diseases	External causes
2011	13.5	7.5	2.1	0.5	1.4	–	–	–	–	–
2015	13.1	6.4	2.1	0.5	1.2	–	–	–	–	–
2018	12.3	5.8	2.0	0.42	0.98	4.7	2.8	0.5	0.2	0.3
2020	14.6	6.4	1.99	0.66	0.95	5.1	3.1	0.41	0.4	0.3

* $p < 0.05$; $t > 2$

Table 4 Dynamics of primary disease incidence in the population of the Russian Federation according to major disease classes over the period from 2011 until 2020 (per 1000 people)

Year	Number of disease incidence cases according to disease classes									
	Russian Federation*					Uzbekistan *				
	Infectious and parasitic diseases	Respiratory system diseases	Circulatory system diseases	Digestive system diseases	Traumas and food poisoning	Infectious and parasitic diseases	Respiratory diseases	Circulatory diseases	Digestive system diseases	Traumas and food poisoning
2011	33	324	26	34	92	12	120	16	56	33
2015	28	338	31	35	90	12	164	23	71	31
2020	20	371	29	26	81	21	134	29	103	28

* $p < 0.05$; $t > 2$

Table 5 Comparative characteristic of primary qualification for disability rates in the Russian Federation and Uzbekistan over the period of time from 2010 to 2020 (number of cases per 10,000 people)

Year	Russian Federation*	Uzbekistan*
2010	77	1
2015	59	9
2020	56	22

* $p < 0.05$; $t > 2$

Table 6 The dynamics of primary qualification for disability in adult population of the Russian Federation according to major disease classes over the period of time from 2010 to 2020 (number of cases per 10,000 people)

Disease class	Russian Federation*			Uzbekistan*		
	2017	2019	2020	2017	2019	2020
Tuberculosis	1.3	1.0	0.8	0.7	0.3	0.6
Malignant tumours	18.3	19	16.9	2.7	2.0	3.6
Mental illnesses	2.8	3,0	2.2	2.4	1.0	1.6
Circulatory system diseases	17.1	16	14.4	3.1	1	2.1
Musculoskeletal system diseases	3.5	3	2.9	1.6	1	1.9
Trauma consequences	2.0	2	1.5	1.6	0,4	0.7

* $p < 0.05$; $t > 2$

4 Conclusion

In order to study the trends in the health status of the population of the Russian Federation and the Republic of Uzbekistan, we carried out the monitoring of birth rate, mortality, morbidity and primary qualification for disability in the Russian Federation and the Republic of Uzbekistan based on data obtained from the official websites of the WHO, the CIS Interstate Statistical Committee, the Federal State Statistics Service of the Russian Federation and the Republic of Uzbekistan for the period from 2010 to 2020. The simulations and differences shown were conditioned by the type of population age structure in these countries (regressive or progressive). According to WHO data, over the last two decades, despite an increase in the proportion of deaths due to noncommunicable diseases in the total mortality structure, the overall level of premature mortality from these causes has decreased. Since 2010, however, progress in this area has slowed [16].

In order to reduce the high mortality and morbidity caused by noncommunicable diseases, the governments of both countries have adopted normative legal acts and regulations aimed at strengthening measures for the prevention, treatment and control of noncommunicable diseases and related risk factors, and reducing premature death and morbidity in the population. Both countries are striving to achieve normative target values within specified time frames [17, 18].

National health refers to a set of indicators that reflect the level of socio-economic development of a society. Despite the tendency to increase the effectiveness of the standard medical examination of the population, there is a low level of preventive measures demonstrated in the course of primary medical services provided to patients with socially significant diseases, along with the need to optimise its organisation [19]. The Sar-Cov-19 pandemic introduced its adjustments in the organisational mechanism of providing medical services to patients with socially significant diseases. Patients with non-infectious diseases have become more vulnerable and at greater risk, not only because of the higher risk of contracting the virus, but also because of the significant medical resources allocated to COVID-19 patients. Therefore, the prevention of risk factors associated with non-infectious diseases is becoming increasingly important in the current context.

Similarly to the demographic situation in Russia and Uzbekistan—regressive and progressive—the health characteristics of the population are clear indicators of the effectiveness of national health systems.

Acknowledgements The study was carried out within the framework of the advanced research project № INTL–SARGMU–SAMGMI-2021-05 dated 02.08.2021. (№ НИОКТР 121090200094-8 от 01.09.2021).

References

1. WHO: Top 10 causes of death. https://www.who.int/news-room/fact-sheets/detail/the-top-10-causes-of-death
2. Townsend, N., Wilson, L., Bhatnagar, P., et al.: Cardiovascular disease in Europe: epidemiological update 2016. Eur. Heart J. **37**(42), 3232–3245 (2016)
3. Movsisyan, N., Vinciguerra, M., Medina-Inojosa, J., et al.: Cardiovascular diseases in central and Eastern Europe: a call for more surveillance and evidence-based health promotion. Ann. Glob. Health **86**(1), 21 (2020)
4. Kesteloot, H., Sans, S., Kromhout, D.: Dynamics of cardiovascular and all-cause mortality in Western and Eastern Europe between 1970 and 2000. Eur. Heart J. **27**(1), 107–113 (2006)
5. Wilkins, E., Wickramasinghe, K., Bhatnagar, P., et al.: European Cardiovascular Disease Statistics 2017. European Heart Network Brussels (2017)
6. Passport of the national project "Healthcare" (approved by the Presidium of the Council under the President of the Russian Federation for Strategic Development and National Projects, protocol dated December 24, 2018 N 16). https://www.consultant.ru/document/cons_doc_LAW_319209/. Accessed 25 Jan 2022
7. Passport of the national project "Demography" (approved by the Presidium of the Council under the President of the Russian Federation for Strategic Development and National Projects, protocol dated December 24, 2018 N 16). https://www.consultant.ru/document/cons_doc_LAW_317388/. Accessed 25 Jan 2022
8. Ibragimova, N.M.: Major results of the implementation of national goals and objectives of sustainable development of Uzbekistan. Econ.: Anal. Forecasts **3**(14), 22–33 (2021)
9. Lee, R., Mason, A.: Fertility, human capital, and economic growth over the demographic transition. Eur. J. Popul. **26**(2): 159–182

10. Drapkina, O.M., Maslennikova, G.Y., Shepel, R.N., et al.: Priority areas for the prevention of non-communicable diseases on the agenda of the 75th World Health Assembly: plans for the future. Prevent. Med. **25**(6), 7–11 (2022). https://doi.org/10.17116/profmed202250617
11. Core health indicators in the WHO European region. https://www.euro.who.int/ru/data-and-evidence/evidence-resources/core-health-indicators-in-the-who-european-region/. Accessed 15 Dec 2021
12. Monitoring of population's quality of life indicators in the Commonwealth of Independent States' countries. http://www.cisstat.com/. Accessed 15 Dec 2021
13. Federal State Statistics Service of the Russian Federation. https://rosstat.gov.ru/incomparisons/. Accessed 15 Dec 2021
14. State Statistics Committee of the Republic of Uzbekistan. https://stat.uz/ru/. Accessed 15 Dec 2021
15. Paniotto, V.I., Maksimenko, V.S.: Quantitative methods in sociological studies. Kiev, 170pp. (2003)
16. World health statistics 2020: monitoring health for the SDGs, sustainable development goals. World Health Organization, Geneva (2020). https://apps.who.int/iris/bitstream/handle/10665/332070/9789240011977-rus.pdf?sequence=32&isAllowed=y/. Accessed 15 Dec 2021
17. Resolution of the Government of the Russian Federation dated December 26, 2017, № 1640 "On the approval of the State Programme of the Russian Federation "Healthcare Development" (as amended on July 24, 2021). https://base.garant.ru/71848440/. Accessed 15 Dec 2021
18. Resolution of the Government of the Republic of Uzbekistan dated 18.12.2018 "On prevention measures for non-infectious diseases, promotion of healthy lifestyle and increase of the population's physical activity level". https://lex.uz/docs/4111360/. Accessed 15 Dec 2021
19. Sazanova, G.Y.: Peculiarities of morbidity, disability and mortality of the population of Saratov region / G. Yu. Sazanova. Kuban Sci. Med. Bull. **1**(130), 144–146 (2012)

Digital Platform for the Treatment of Multiple Sclerosis with SSRIs

Iuliia Zhelnina, Tatiana Baidina, Alexander Zhelnin, Tatiana Trushnikova, and Julia Akintseva

Abstract Serotonin levels in blood serum and peripheral blood platelets were investigated in patients with multiple sclerosis (MS) (71 and 60 patients, respectively) using enzyme-linked immunosorbent assay. The mean serum serotonin concentration in patients with MS (210.11; 124–334.6 ng/ml) was lower ($p = 0.014$) than in the control group (269.48; 120–354.6 ng/ml). The mean value of platelet serotonin was 206.1 (124–334.6) ng/109, which was significantly lower ($p = 0.000$) than in the control group (641.7; 611.5–712.9 ng/109). A significant decrease in this neurotransmitter—its transport and free binding—was found in patients with MS. Comparative and correlational analysis showed that the level of serotonin was negatively correlated with many characteristics of the disease (its duration, severity, severity of individual neurological syndromes), which suggests the necessity of correcting the deficiency of this neurotransmitter with drugs that can affect the activity of the serotonergic system (SSRIs). The necessity of this correction can also be justified by the data available in the literature on the effect of serotonin on immune processes, therefore the appointment of serotonergic therapy is dictated by the possibility of pathogenetic influence on the autoimmune process.

Keywords Multiple sclerosis · Serum serotonin · Platelet serotonin

1 Introduction

Multiple sclerosis (MS) is a debilitating disease of the nervous system that affects more than two million people worldwide [1]. The fight against this disease is being waged in three directions: managing exacerbations, preventing exacerbations and influencing the course of the immunopathological process, and developing drugs

I. Zhelnina
Perm Regional Clinical Hospital, Perm, Russian Federation

T. Baidina (✉) · A. Zhelnin · T. Trushnikova · J. Akintseva
E.A. Vagner Perm State Medical University, Perm, Russian Federation
e-mail: tatiana_baidina@mail.ru

T. Antipova (ed.), *Digital Technology Platforms and Deployment*, Information Systems Engineering and Management 36, https://doi.org/10.1007/978-3-031-86547-3_8

that affect individual symptoms of the disease. In this context, some neurotransmitters that may be involved in the development of disease symptoms could be therapeutic targets. The immunomodulatory, neurotrophic and neuroprotective properties of neurotransmitters, especially serotonin, are discussed [2, 3]. There is information in the literature about decreased levels of serotonin in various body tissues and disruption of its transport in MS patients. In particular, the study of metabolites of biogenic amines, including serotonin (5-hydroxyindoleacetic acid), in the cerebrospinal fluid of MS patients (39—relapsing–remitting type, 26—primary progressive) was conducted by Markianos M. et.al. (2009 [4]). The control group consisted of 50 healthy subjects. The clinical status of the patients was assessed using the Expanded Disability Status Scale (EDSS) and the Multiple Sclerosis Severity Score (MSSS). Levels of 5-hydroxyindoleacetic acid were significantly reduced in both groups of MS patients. In patients with a remitting course, this serotonin metabolite in the cerebrospinal fluid was negatively correlated with EDSS and MSSS.

Hesse et al. (2014) [5] used positron emission tomography (PET) to investigate central serotonergic mechanisms. The authors performed PET using specific serotonin transporters to determine their distribution in the brain of 23 patients with MS and 22 healthy volunteers. The study showed that patients with MS had decreased activity of serotonin transporters in the cingulate cortex, thalamus and insula, and increased activity in the fronto-orbital cortex.

Differences were found between patients with relapsing–remitting and primary progressive forms of the disease. In the former, a decrease in serotonergic transmission was found mainly in the hippocampus, and in the latter in the prefrontal cortex. A positive correlation was found between the activity of serotonin transporters and the severity of depression and fatigue. The results of the study allowed us to conclude that patients with MS have impaired serotonin neurotransmission in the limbic system and frontal cortex, which appears to contribute to the development of psychiatric symptoms of MS.

One of the approaches to indirectly assess neurotransmitter processes in the brain is to study the effect of their correctors on various pathophysiological and clinical features of the disease. Selective serotoninergic reuptake inhibitors (SSRIs) are used as such correctors for the serotoninergic system. Such studies are being conducted in both experimental and clinical settings.

Studies in animal models of MS demonstrate the presence of serotoninergic mechanisms in the pathogenesis of the disease and the anti-inflammatory and neurotrophic effects of their pharmacological correction. In particular, in the experimental model of autoimmune encephalomyelitis, it has been shown that fluoxetine administration induces remission in rats with reduction of the histologically defined inflammatory lesion and suppression of IFN-γ production [6]. The authors suggest that prophylactic use of fluoxetine may have a neuroprotective effect related to the anti-inflammatory mechanisms of SSRIs.

The view that fluvoxamine has a beneficial effect on the mechanisms of brain damage in MS is supported by an experimental study by Ghareghani M. et al. (2017), which showed that this SSRI stimulates neurogenesis of oligodendroglia in stem cell culture and prevents inflammation and demyelination in animal models of MS [6].

The potential immunosuppressive effect of another SSRI, sertraline, in experimental autoimmune encephalomyelitis as a model of MS was investigated by Taler et al. (2011) [7]. Mice were treated with sertraline 5 mg/kg or dexamethasone 1 mg/kg 3 times daily for 3 weeks. Sertraline moderately improved neurological status and reduced disease-related weight loss. In addition to the clinical data, the authors analysed the sanogenetic mechanisms of this improvement. They showed that sertraline significantly reduced the production of pro-inflammatory cytokines in mice with autoimmune encephalomyelitis. The authors suggest that this drug could be used for adjuvant therapy in MS.

In addition to experimental studies, there are clinical studies of the effects of SSRIs on people with MS. The effects of fluoxetine on the human brain in MS have been studied using proton magnetic resonance spectroscopy, which provides non-invasive information about the chemical composition of the tissue being studied, and diffusion tensor MRI [8]. Fluoxetine was found to have a stimulating effect on gluconeogenesis in astrocytes, which provides energy processes in axons. Such an effect, the authors believe, may reduce the processes of axonal degeneration in patients with MS.

In a preliminary pilot study, 15 patients with MS underwent diffusion tensor MRI and magnetic resonance spectroscopy to determine the diffusion properties of brain tissue (fractional anisotropy, diffusion coefficient) and the levels of metabolites (choline, creatine, N-acetylaspartate) in grey matter, visually unaltered white matter and foci of white matter damage.

The studies were repeated after one week of fluoxetine (20 mg/day) and the following week of high-dose fluoxetine (40 mg/day). A decrease in choline was observed in grey matter after the first and second week of treatment (significant in the subgroup of 11 patients with relapsing–remitting and secondary progressive multiple sclerosis). In white matter, diffusion coefficient increased after the first week of treatment and N-acetylaspartate increased after the second week (both changes were statistically significant). Based on these results, the authors conclude that fluoxetine has neuroprotective properties in multiple sclerosis.

Another neuroimaging study on the effect of fluoxetine on the course of MS was carried out by Mostert et al. (2008) [9]. The double-blind, placebo-controlled study included 40 patients without depression with relapsing–remitting and secondary progressive multiple sclerosis. Patients received 20 mg of fluoxetine or placebo daily for 24 weeks. The formation of new gadolinium-enhancing lesions on MRI was assessed after 4, 8, 16 and 24 weeks of treatment. Nineteen patients completed the study. The mean number of new lesions over 24 weeks was 1.84 (2.9) in the fluoxetine group and 5.16 (8.6) in the placebo group ($p = 0.15$). The proportion of images showing new lesions was 25% in the fluoxetine group versus 41% in the placebo group ($p = 0.04$). During the last 16 weeks of treatment, the mean number of lesions was 1.21 (2.6) in the fluoxetine group and 3.16 (5.3) in the placebo group ($p = 0.05$). The proportion of patients with no new lesions was 63% in the fluoxetine group versus 26% in the placebo group ($p = 0.02$). This study shows that fluoxetine reduces the formation of new lesions in MS, but the authors acknowledge that this phenomenon requires further study.

The effect of SSRIs on non-affective symptoms of MS was conducted by Di Rezze S. (2012). The authors conducted a pilot study on the effect of duloxetine on overactive bladder in MS. 23 patients with MS (13 with a relapsing–remitting course, 10 with a secondary progressive course) took duloxetine or placebo for 8 weeks. The researchers found a significant improvement in bladder function, in the form of a decrease in residual urine volume and an increase in maximum bladder volume in the duloxetine group [10].

However, not all studies confirm the beneficial effect of SSRIs on the course of MS. In particular, Zhang T. et al. (2016) found no evidence that the use of SSRIs can slow disease progression [11].

A meta-analysis of the effect of SSRIs on the course of MS was performed in 2014 [12]. It only included data from trials in people with MS without depression. The author was only able to identify three trials that met all the requirements of evidence-based medicine. The conclusion of the meta-analysis was a statement of the need for further research into the possibility of introducing SSRIs into clinical practice for the management of patients with MS in order to modify the course of the disease.

2 Data and Methodology

Serum serotonin was measured in 71 patients with relapsing–remitting and secondary progressive MS using the "Enzyme-linked immunosorbent assay for the quantitative determination of serotonin in serum, plasma, platelets, urine, tissue homogenates and cell culture supernatants" ELISA (IBL, Hamburg). Given the impairment of blood–brain barrier function in patients with MS, it is thought that serum levels correlate with those in the central nervous system. Platelet serotonin levels were measured in 60 patients with MS (including relapsing–remitting and secondary progressive forms). The control groups consisted of 38 healthy individuals (for serum serotonin) and 12 (for platelet serotonin). All patients were diagnosed with definite MS according to the W.I. McDonald criteria [13]. The total disability score according to the Expanded Disability Status Scale (EDSS) [14] was 4.00 ± 1.33 points. In all patients, the diagnosis was confirmed by MRI. The example of the MRI of the patient with MS is shown below (Figs. 1, 2, 3).

In addition to assessing the severity of the neurological deficit according to the EDSS scale, neuropsychological tests were used to assess the presence of emotional disorders: Hospital Anxiety and Depression Scale—HADS [15], State-trial Anxiety Inventory—STAI [16], Beck Scale [17], to assess fatigue syndrome—Fatigue Severity Scale (FSS) [18] and Modified Fatigue Impact Scale (MFIS) [18]. To assess quality of life, the EQ-D5 questionnaire [19] and the MS-specific Functional Assessment of Multiple Sclerosis (FAMS) questionnaire [20] were used.

Statistical data processing was performed using the STATISTICA 10.0 software package (StatSoftInc., USA). Descriptive characteristics of quantitative features are

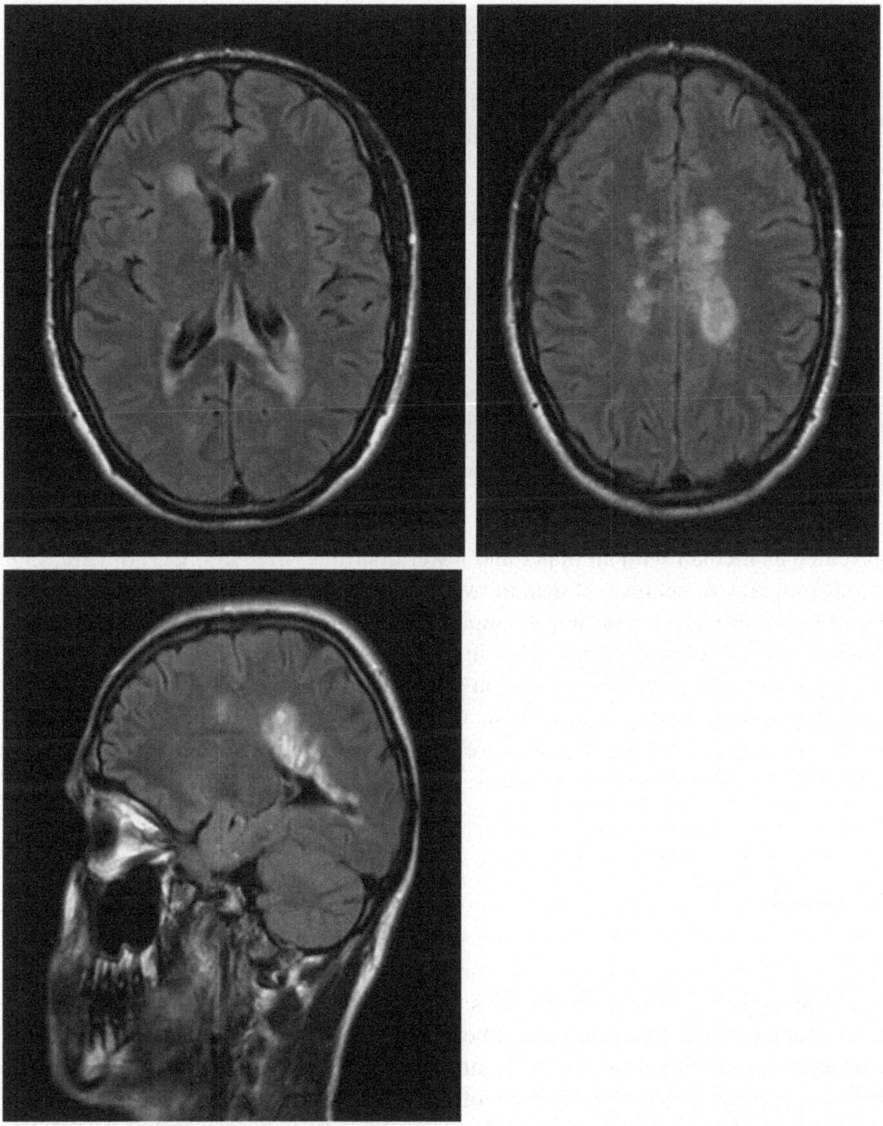

Fig. 1 MRI of the brain, T2-FLAIR modes In a series of MRI tomograms in axial and sagittal projections there are multiple foci of increased signal of ovoid and irregular shape in the deep and periventricular white matter of the cerebral hemispheres on both sides and in the corpus callosum with signs of perifocal edema

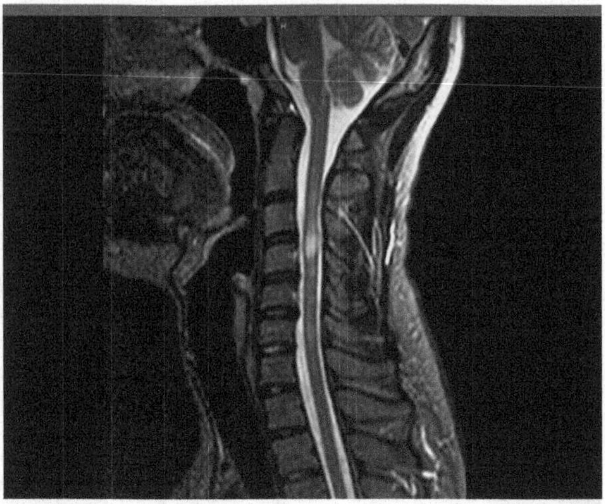

Fig. 2 MRI of the spinal cord. On the sagittal tomogram in T2 mode, there are different-sized foci of demyelination C1-C7, up to 16 mm in length, up to 5.3 mm in thickness without edema

presented as median with an upper and lower quartile—Me (LQ; UQ) or (Me; LQ-UQ). Comparative analysis of data in two independent groups was performed using the Mann–Whitney U criterion. Comparison of frequencies of binary qualitative characteristics in two independent groups was performed using the $\chi 2$ criterion and Fisher's exact criterion. The Wilcoxon signed rank test was used to compare two dependent groups. Correlation analysis was performed using Spearman's rank coefficient (R). The critical significance level for testing statistical hypotheses was set at 0.05. If the p-value was less than 0.001, the value was reported as $p = 0.000$.

3 Results

The mean serum serotonin concentration in MS patients (210.11; 124–334.6 ng/ml) was lower ($p = 0.014$) than in controls (269.48; 120–354.6 ng/ml). Humoral serotonin deficiency was characteristic of both sexes. In all age groups, serum serotonin concentration was lower than in the healthy group, and this decrease was particularly pronounced in patients over 50 years of age (56.1; 20–84.6 ng/ml). Serotonin was particularly low in severe cases of the disease (57.51; 30.1–76.2 ng/ml), the reliability of the differences with mild disease $p = 0.033$, with moderate disease—$p = 0.023$. Serotonin deficiency increased with the duration of the disease: with a history of MS of up to 10 years—226.3; 120–314.6 ng/ml, 10 years and more—144.0; 101–203 ng/ml ($p = 0.029$), the relationship of serotonin with the duration of the disease was also confirmed by correlation analysis ($R = -0.24$; $p = 0.040$).

A correlation analysis of the relationship between serotonin concentration and the degree of damage to the functional systems was carried out. The results are shown in Table 1.

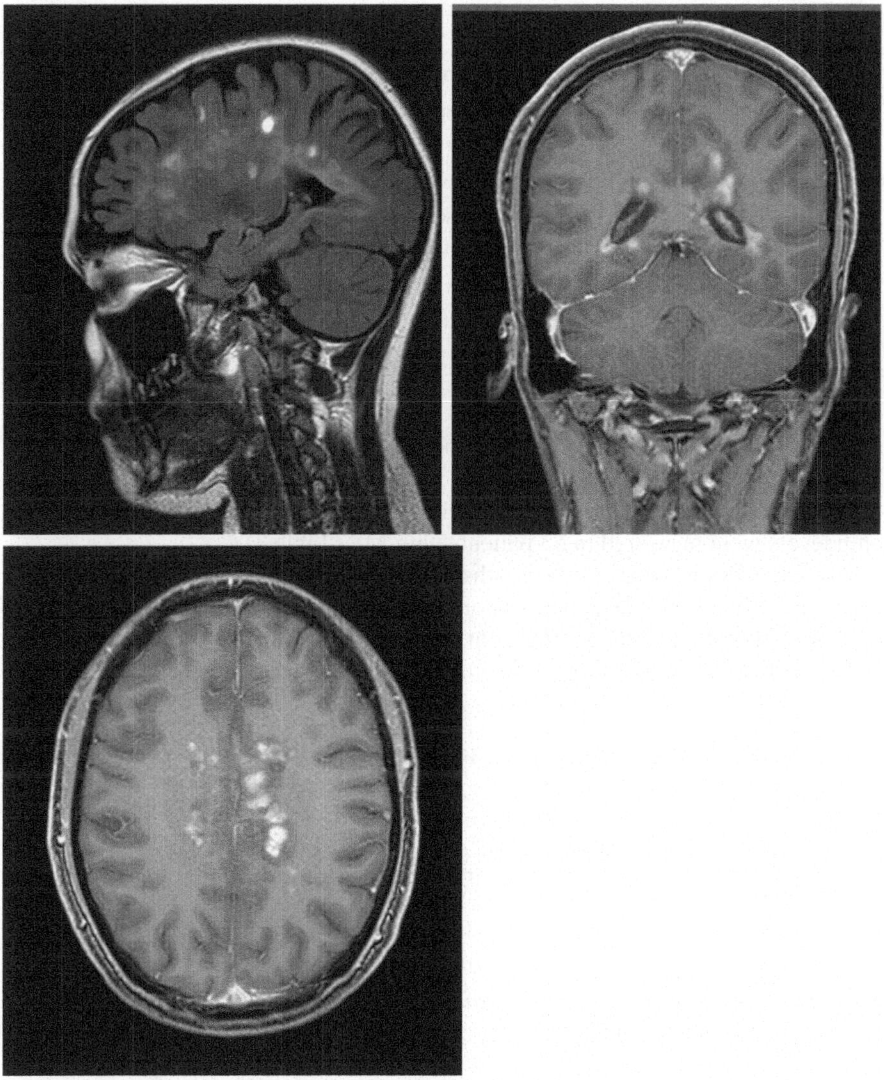

Fig. 3 MRI of the brain with contrast enhancement. On the sagittal section in FLAIR mode in the frontal and axial projection in T1 mode, massive accumulation of contrast was noted, which confirms the violation of the permeability of blood–brain barrier and indicates the activity of the disease at the present time

An inverse relationship has been found between serotonin levels and pelvic and visual disturbances and spasticity.

A comparison has been made between humoral serotonin and the severity of emotional disturbances in people with MS. It was found that anxiety disorders were

Table 1 Relationship between serum serotonin concentration in patients with multiple sclerosis and the severity of impairment of various functions according to EDSS

Indicators	R	p
Serotonin and pyramidal function	−0.08	0.48
Serotonin and spasticity	−0.10	0.04
Serotonin and cerebellar function	−0.13	0.25
Serotonin and brainstem functions	−0.05	0.67
Serotonin and sensory functions	−0.14	0.25
Serotonin and pelvic functions	−0.20	0.08
Serotonin and visual function	−0.25	0.037
Serotonin and thinking function	−0.01	0.91
Serotonin and EDSS	−0.01	0.90

more pronounced in patients with low blood serotonin levels, as would be expected (Table 2).

During immunomodulatory therapy with interferon-beta, peripheral blood serotonin levels were measured in 53 patients with MS (Table 3).

The Table 3 shows that after 3 months of treatment with interferon-beta, serotonin levels in the blood of MS patients approached control levels and remained at this level until the 6th month of treatment, when they again fell significantly below control levels. After one year of treatment with interferon-beta, serum serotonin

Table 2 Level of anxiety and depression in patients with multiple sclerosis with different concentrations of serotonin in the blood serum

Scales	Serotonin concentration		p^{M-W}
	<50 ng/ml (n = 10)	>50 ng/ml (n = 61)	
HADS, depression	7.87 ± 3.52	5.60 ± 3.42	0.086
HADS, anxiety	11.75 ± 4.62	7.44 ± 3.33	0.003
Spielberger, trait anxiety	57.12 ± 9.07	48.91 ± 8.27	0.007
Spielberger, reactive anxiety	55.75 ± 10.47	44.95 ± 8.41	0.014

Table 3 Concentration of serotonin in the blood serum of patients with multiple sclerosis during immunomodulatory therapy

Observation period	n	Serotonin (ng/ml)	p_{base}	$P_{control}$
Control value	28	269.48 ± 98.8		
Background state	71	210.11 ± 134.2		0.014
3 months	22	264.16 ± 149.1	0.111	0.880
6 months	16	221.93 ± 154.2	0.757	0.218
9 months	12	182.24 ± 132.8	0.506	0.026
Year	14	170.36 ± 116.3	0.304 $p_{3-12\ months} = 0.050$	0.001

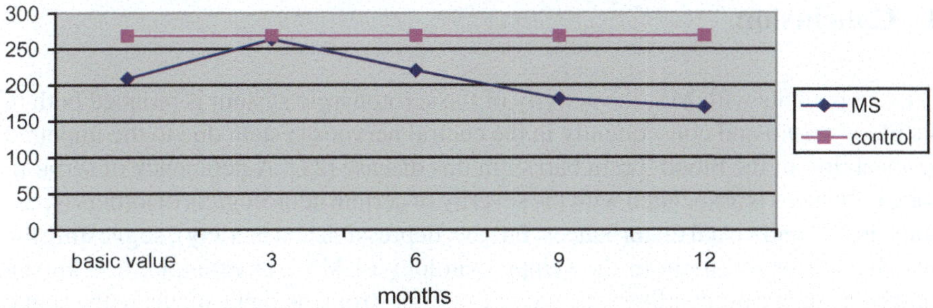

Fig. 4 Dynamics of serotonin (ng/ml) in peripheral blood during interferon-beta therapy

levels remained low. Thus, the short-term increase in serotonin was replaced by a return to the initial low levels (Fig. 4).

The transport link of the serotonergic system, represented by platelet serotonin, was studied in 60 patients with MS and 12 healthy individuals. In the main group it ranged from 0 to 758.58 ng/109. The mean serotonin level was 206.1 (124–334.6) ng/109, which was significantly lower ($p = 0.000$) than in the control group (641.7; 611.5–712.9 ng/109). Platelet serotonin deficiency was characteristic of both men and women. Platelet serotonin levels in MS patients did not depend on age ($R = -0.14$, $p = 0.257$). The duration of MS was not associated with platelet serotonin levels ($R = -0.04$, $p = 0.748$). Patients with low platelet serotonin levels ($\leq$100 ng/109) were identified. This group included 11 patients with MS (8 females and 3 males). Patients with low platelet serotonin levels did not differ in gender composition from the group with platelet serotonin levels > 100 ng/109 ($\chi 2 = 0.00$; $p = 0.959$), but were significantly older ($p = 0.022$), which may reflect a physiological depletion of the serotonergic system with age. Patients with serotonin levels $\leq$ 100 ng/109 had a higher EDSS disability score (4; 3.5–6) than patients in the second group (3.5; 2–4.5), $p = 0.05$. A correlation analysis was performed within a group of patients with low platelet serotonin levels ($\leq$100 ng/109). It was found that in these patients there was an inverse correlation between platelet serotonin and the sensory ($R = -0.28$, $p = 0.029$) and pelvic ($R = -0.26$, $p = 0.048$) scores, with the total FSS score ($R = -0.26$, $p = 0.044$), with the level of depression ($R = -0.29$, $p = 0.022$) and with the EQ-5D/mobility, self-care and total quality of life subscales.

Thus, low levels of serotonin in platelets are associated with more pronounced neurological deficits, emotional disturbances, particularly depression, fatigue syndrome, and quality of life in MS patients. The established patterns once again demonstrate the involvement of the serotonergic system in the pathogenesis of MS and the mechanisms of its symptom formation. During immunomodulatory therapy, the concentration of platelet serotonin was measured in 30 patients receiving interferon-beta. After 6 months of taking interferon-beta, this indicator increased significantly ($p = 0.017$), but without any relation to the dynamics of the EDSS.

4 Conclusion

Thus, in patients with MS, the activity of the serotonergic system is reduced both in the blood serum and consequently in the central nervous system due to the impaired permeability of the blood–brain barrier in this disease [21]. A deficiency of serotonergic influences is associated with the severity of certain neurological disorders (spasticity, pelvic and visual disturbances, fatigue, depression and anxiety), suggesting the involvement of serotonin in the symptomatology of MS (for example, it is known that there are two descending serotonergic systems from the raphe nuclei to the spinal cord, which are part of the pyramidal tract [22]). Many of these symptoms determine the quality of life of people with MS [23]. Drug activation of the serotonergic system can be used to treat MS symptoms and significantly improve the condition.

The established dependence of serotonin deficiency on the severity of the neurological deficit with the severity of the disease confirms the involvement of serotonin in the pathogenesis of the disease, i.e. in the course of autoimmune processes. In this regard, drugs such as SSRIs can be classified as options for targeted therapy in MS [24]. Adequate treatment of MS in this way, in addition to appropriate organisational medical measures [25], can be provided by a digital assessment of the brain's serotonergic system.

References

1. San Hernandez, A., Singh, C., Valero, D., et al.: Multiple sclerosis and serotonin: potential therapeutic applications. Cureus **12**(11), 2–8 (2020)
2. Sacramento, P., Monteiro, C., Dias, A., et al.: Serotonin decreases the production of Th1/Th17 cytokines and elevates the frequency of regulatory CD4 + T-cell subsets in multiple sclerosis patients. Eur. J. Immunol.Immunol. **48**, 1376–1388 (2018)
3. Boyko, A., Melnikov, M., Zhetishev, R., Pashenkov, M., et al.: The role of biogenic amines in the regulation of interaction between the immune and nervous systems in multiple sclerosis. NeuroImmunoModulation **23**, 217–223 (2016)
4. Markianos, M., Koutsis, G., Evangelopoulos, M., et al.: Relationship of CSF neurotransmitter metabolite levels to disease severity and disability in multiple sclerosis. Neurochemistry **108**(1), 158–164 (2009)
5. Hesse, S., Moeller, F., Petroff, D., et al.: Altered serotonin transporter availability in patients with multiple sclerosis. Eur. J. Nucl. Med. Mol. ImagingNucl. Med. Mol. Imaging **41**(8), 1640–1641 (2014)
6. Ghareghani, M., Zibara, K., Sadeghi, H.: Fluvoxamine stimulates oligodendrogenesis of cultured neural stem cells and attenuates inflammation and demyelination in an animal model of multiple sclerosis. Sci. Rep. **7**(1), 4923 (2017)
7. Taler, M., Gil-Ad, I., Korob, I., et al.: The immunomodulatory effect of the antidepressant sertraline in an experimental autoimmune encephalomyelitis mouse model of multiple sclerosis. NeuroImmunoModulation **18**(2), 117–122 (2011)
8. Sijens, P., Mostert, J., Irwan, R., et al.: J. Psychiatry Res. **164**(3), 274–282 (2008)
9. Mostert, J., Admiraal-Behloul, F., Hoogduin, J., et al.: Effects of fluoxetine on disease activity in relapsing multiple sclerosis: a double-blind, placebo-controlled, exploratory study. J. Neurol. Neurosurg. PsychiatryNeurosurg. Psychiatry **79**(9), 1027–1031 (2008)

10. Di Rezze, S., Frasca, V., Inghilleri, M., et al.: Duloxetine for the treatment of overactive bladder syndrome in multiple sclerosis: a pilot study. Clin. Neuropharmacol.. Neuropharmacol. **35**(5), 231–234 (2012)
11. Zhang, T., Kingwell, E., De Jong, H., et al.: Association between the use of selective serotonin reuptake inhibitors and multiple sclerosis disability progression. Pharmacoepidemiol. Drug Saf.. Drug Saf. **25**(10), 1150–1159 (2016)
12. Foley, P., Lawler, A., Chandran, S., et al.: Potential disease-modifying effects of selective serotonin reuptake inhibitors in multiple sclerosis: systematic review and meta-analysis. J. Neurol. Neurosurg. PsychiatryNeurosurg. Psychiatry **85**(6), 709–710 (2014)
13. McDonald, W., Compston, A., Edan, G.: Recommended diagnostic criteria for multiple sclerosis guidelines from the International Panel on the diagnosis of multiple sclerosis. Ann. Neurol. **50**(1), 121–127 (2001)
14. Kurtzke, J.: Rating neurologic impairment in multiple sclerosis: an expanded disability status scale (EDSS). Neurology **33**(11), 1444–1452 (1983)
15. Zigmond, A., Snaith, R.: The hospital anxiety and depression scale. Acta Psychiatr. Scand. Psychiatr. Scand. **67**(6), 1–4 (1983)
16. Spilberger, C., Gorsuch, K., Palo Alto, C.: Manual for the state-trial anxiety inventory. Consulting Psychologist Press (1970)
17. Beck, A.T., Ward, C., Mendelsohn, M.: An inventory for measuring depression. Arch. Gen. Psychiatry **4**, 561–571 (1961)
18. Dittner, A., Wessely, S., Brown, R.: The assessment of fatigue: a practical guide for clinicians and researchers. J. Psychosom. Res.Psychosom. Res. **56**(2), 57–170 (2004)
19. EuroQol, Group: EuroQol - a new facility for the measurement of health-related quality of life. Health Policy **16**(3), 199–208 (1990)
20. Fischer, J., LaRocca, N., Miller, D.: Recent development in assessment of quality of life in multiple sclerosis. Mult. Scler. **5**(4), 251–25 (1999)
21. Konysova, A. Serotonin metabolism in multiple sclerosis and retrobulbar neuritis (clinical and biochemical study) Cand. Sci. (Medicine). 5 (10), 30 (1995).
22. Zavalishin, I.A., Osadchikh, A.I., Vlasov, Y.V.: Upper motor neuron syndrome. Samara branch of the Literary Fund, 440p. (2005)
23. Akintseva, Y., Trushnikova, T., Baidina, T.: Impact of fatigue syndrome on the quality of life of patients with multiple sclerosis. Neurol. Bull. **XLII**(1), 132 (2010)
24. Baidina, T., Kuklina, E., Selyanina, N., et al.: Study of the pathogenesis of multiple sclerosis as the basis for its targeted therapy. Perm Med. J. **35**(1), 27–32 (2018)
25. Antipova, T., Zhelnin, A., Zhelnina, I.: Briefs in assessing the adequacy of health care facilities' fixed assets. J. Digit. Sci. **4**(1), 85–91 (2022)

Configuration of a Post-covid Care Center

Lidiia Soprun, Nataliia Gavrilova, Pavel Kislyy, and Igor Akulin

Abstract Post-covid syndrome was defined as the presence of symptoms and/or signs of damage to various organ systems that develop during or after a previous COVID-19 infection persist for more than 12 weeks and cannot be explained by an alternative diagnosis. (A Quick Guide to COVID-19: Managing the Long-term Effects of COVID-19 (NG188). National Institute of Excellence in Healthcare and Medical Care (UK); 2020.) Although the literature on symptoms after COVID is still in the early stages. A variety of symptoms affecting various systems are reported: neurocognitive post-COVID (brain fog, dizziness, loss of attention, confusion), vegetative post-COVID (chest pain, tachycardia, palpitations), gastrointestinal post-COVID (diarrhea, abdominal pain, vomiting), respiratory post-COVID (general fatigue, shortness of breath, cough, sore throat), musculoskeletal post-COVID (myalgia, arthralgia), related to psychology after COVID (post-traumatic stress disorder, anxiety, depression, insomnia), and other manifestations (agesia, anosmia, parosmia, skin rashes. In this regard, the opening of centers for post-covid syndrome becomes relevant. This paper presents the possibilities of organizing centers, departments dealing with postcovid syndrome.

Keywords COVID-19 · Post-covid syndrome · SARS-CoV2 · Telemedicine technologies · Healthcare organization · Health care safety management

L. Soprun (✉) · N. Gavrilova · I. Akulin
Saint-Petersburg State University, Saint-Petersburg, Russia
e-mail: lidas7@yandex.ru

L. Soprun · N. Gavrilova · P. Kislyy
Saint-Petersburg State University Hospital, Saint-Petersburg, Russia

P. Kislyy
Gatchina Clinical Interdistrict Hospital, Saint-Petersburg, Russia

T. Antipova (ed.), *Digital Technology Platforms and Deployment*, Information Systems Engineering and Management 36, https://doi.org/10.1007/978-3-031-86547-3_9

1 Introduction

The 2019 coronavirus disease pandemic (hereinafter COVID-19) was caused by severe acute respiratory syndrome coronavirus 2 (SARS-CoV-2) [1, 2]. Although the main acute clinical manifestations of the disease are associated with the respiratory tract, neurological symptoms have been reported, including loss of taste and smell, headache, dizziness, numbness and neuropathic pain. According to data from China, among 44,415 COVID-19 patients, 81% had mild or moderate diseases, 14% had severe diseases and 5% had critical diseases. In the current study, 83% of the patients had mild to moderate disease (without clinical signs of pneumonia and mild pneumonia that did not require oxygen support). 10% had a severe illness (the need for oxygen therapy), and 8% had a critical illness (requiring hospitalization in the intensive care unit due to respiratory failure), which indicates that our study population is comparable to other large populations in terms of the severity of the disease [3–5].

A new coronavirus infection (COVID-19) was added to the List of diseases that pose a danger to others (approved by Decree of the Government of the Russian Federation No. 715 of December 1, 2004) by Decree of the Government of the Russian Federation No. 66 of 31.01.2020—item B 34.2. Subsequently, this served as the basis for the creation of anti-epidemic measures. One of the most important anti-epidemic measures is the introduction of new regulatory documents and the formation of a legislative base of different levels of regulation. The regulatory legal act on the territory of the Russian Federation regulating the requirements for reducing the risks of the development and spread of a new coronavirus infection was the Resolution of the Chief State Sanitary Doctor of the Russian Federation No. 2 of 24.01.2020. In this document, for the first time in the territory of the Russian Federation, the following requirements were established

- to ensure the readiness of medical organizations to receive infected patients;
- carrying out work on informing the population;
- creation of a stock of antiviral drugs, disinfectants and PPE in medical organizations, organizations of school, preschool education, nutrition organizations;
- strengthening of sanitary and quarantine control at checkpoints across the state border;
- monitoring of compliance with the disinfection regime at transport hubs and in crowded places;
- conducting in-depth molecular genetics and virological studies of biomaterial from patients suspected of a new infection.

Before the outbreak of a pandemic of a new coronavirus infection, the Decree of the President of the Russian Federation No. 254 dated June 6, 2019 "On the Strategy for the development of healthcare in the Russian Federation for the period up to 2025" was developed, possible epidemiological risks were described:

* the risk of complications of the epidemiological situation against the background of the unfavorable situation in foreign countries for a number of new and dangerous infectious diseases.
* the risk of new infections caused by unknown pathogens, the introduction of rare or previously unknown infectious and parasitic diseases on the territory of the Russian Federation, the occurrence and spread of native focal infections, spontaneous infection with infectious agents, the return of disappeared infections, the overcoming of interspecific barriers by microorganisms.

Thus, the developed regulatory and legal documents provide the necessary level of legislative framework and contribute to reducing the risks of the development and spread of a new coronavirus infection.

Encoding statistical information in the presence of dosinia or established diagnosis of COVID-19 are carried out in accordance with the following procedure defined in ICD 10.

U07.1—Coronavirus infection COVID-19, the virus confirmed by laboratory testing regardless of the severity of clinical signs or symptoms.

U07.2—Coronavirus infection COVID-19, virus not identified (COVID-19 diagnosed clinically or epidemiologically).

Z03.8—Surveillance in case of suspected coronavirus infection.

Z22.8—Carrier of the causative agent of coronavirus infection.

Z20.8—Contact with a patient with a coronavirus infection.

Z11.5—Screening examination to detect coronavirus infection.

B34.2—Unspecified Coronavirus infection (except COVID-19).

B33.8—Specified Coronavirus infection (except COVID-19).

Z29.0—Isolation.

U08.9—Personal history of COVID-19.

U09.9—Condition after COVID-19.

U11.9—The need for immunization against COVID-19.

U12.9—Vaccine against COVID-19, which caused an adverse reaction.

Post-covid syndrome (U09.9—The condition after COVID-19) has been defined by the World Health Organization (WHO) as a condition characterized by the presence of symptoms of damage to various organ systems that develop within three months after a proven episode of a new coronavirus infection, persist for at least eight weeks and cannot be explained by an alternative diagnosis (WHO Delphi consensus on the post-COVID-19 complication. October, 6th, 2021). According to this concept, patients suffering from the consequences of a new coronavirus infection may have two types of complications:

(1) Respiratory failure and other complications developing against the background of a severe course of a new coronavirus infection. This type of complications should not be considered exclusively within the framework of the post-covid syndrome, due to its nonspecific nature, typical for severe respiratory viral infections.
(2) The actual post-covid syndrome should be considered immunopathological and/or probably immunopathological complications from the cardiovascular,

nervous, respiratory, immune systems, as well as gastrointestinal tract organs that meet the WHO definition.

These symptoms have been approved by WHO experts as possible complications of coronavirus infection, however, no pathophysiological concept combining these symptoms into simp-tom complexes has been proposed at the moment, despite the fact that this topic is being actively studied all over the world. Due to the spread of a new coronavirus infection in the territory of the Russian Federation, the number of postcovid syndrome among the working-age population is increasing [2, 6, 7]. That is why there was a need to organize a postcovid center on the basis of St. Petersburg State University.

The purpose of this work was to analyze the proposal for the organization of admission of patients with postcovid syndrome. Patients who, after being cured of infection, developed subacute and chronic health disorders of an immunopathological or presumably autoimmunological nature. Tasks of creating a center:

* creation of a general patient register,
* studying the etiology and pathogenesis of these disorders,
* development and application of methods of their treatment,
* implementation of the subsequent stages of their rehabilitation and long-term follow-up,
* creation and implementation of protocols for diagnosis, treatment, rehabilitation and dynamic follow-up at the stages of outpatient, inpatient and social care for patients with these forms of pathology,
* development of an individualized treatment plan for each patient in a short time within the framework of outpatient consultations, a dispensary examination by a team of specialists and a one-day hospital,
* development of telemedicine consultation techniques for working with patients in a timely manner.

2 Materials and Methods

The results of the analysis of the Center's work were used. In order to determine the tasks of activating the development of the Center, the methodology of SWOT analysis of the formation of strategic tasks was applied. Applied methods of content analysis, analytical, SWOT analysis.

A Center for the study and treatment of postcovid syndrome has been operating on the basis of St. Petersburg State University since September 2021. The scientific concept developed by a team of authors [8–13] formed the basis for creating a plan of diagnostic and therapeutic measures.

When creating a Center for patients with postcovid syndrome, the following tasks were set and implemented:

- Formation of a general register of patients after recovery from NCVI (with subacute and chronic health disorders);

- Study of the etiology and pathogenesis of postcovid syndrome;
- Development and application of postcovid syndrome treatment methods;
- Implementation of the subsequent stages of rehabilitation and long-term follow-up of patients with postcovid syndrome;
- Implementation of diagnostic, treatment, rehabilitation and dynamic monitoring protocols at the stages of providing social and medical care to patients in outpatient and inpatient settings;
- Development of a treatment plan for each patient in a short time within the framework of outpatient consultations and dispensary examination by a team of specialist doctors, treatment in a "one-day hospital";
- Development of telemedicine consultation techniques for remote work with patients.

3 Results. Organization of the Center's Work

If the attending physician suspects that the patient has post-CoV syndrome, consultation at the centre is recommended. In case of negative tests for the SARS-CoV2 virus, the attending physician refers the patient or a package of documents for the selection committee to the designated centre.

The selection committee coordinates the outpatient consultation at the centre and prepares the initial documentation. If there are indications, after an outpatient consultation, the patient may be recommended to spend a day in hospital for further examination and treatment. The centre's attending physician invites the patient for further treatment and rehabilitation. If necessary, a pharmacy is set up and a health check is carried out, which increases the patients' awareness of the disease and their motivation for treatment and rehabilitation. At present, the market for medical services is actively developing. This is due to government support, which increases the attractiveness of the industry for investment. In particular, rehabilitation centres are in high demand by the population, especially against the background of the development and spread of the pandemic caused by the new COVID-19 coronavirus infection and the emergence of a new "post-COVID syndrome". This business plan has been drawn up for the opening of a rehabilitation centre and a day hospital, designed for 2 wards. The working hours of the centre: Monday–Friday from 08:00 to 20:00; Saturday—from 09:00 to 15:00; Sunday—only by appointment. With such a number of rooms, the payback period is 18 months and the break-even point is 4 months. For the opening of the centre, it is possible to transform the outpatient department, which has offices for the reception of neurological patients in the structure, into an economic unit. On average, about 80 people per month will undergo a rehabilitation course. The cost of a 3–4 day course will be 25,000 m.u.

As a result, the operational indicators of the project will be as follows:

- Total investment—3,804,000 m.u.
- Monthly expenses—1,580,833 m.u.
- Monthly profit—257,840 m.u.

- Payback period—18 months
- The break-even point is 3 months.
- Return on sales—19%.

Description of the market

In order to create a competent advertising company and marketing strategy, it is necessary to determine the target group. The main customers of the centre are citizens who have postcovid violations and who are sent in accordance with the regional roadmap—routing. At the same time, the gender ratio will be approximately 40% for men: 60% for women. When creating a profile of a potential client, we take into account the following factors: gender, age group, profession, lifestyle, marital status and presence of children. The team of doctors is 100% professional and guarantees an accurate diagnosis. We offer a high level of service, from the initial consultation to the client's departure, with exceptional quality and courtesy.

* Speed of work: organisation of all complex examination activities in the shortest possible time;
* Convenience of the Centre's location: availability of parking and public transport;
* Optimal cost of services, competitive prices.

It should also be noted that the success of this business is based on the provision of quality services. In order to attract patients to the centre, it will first be necessary to carry out a powerful PR campaign and to obtain recommendations from industry experts. In addition, the first clients can be offered favourable conditions for completing the course, additional procedures or discounts.

The main promotional channels include:

Roadmap/Routing.

Promoting the centre on social networks.

Participation in exhibitions and professional events.

Presentation of the centre's programme to industry professionals.

Contextual advertising

These promotional channels are not exhaustive. You can also actively promote the center in your region using radio or television, but then advertising costs will increase dramatically.

SWOT analysis of the medical center

Strengths

- Highly qualified staff: Having a team of professionals with academic qualifications is an undeniable advantage that enhances patient confidence and improves the quality of services provided.
- Narrow specialization: Allows you to focus on a specific area of medicine, study it in depth and provide more effective treatment and advice.

- Full range of services: We provide a comprehensive approach to diagnosis and treatment, which is convenient for patients and ensures the effectiveness of medical care.
- Comfortable environment: Our clinic's location and interior are designed to promote relaxation and psychological well-being, creating a positive experience for our patients.
- Positive reputation: The well-established reputation of the medical center helps to retain regular customers and attract new ones.
- Professional development: Regular training of specialists guarantees the use of the latest techniques and approaches in treatment.
- Discount system: Stimulates repeated requests and builds customer loyalty.
- Telemedicine: The introduction of telemedicine technologies expands the availability of services and increases convenience for patients.

Weaknesses

* Narrow specialization: While it may limit the number of patients interested in receiving treatment for other conditions, it also allows for a more focused approach to patient care.
* Reputation: Any negative feedback or incidents can seriously damage a clinic's reputation, leading to a loss of customers. Maintaining a positive reputation is essential for attracting and retaining patients.
* Cost: Due to the high level of expertise required for specialized treatments, the cost of services may be higher. This can make certain treatments inaccessible for some patients.
* Technology: Telemedicine requires a reliable internet connection and appropriate equipment from both the clinic and the patient. This may present a challenge in some areas.
* Growth Opportunities: Expanding services within a chosen area of specialization or exploring related fields of medicine can help a clinic grow and maintain a competitive edge.
* Marketing: Using online platforms such as blogging and social media can help to attract patients and increase the visibility of the clinic.
* Partnership with insurance companies: The conclusion of contracts with insurance companies can help to expand the customer base and provide more coverage to patients.
* Telemedicine development: Expanding telemedicine services such as online consultations and remote patient monitoring can improve patient care and convenience.
* Competition: High competition from other clinics and medical centers can be a threat.
* Economic instability: Economic instability can lead to a decrease in demand for medical services.
* Legislative changes: Changes to the regulatory framework can require additional financial and organizational efforts.

* Technology development: Advances in medical technology may require investments in equipment and training. Conducting a SWOT analysis can help you develop an effective strategy for the growth of your medical center. This analysis takes into account the strengths, weaknesses, opportunities, and threats of your company. It can also help you create a website for your company that allows people to schedule online consultations. The director/head of the center is accountable for the efficient functioning of the center. This includes assuming responsibility for the risks involved in the work of the center as a whole and for the doctors in particular.

The personnel of your medical facility, including doctors, must possess a diploma in medical education and have at least three years of professional experience. Additionally, they should have received positive feedback from patients. To ensure their success, the doctors at your facility must continuously enhance their skills. To support this, you must provide them with the necessary resources, including financial incentives and access to specialized literature.

Work schedule: individual.

Nursing staff (nurses).

The nurse is the person who interacts most closely with patients, so it is essential that they are competent, polite, and patient with every patient. Experience and positive references from a previous position are required before entering into employment. The average number of medical staff is determined by the ratio of 1 doctor to 2 nurses in the center.

Junior medical staff (nurse).

Ensures cleanliness and order in all rooms of the center.

Registry (administrators).

In the early days of the center's launch, these personnel have a wide range of tasks: answering phone calls, taking patient records, informing doctors about appointments, and greeting patients. Therefore, it is crucial to closely monitor their workload: while administrators can initially assist with the cashier duties, once the center reaches its full capacity, all these responsibilities should be shared among three individuals: the administrator, the cashier, and the receptionist.

Cash register (cashier). Acceptance of funds from patients based on documents, accounting for funds and collection.

The Table 1 shows the staffing of the medical center at the initial stage of work.

Financial plan. Investments in the opening of a medical center (Table 1, 2).

More than 80% of the costs of opening a medical center fall on the purchase of medical equipment, as well as on the repair and decoration of the premises (Table 3).

It should be borne in mind that for the first three months you will have to invest money in obtaining a license, paying wages and advertising the center (Table 4).

The average receipt for a visit to a medical center consists of two indicators: the average cost of visiting a specialist, as well as the average cost of taking an analysis. A comprehensive sales strategy for a two-year period, factoring in seasonal variations, evaluating the profitability of investments, and anticipating economic indicators, is outlined in the financial model.

Table 1 The medical center at the initial stage of work

	Salary	Number	Amount
Director	100,000	1	100,000
Doctor	100,000	4	400,000
Medical nurse	50,000	2	100,000
Kliner	18,000	2	36,000
Administrator	18,000	2	36,000
Cashier	20,000	1	20,000
Inshurance			207 600
Total			899 600

Table 2 The total amount of investments in the opening and initial operation of the center will be approximately 5,000,000 m.u

Investments for opening	Amount
Registration, including obtaining all permissions	200 000
Commissioning of all equipment	350 000
Software Installation	20 000
Purchasing a form for staff	50 000
Purchase of office equipment	150 000
Purchase of office furniture	25 000
Repair	300 000
Initial purchase of medicines	100 000
Promotional materials	100 000
Rent for the time of repair	30 000
Purchase of equipment	3 423 203
Other	30 000
Total	**4 778 203**
Monthly expenses	
PHOTOS (including deductions)	899 600
Rent (15 sq.m.)	15 000
Depreciation	73 061
Utilities	25 000
Advertisement	20 000
Purchase of medicines	30 000
Other expenses	30 000
Total	**1 092 661**

Potential risks. The field of healthcare services is inherently risky from a business perspective. However, by meticulously assessing all potential risks, you can safeguard your business against their adverse effects. Internal hazards. Preparing for the launch of the facility. The facility, its interior, maintenance, equipment, and staff—all must

Table 3 Equipment for the medical center

Name	Quantity	Price per 1 pc	Total
TSA II thermal stimulation system (MEDOC, USA)	1	390 000	390 000
Stabilometric platform	1	1 320 000	1 320 000
Automated doctor's workplace (4 sets)	4	27 000	108 000
UV bactericidal irradiator wall mounted	1	4 250	4 250
Filing cabinet	1	16 400	16 400
Medical sofa	1	5 329	5 329
Doctor's chair	4	6 020	24 080
Doctor's desk	4	14 100	56 400
Medical couch	4	6 625	26 500
Infusion equipment for infusion of polyclonal donor human immunoglobulins	300	50	15 000
Electric bed adaptive	4	36 210	144 840
Drying and sterilization cabinet	0	0	0
Filing cabinet	1	6 280	6 280
Medical sofa	1	5 239	5 239
Doctor's chair	1	6 020	6 020
Doctor's desk	1	10 800	10 800
Medical couch	1	6 625	6 625
Ultraviolet bactericidal irradiator	1	4 250	4 250
A set of odorous substances for the study of the functions of the olfactory analyzer	2	48 000	96 000
Refrigerator for storing immunoglobulins with labeled shelves	1	30 900	30 900
Cabinet for tools and medicines	1	13 600	13 600
Tonometer	2	1 890	3 780
Thermometer	5	280	1 400
Equipment for training with biofeedback according to EMG parameters (BOS)	2	460 100	920 200
Disposable syringes	1	10 000	10 000
Cabinet with a set for antishock therapy	1	8 000	8 000
Medical couch	2	6 625	13 250
Bixes with sterile material	0	12 000	0
Filing cabinet	1	14 250	14 250
Containers with disinfectant solutions	1	5 000	5 000

(continued)

Table 3 (continued)

Name	Quantity	Price per 1 pc	Total
Germicidal lamps	2	2 390	4 780
Means of antishock therapy	1	7 000	7 000
Steam autoclave	1	21 990	21 990
Ultrasonic bath for pre-sterilization cleaning of instruments	1	13 800	13 800
Dry—burning cabinet	1	20 000	20 000
Germicidal cabinet	1	13 800	13 800
Water distiller	1	14 990	14 990
Bactericidal irradiator	1	4 250	4 250
Consumables	1	30 000	30 000
Disposable materials	1	20 000	20 000
	0	32 000	0
Medical sofa	0	12 600	0
Cooler	1	3 200	3 200
Shoe rack	2	1 500	3 000
TV	0	20 000	0
Total			3 423 203

Indicator	**Value**
Discount rate, %	17.00
Discounted payback period—DPB, months	9
Net discounted income—NPV, m.u.	25,497,096
Profitability index—PI	6,50
Internal rate of return—IRR,%	51.32
Return on sales, %	176%
PV1 year 1	10,379,541
PV1 year 2	19755,758
Income 1	12144,063
Income 2	27043,657

adhere to stringent regulations and specifications. This implies that it may take you several months to locate the ideal alternatives.

The expertise and competence of medical professionals. A precise diagnosis and a well-designed treatment plan are essential for turning a one-time patient into a loyal customer of your clinic. Therefore, when hiring staff, it is crucial to thoroughly review their resumes and seek references from their previous employers.

External threats. Obtaining a license for medical practice. To secure a license, you must possess suitable premises, trained staff, and certified equipment. The process

Table 4 Current expenses

Expenses	1 month	2 month	3 month	4 month	5…10 month	11 month	12 month
Rent	15 000	15 000	15 000	15 000		15 000	15 000
Utility costs	25 000	25 000	25 000	25 000		25 000	25 000
Depreciation of equipment	73 061	73 061	73 061	73 061		73 061	73 061
Purchase of medicines	3 600	7 200	10 800	14 400		36 000	36 000
Advertisement	20 000	20 000	20 000	20 000		20 000	20 000
Other expenses	30 000	30 000	30 000	30 000		30 000	30 000
Salary (including insurance premiums)	899 600	899 600	899 600	899 600		899 600	899 600
Total	**1 066 261**	**1 069 861**	**1 073 461**	**1 077 061**		**1 098 661**	**1 098 661**

involves gathering all required paperwork, but this process may be prolonged if you are denied for any reason, which must be promptly addressed.

The pricing strategy for the services offered by the center should consider the needs and perspectives of not only the business owner, but also the patients and even the medical professionals. It is crucial to take into account and support the interests of your staff, who may not want to accept a lower compensation.

Reputation is a complex and multifaceted aspect that can significantly affect a medical center's success. It encompasses the collective perception and opinions of patients, staff, and the wider community about the center's quality of care, professionalism, and overall experience.

Here's a breakdown of key aspects regarding a medical center's reputation:

Factors Influencing Reputation;

Patient Experience;

Quality of Care: This is paramount. Effective treatments, accurate diagnoses, and successful outcomes build a strong reputation. Conversely, medical errors, misdiagnoses, and poor outcomes can severely damage it.

Communication and Bedside Manner: Clear, empathetic communication from doctors and staff is crucial. Patients value feeling heard, understood, and respected.

Wait Times: Long wait times for appointments, procedures, or even in the waiting room can frustrate patients and negatively impact their perception of the center.

Facility Cleanliness and Comfort: A clean, well-maintained, and welcoming environment contributes positively to patient experience.

Staff Expertise and Conduct: Professionalism and Qualifications: A highly qualified and experienced medical team with a demonstrable record of accomplishment instills confidence in patients.

Staff Attitude: Friendly, helpful, and compassionate staff throughout the center (reception, nurses, and technicians) contribute to a positive atmosphere and enhance reputation. Community Engagement and Outreach: Local Involvement: Participating in health fairs, offering educational workshops, or supporting community initiatives can foster goodwill and a positive image within the community.

Online Presence and Reviews: Online Reviews: Platforms like Google Reviews, specialized healthcare review sites, and social media significantly influence how potential patients perceive the center. Website Content: An informative, up-to-date website that is easy to navigate and provides essential information about services, doctors, and contact details is essential.

Transparency and Ethical Practices: Open Communication: Being transparent about treatment options, potential risks, and costs builds trust with patients.

Ethical Billing Practices: Fair and transparent billing practices are essential to avoid negative perceptions and potential reputational damage.

Managing and Improving Reputation: Active Monitoring: Regularly track online reviews, social media mentions, and patient feedback to understand public perception and address any emerging issues promptly. Responding to Feedback: Acknowledge and address both positive and negative feedback professionally and constructively. Show empathy and a willingness to resolve issues. Encouraging Patient Reviews: Make it easy for satisfied patients to share their positive experiences through online platforms and testimonials. Internal Communication and Training: Foster a culture of patient-centered care by providing staff with ongoing training on communication, empathy, and professionalism.

Remember, a strong reputation is an invaluable asset for a medical center, attracting new patients, retaining existing ones, and ultimately contributing to its long-term success. In the Russian Federation, the following tariffs have been introduced into the programs of the federal and some territorial mandatory medical insurance funds: Medical rehabilitation after a coronavirus infection COVID-19 (2 points on the SRM), medical rehabilitation after a coronavirus infection COVID-19 (3 points on the SRM).

4 Expected Results

1. The opening of a post-ovoid center, contributes to the provision of high-quality medical care in one institution
2. Obtaining new knowledge about the etiology and pathogenesis of autoimmune diseases of viral genesis, clarifying their clinical course and diagnostic features.
3. Clarification of the etiology, pathogenesis and clinical features of the cluster of autoimmunity-associated syndromes, the clinical pathophysiology of which is currently insufficiently studied.
4. Development of new methods and means of treatment and prevention of post-covid disorders.

5. Creation of the first in the world practice and the only center in Russia to help patients with post-COVID health disorders and COVID-associated autoimmune diseases, uniting a team of specialists of various profiles around a translational medical task, accumulation of unique clinical experience.
6. Close cooperation and exchange of information with the patient community, receiving feedback in the process of treatment, medical and social effect.
7. Restoration to work of patients of working age suffering from autoimmune complications against the background of COVID-19 virus infection, psychological and medical care, prevention of depressive-phobic behavioral disorders and suicidal tendencies.
8. Conducting clinical trials on the basis of the Center within the framework of grants and state assignments, orders of local authorities, the creation of a unified register and database of these patients.
9. Strengthening domestic and international cooperation in the study, diagnosis, prevention and treatment of autoimmune diseases.
10. Professional development of specialists within the framework of schools and telemedicine conferences
11. Improving the medical and economic efficiency and focus of the work of commercial rehabilitation units of clinics involved in the project and affiliated with its participants.

5 Discussion

As a result of the functioning of this center, it is already possible to draw conclusions about the pathophysiological link of the postcovid syndrome. Pathophysiological concept of post-covid syndrome.

The scientific concept developed by a team of authors [1, 2] formed the basis for creating a plan of diagnostic and therapeutic measures at the Center for the Study and Treatment of Post-Covid Syndrome, established at St. Petersburg State University in September 2021. It can be briefly presented in the form of the following theses:

In patients predisposed to the development of autoimmune diseases, an autoreactive immunological process develops against the background of a COVID-19 infection, with the possible production of autoantibodies and proinflammatory cytokines. These disorders lead to damage to the terminal parts of the nerve fibers, contributing to the development of sensory and autonomic poly-neuropathy. This type of polyneuropathy cannot be detected during routine examinations due to the small diameter of the damaged fibers. Vegetative polyneuropathy leads to the development of microcirculation disorders. In addition to the disruption of autonomic regulation, the direct impact of a new coronavirus infection on the vascular wall with the development of endothelial dysfunction, described in the literature, may also be important here. Vascular damage at the microlevel leads to deregulation of the cardiovascular system as a whole with the development of constant fatigue, tachycardia and other symptoms. Symptoms from other organs and systems may also occur [14–16]. Various manifestations on the part of the central nervous system can be explained both by the direct effect of the virus and the influence of autoantibodies, as well as by the disruption of autoregulation of cerebral circulation. Due to the fact that thin nerve fibers can be a source of neuropeptides that regulate (and lower) inflammation and the intensity of immunopathological, including autoreactive, processes [2, 17, 18], their involvement in post-covid polyneuropathy can lead to a vicious circle in pathogenesis, fixing neuroimmune manifestations of inflammation. Thus, the lesion of small diameter fibers may be one of the leading manifestations of the post-covid syndrome, pathophysiologically justifying a significant number of sensory and vegetative complaints recorded in these patients [8–11, 19]. The main issue here may be a development of the complicated pathological processes, involving damage to small nerve fibers and the development of micro-, microcirculatory, and respiratory disorders in combination with autoimmune dysfunction, the development of hypercoagulability, and immunological inflammation contributes to the progression of multiple organ dysfunctions in these patients.

6 Conclusions

The chapter considers a model of post-Covid syndrome as a particular post-viral immunological complication. Currently, the search for new organizational models of medical care for patients with post-Covid syndrome continues to be relevant. This chapter presents the experience of the first medical center for the provision of medical care to patients with post-Covid syndrome in St. Petersburg. The main principle of successful therapy for post-COVID syndrome is a holistic, multidisciplinary approach involving a large team of medical specialists. Therefore, treatment options include corrections of all of the abovementioned conditions that require further extensive research.

References

1. Varela, N., Chorny, A., Gonzalez-Rey, E., Delgado, M.: Tuning inflammation with anti-inflammatory neuropeptides. Expert. Opin. Biol. Ther. **7**(4):461–478 (2007). https://doi.org/10.1517/14712598.7.4.461; Souza-Moreira, L., Campos-Salinas, J., Caro, M, Gonzalez-Rey, E.: Neuropeptides as pleiotropic modulators of the immune response. Neuroendocrinology **94**(2), 89–100 (2011). https://doi.org/10.1159/000328636
2. Gavrilova, N., Soprun, L., Lukashenko, M., Ryabkova, V., Fedotkina, T.V., Churilov, L.P., Shoenfeld, Y.: New clinical phenotype of the post-covid syndrome: fibromyalgia and joint hypermobility condition. Pathophysiology **29**, 24–29 (2022). https://doi.org/10.3390/pathophysiology29010003
3. US CDC confirmed 2019-nCoV cases globally. https://www.cdc.gov/coronavirus/2019-ncov/locations-confirmed-cases.htmlmap
4. 2019 novel coronavirus, Wuhan, China. CDC (2020). https://www.cdc.gov/coronavirus/2019-ncov/about/index.html

5. Novel Coronavirus (2019-nCoV) Situation report – 22. WHO (2020). https://www.who.int/docs/default-source/coronaviruse/situation-reports/20200211-sitrep-22-ncov.pdf
6. Lukashenko, M.V., Gavrilova, N.Y., Bregovskaya, A.V., Soprun, L.A., Churilov, L.P., Petropoulos, I.N., Malik, R.A., Shoenfeld, Y.: Corneal confocal microscopy in the diagnosis of small fiber neuropathy: faster, easier, and more efficient than skin biopsy? Pathophysiology **29**, 1–8 (2022). https://doi.org/10.3390/pathophysiology29010001
7. Liu, X., Treister, R., Lang, M., Oaklander, A.L.: IVIg for apparently autoimmune small-fiber polyneuropathy: first analysis of efficacy and safety. Ther. Adv. Neurol. Disord.. Adv. Neurol. Disord. **11**, 175628561774448 (2018). https://doi.org/10.1177/1756285617744484
8. Shouman, K., Vanichkachorn, G., Cheshire, W.P., Suarez, M.D., Shelly, S., Lamotte, G.J., Sandroni, P., Benarroch, E.E., Berini, S.E., Cutsforth-Gregory, J.K., Coon, E.A., Mauermann, M.L., Low, P.A., Singer, W.: Autonomic dysfunction following COVID-19 infection: an early experience. Clin. Auton. Res.. Auton. Res. **31**(3), 385–394 (2021). https://doi.org/10.1007/s10286-021-00803-8
9. Finsterer, J.: SARS-CoV-2 vaccinations may not only be complicated by GBS but also by distal small fibre neuropathy. J. Neuroimmunol.Neuroimmunol. **15**(360), 577703 (2021). https://doi.org/10.1016/j.jneuroim.2021.577703
10. Aranyó, J., Bazan, V., Lladós, G., Dominguez, M.J., Bisbal, F., Massanella, M., Sarrias, A., Adeliño, R., Riverola, A., Paredes, R., Clotet, B., Bayés-Genís, A., Mateu, L., Villuendas, R.: Inappropriate sinus tachycardia in post-COVID-19 syndrome. Sci. Rep. **12**(1), 298 (2022). https://doi.org/10.1038/s41598-021-03831-6
11. Soprun, L., Gavrilova, N., Ryabkova, V., Yastrebov, S.: A model of clinical symptoms in patients with chronic fatigue syndrome with postCOVID. JDAH **4**(2), 22–30 (2023). https://doi.org/10.33847/2712-8149.4.2_3
12. Drias, Y., Drias, H.: Sentiment evolution analysis and association rule mining for COVID-19 tweets. J. Digit. Art Humanit. **2**(2), 3–21 (2021). https://doi.org/10.33847/2712-8148.2.2_1
13. Jiang, J., Srinivasan, K.: Comparing pregnancy and childbirth-related hospital visits in Arizona before and during COVID-19 using network analysis. J. Digit. Sci. **3**(2), 37–52 (2021). https://doi.org/10.33847/2686-8296.3.2_2
14. Culver, D.A., Dahan, A., Bajorunas, D., Jeziorska, M., van Velzen, M., Aarts, L.P.H.J., Tavee, J., Tannemaat, M.R., Dunne, A.N., Kirk, R.I., et al.: Cibinetide improves corneal nerve fiber abundance in patients with sarcoidosis-associated small nerve fiber loss and neuropathic pain. Investig. Ophthalmol. Vis. Sci.. Ophthalmol. Vis. Sci. **58**, BIO52–BIO60 (2017). https://doi.org/10.1167/iovs.16-21291
15. Gorbalenya, A.G., et al.: The species severe acute respiratory syndrome-related coronavirus: classifying 2019-nCoV and naming it SARS-CoV-2. Nat. Microbiol.Microbiol. **5**, 536–544 (2020). https://doi.org/10.1038/s41564-020-0695-z
16. Gemignani, F.: Small fiber neuropathy and SARS-CoV-2 infection. Another piece in the long COVID puzzle? Muscle Nerve, 11 Jan 2022 (2022). https://doi.org/10.1002/mus.27495
17. Barros, A., Queiruga-Piñeiro, J., Lozano-Sanroma, J., Alcalde, I., Gallar, J., Fernández-Vega Cueto, L., Alfonso, J.F., Quirós, L.M., Merayo-Lloves, J.: Small fiber neuropathy in the cornea of Covid-19 patients associated with the generation of ocular surface disease. Ocul. Surf. **23**, 40–48. https://doi.org/10.1016/j.jtos.2021.10.010. Epub 2021 Nov 12
18. Waheed, W., Carey, M.E., Tandan, S.R., Tandan, R.: Post COVID-19 vaccine small fiber neuropathy. Muscle Nerve **64**(1), E1–E2 (2021). https://doi.org/10.1002/mus.27251
19. Geographical distribution of 2019-nCov cases globally, European Centre for disease prevention and control (2020). https://www.ecdc.europa.eu/en/geographical-distribution-2019-ncov-cases

Artificial Intelligence Deployment

AI-Optimized OWC/RF Networks in Urban Areas

Olimpiu Stoicuta, Monic a Leba, Andreea Ionica, and Remus Sibişanu

Abstract This paper proposes an innovative approach to enhancing energy efficiency in Optical Wireless Communication (OWC) networks by integrating Artificial Intelligence (AI)-driven green management strategies. The increasing demand for sustainable communication technologies is determined by a combination of environmental, regulatory, economic, technological, and social factors. As the global digital infrastructure continues to expand, the adoption of sustainable practices in communication technologies is essential for mitigating environmental impacts and ensuring long-term viability. By leveraging machine learning algorithms and advanced data analytics, the study aims to develop sustainable and efficient management practices of the OWC networks with the aim to reduce the environmental impact while maintaining high performance and reliability of the communication link. This paper explores the various ways AI is applied to optimize OWC networks, highlighting key areas and real-world applications. We propose the use of a Machine Learning (ML)-based Random Forest algorithm to implement a soft-switching strategy in urban OWC/RF networks with the aim to enhance the optical network efficiency by dynamically adjustment of transmission parameters or switching to RF depending on the real-time conditions.

Keywords Random forest algorithm · ML · Hybrid OWC/RF

1 Introduction

The development and wide application of wireless communication technologies have brought about a new era of connectivity, with an important impact and a number of benefits for work, life, and society. However, as we continue to enjoy the accessibility and benefits of these technologies, we must also consider their environmental impact. The urgent need for sustainable communication technologies is becoming

O. Stoicuta (✉) · M. Leba · A. Ionica · R. Sibişanu
University of Petrosani, Petrosani, HD, Romania
e-mail: stoicuta_olimpiu@yahoo.com

T. Antipova (ed.), *Digital Technology Platforms and Deployment*, Information Systems Engineering and Management 36, https://doi.org/10.1007/978-3-031-86547-3_10

increasingly obvious, as industries and consumers alike seek solutions to combat energy consumption, greenhouse gas emissions, and electronic waste.

OWC has already a number of a cutting-edge wireless communication technologies based on light, that harnesses the power of light to transmit data wirelessly. It differs significantly from traditional RF communication, as it leverages light waves across the visible, infrared, or ultraviolet spectrum, rather than relying on the electromagnetic waves used by RF systems. OWC includes a variety of technologies, including Infrared (IR) communication, which utilizes infrared light for data transmission, Visible Light Communication (VLC), which utilizes the visible spectrum, Li-Fi which is a full duplex indoor network communication using both visible light (LEDs) and IR spectrum, Free Space Optics (FSO), which relies on infrared or laser light for communication, Optical Camera Communication (OCC) that leverages camera sensors in smartphones, tablets, and other devices to receive data transmitted by light sources [1].

The main benefits of OWC are energy efficiency, high bandwidth and data rates, security and interference resistance, spectrum efficiency, cost-effectiveness, flexibility, and scalability. The ever-growing global demand for communication services, especially wireless communication, has brought an increased awareness regarding the environmental consequences, associated with the infrastructure and systems required to support these services [2]. In response, integrating green management practices into communication networks has become crucial, as it helps minimize energy consumption, reduce carbon emissions, and foster sustainable development. These practices enable communication networks to function more efficiently while mitigating their environmental impact. AI has revolutionized numerous industries by enabling, for example, intelligent data analysis, pattern recognition, and data-driven decision-making. In the domain of energy and resource management, AI has proven to be a valuable asset, with its ability to optimize efficiency, reduce costs, and promote sustainability. Across various sectors, AI is being leveraged to improve energy and resource management, from smart building systems to intelligent grid management and more [3].

This paper comprehensively covers the role of ML in optimizing energy efficiency and resource management for integration of OWC technologies in green management practices. It also provides a comprehensive approach to using a Random Forest algorithm (RFA) for soft-switching in urban OWC/RF networks, ensuring high performance and energy efficiency within challenging environmental conditions.

2 Sustainable Management in OWC Networks

As the global demand for communication services continues to rise, the environmental impact of communication networks has become a significant concern. Integrating green management practices into communication networks is essential to reduce energy consumption, minimize carbon emissions, and promote sustainable development. These practices ensure that communication networks operate efficiently while mitigating their environmental footprint.

2.1 Current Advancements Benefits and Applications of OWC

The last two decades, OWC technologies and applications have made significant steps, especially in areas as Indoor Positioning Systems (IPS), indoor optical networks (Li-Fi), outdoor wireless communication between buildings, vehicles [4], unmanned aerial vehicle (UAV), optical ground terminal and drones across turbulent atmospheres, and underwater communication [5, 6]. Internet of Underwater Things (IoUT) is a scientific development that aims to revolutionize research, business, and underwater military applications. It plays a crucial role in sea species tracking, water contamination monitoring, oil and gas production, natural hazard control, maritime security, and naval military activities.

Is also an essential feature of upcoming 5G and 6G networking systems, since the fifth (5G) and sixth (6G) generation networks [7, 8] are expected to surpass current fourth-generation systems. Challenges include coverage performance, high-ability networking, low latency, high protection, low power usage, strong knowledge, and stable networking [9]. Innovations like OWC are used to address these challenges in 5G networks. OWC combines the advantages of fiber optics (high bandwidth) with wireless technologies (flexibility). It offers wide spectrum, high wireless data rates and security, low latency and cost, as well as low energy consumption [10].

Current OWC research emphasizes advances in optical receiver (oRx) design to improve signal-to-noise ratio (SNR), efficient pointing and tracking (APT) units for better beam pointing, and study of atmospheric turbulence to optimize the light wave performance [11].

VLC offers dual functionality by providing both illumination and data communication by using visible light emitted by LED bulbs to transmit data. FSO is ideal for outdoor point-to-point communication, such as connecting buildings or towers and uses infrared or laser light to transmit data over long distances through the air [12]. IRC is commonly used in remote controls and certain types of sensor networks, and employs infrared light for short-range communication.

OWC technologies have a broader spectrum availability compared to RF communication, which is highly regulated and congested. This allows for higher data rates and the potential for ultra-high-speed communication. They also have higher data throughput. VLC, for instance, can achieve data rates exceeding 10 Gbps, making it suitable for applications requiring high bandwidth, such as HD video streaming and virtual reality.

LEDs used in VLC are highly energy-efficient. They can modulate data onto light without significant additional power consumption. OWC systems can integrate energy harvesting technologies, such as solar panels, to further reduce their environmental footprint.

Since light waves do not penetrate walls, which limits the propagation of signals outside the intended area, enhancing security and reducing the risk of eavesdropping. OWC systems are immune to electromagnetic interference, making them ideal for environments with high RF noise, such as industrial settings or densely populated urban areas. The use of light waves alleviates congestion in the RF spectrum, allowing for more efficient use of available bandwidth and reducing the need for frequency management and licensing. OWC can be used alongside RF communication to enhance overall network capacity and reliability, particularly in heterogeneous network environments.

In certain applications, such as FSO for building-to-building communication, OWC can reduce the need for expensive fiber optic cables. VLC systems that combine lighting and data transmission offer cost savings by utilizing existing lighting infrastructure for communication purposes. OWC systems can be rapidly deployed in various settings without the need for extensive infrastructure changes, making them ideal for both temporary and permanent installations. OWC can be scaled to different application sizes, from small indoor networks using VLC to extensive outdoor networks using FSO [13].

As an indoor communication, smart lighting systems and secure communications are prevalent application. VLC can be integrated into smart lighting systems to provide both illumination and high-speed internet access within homes, offices, and public spaces, and OWC is ideal for secure, short-range communications within buildings, such as in healthcare facilities and financial institutions.

Outdoor communication, on the other hand, refers to urban networks and rural connectivity. OWC can enhance urban communication networks by providing high-speed data links between buildings, traffic signals, vehicles and public Wi-Fi hotspots. FSO can also bridge the digital divide by offering high-speed internet connections in rural and remote areas where fiber optic infrastructure is impractical.

Industrial and military applications are developed in harsh environments providing secure communications [14, 15]. OWC systems can operate effectively in environments with high RF interference or electromagnetic sensitivity, such as factories and military bases. The inherent security of OWC makes it suitable for sensitive communication in military operations and secure facilities.

2.2 Green Management in Urban Networks

The are a number of key areas of sustainable communication technologies naming energy-efficient network design, renewable energy integration, sustainable manufacturing and materials, lifecycle management.

Data centers are significant energy consumers. Implementing energy-efficient cooling systems, renewable energy sources, and optimized server utilization can drastically reduce their carbon footprint. Adaptive protocols that optimize data routing and reduce redundant transmissions contribute to energy savings [16].

Deploying base stations powered by renewable energy sources reduces dependence on fossil fuels and lowers operational costs. Devices that harvest energy from ambient sources, such as solar or kinetic energy, can operate sustainably without relying on traditional power supplies.

Using recyclable and biodegradable materials in the manufacturing of communication devices reduces environmental impact. Ensuring that supply chains adhere to environmental standards and minimize carbon footprints is essential for sustainable technology production.

Designing devices for durability and ease of repair extends their useful life and reduces electronic waste. Implementing efficient recycling programs and encouraging the reuse of components can significantly lower the environmental impact of discarded devices.

3 The Role of AI in OWC Networks

OWC networks have high-speed, secure, and interference-free communication. As the demand for efficient and reliable communication systems grows, integrating Artificial Intelligence (AI) into OWC networks has become crucial for optimizing performance, enhancing energy efficiency, and managing resources effectively.

Key areas where AI are expected to optimize OWC networks are (i) adaptive modulation and coding, (ii) energy efficiency, (iii) resource allocation, (iv) network planning and optimization, (v) security enhancements, and (vi) data analytics and insights.

AI algorithms dynamically adjust modulation and coding schemes based on real-time channel conditions, optimizing data rates and maintaining communication quality. By analyzing historical data, AI can predict future channel conditions, enabling preemptive adjustments to modulation schemes to enhance performance [17].

Advanced AI-driven power control mechanisms offer significant benefits in OWC systems, allowing real-time power adjustments of transmitters to reduce energy consumption while ensuring signal quality remains high. Furthermore, AI algorithms can intelligently manage sleep modes for network components, activating or deactivating them based on real-time usage patterns, conserving energy while maintaining optimal performance. This combination of power management techniques, made possible by AI, promises more energy-efficient and sustainable OWC systems.

AI optimizes the allocation of resources such as bandwidth and transmission power, ensuring efficient use of available resources and reducing bottlenecks. AI-driven load balancing techniques distribute network traffic evenly across OWC links, preventing congestion and enhancing overall network efficiency. Unprecedented capabilities are brought by AI when it comes to optimizing network topology in OWC systems. By analyzing the network's layout, AI can determine the most efficient placement of transmitters and receivers, maximizing coverage while minimizing energy consumption. These algorithms can also detect faults within the network, initiating recovery processes that ensure minimal disruption and maintain network reliability, promoting optimal performance and efficiency. These AI-driven topology optimization techniques pave the way for more robust, resilient, and energy-efficient OWC networks [18].

AI-based security systems monitor OWC networks for unusual activity, detecting and mitigating potential threats in real-time, and can enhance secure authentication mechanisms, ensuring that only authorized devices and users access the network [19].

Moreover, AI leverages its powerful data analysis capabilities to continuously monitor key performance metrics, enabling a real-time understanding of the network's functionality. By analyzing data from various network components, such as transmitters, receivers, and routers, AI systems can predict maintenance needs, enabling proactive measures to be taken and significantly reducing downtime, while also ensuring optimal performance. This approach, which combines big data analysis and predictive maintenance, helps pave the way for more robust and energy-efficient OWC networks [20].

The use of Machine Learning (ML) in Optical Wireless Communication (OWC) technologies allows for predictive maintenance, energy consumption forecasting, dynamic resource allocation, improving communication efficiency, integration of renewable energy sources into OWC, reducing environmental impact. By analyzing data and utilizing ML, mathematical models and predictions can be created regarding equipment failures and maintenance needs within OWC systems, ensuring continuous and efficient operation of these systems. Based on historical data and input data, mathematical models can be created to predict electrical energy consumption, enabling better planning and resource allocation. By analyzing data from various sources and using ML, resources can be dynamically allocated where needed, optimizing energy use and reducing waste. ML can improve the efficiency of OWC systems by optimizing signal processing and reducing interference. This allows for more reliable and energy-efficient communication. ML enables the integration of renewable energy sources into OWC technologies. By predicting energy generation from sources like solar and wind, ML can adjust energy distribution accordingly. By optimizing energy consumption and improving the efficiency of OWC systems, ML contributes to reducing the overall environmental impact and energy consumption [21].

3.1 Areas of Application

Main areas of application of the couple AI&OWC are smart cities, industrial automation, healthcare patient monitoring, transportation. AI-optimized OWC networks provide high-speed, reliable communication for smart city applications, such as intelligent traffic management, public safety, and environmental monitoring, and AI-driven OWC systems enhance energy efficiency in smart city infrastructure, such as street lighting and public Wi-Fi.

The performance of OWC networks in industrial environments, can be significantly enhanced by AI facilitating real-time monitoring and control of manufacturing processes. By optimizing network performance and reducing energy consumption, AI can boost productivity and efficiency. Furthermore, predictive maintenance systems, enabled by OWC networks, can detect potential equipment failures before they occur, reducing costly downtime and extending the lifespan of machinery.

AI enhances the reliability and security of OWC networks in healthcare settings, ensuring fast and secure transmission of medical data between devices and systems, and also allow to an effective real-time patient monitoring, improving response times and patient outcomes.

Figure 1 shows that RF communication has a shorter active window time in winter than FSO, compared with summer, when FSO has a shorter activity time because of high light intensity.

FSO Building-to-Building or Vehicle-to-Everything (V2X) communication in urban areas can be significantly improved by AI, leading to communication availability in certain disaster scenarios, as well as more efficient traffic management, enhanced safety, and overall greater efficiency. Optical and RF wireless communication links can be switched in between depending on the lights intensity predicted in certain areas based on historical data and ML algorithms integrated into the communication system.

By harnessing the power of OWC networks, which provide high-speed and reliable communication, AI can facilitate high-quality in-flight entertainment and connectivity. This advancement will revolutionize the transportation industry, offering passengers a more enjoyable and connected travel experience while also streamlining transport operations [22]. Relevant case studies that are viable examples of reliable OWC networks with AI support are the Google's Project Loon, Li-Fi in office buildings and museums, and Siemens' smart factory. Project Loon, launched by Google in 2013, was a groundbreaking initiative that sought to provide internet

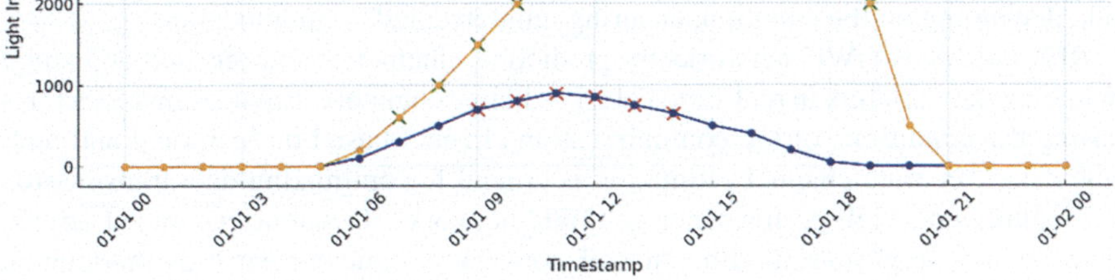

Fig. 1 Hourly light intensity variations and switch to RF communication. Authors elaboration

access to remote and underserved areas using high-altitude balloons. These balloons floated in the stratosphere, twice as high as commercial airplanes, beaming internet connectivity to the ground below. Project Loon proved its potential by providing connectivity in disaster-struck regions, like Peru during flooding and Puerto Rico after a hurricane. Loon balloons were designed to remain aloft for hundreds of days, navigating the stratospheric winds to maintain coverage [23].

Li-Fi in office buildings and museums [24] has implemented AI-optimized VLC systems that can provide high-speed internet access through existing LED lighting infrastructure that enhanced energy efficiency, secure communication, and seamless connectivity for workers [25]. Siemens' Smart Factory implemented AI-driven OWC networks optimize communication between machines and systems in Siemens' smart factories, supporting real-time data exchange and process automation that increases productivity, reduce energy consumption, and improve operational efficiency [26].

4 Soft-Switching Between Wireless Links Using RFA

OWC networks in urban environments face the challenge of maintaining high performance and energy efficiency among variable channel conditions and high interference. A soft-switching strategy, which dynamically adjusts transmission parameters or choose the most reliable link (between optical and RF) based on real-time conditions, can enhance network efficiency [27].

Soft-switching can dynamically adjust the transmission parameters, such as modulation schemes and power levels, to adapt to varying channel conditions. This approach ensures optimal performance and energy efficiency by switching to the most appropriate transmission settings based on current network conditions [28, 29].

The RFA, a robust ML technique known for its accuracy and versatility, is well-suited for implementing soft-switching strategies. It consists of an ensemble of decision trees that operate collectively to improve prediction accuracy and handle complex data patterns [30].

An urban network (Fig. 2) has multiple wireless communication links, including optical links with varying quality and backup RF links.

The Random Forest model, continuously fed with real-time data, predicts that an optical link is about to be less performant due to light intensity or obstructed by a moving vehicle. The model triggers a switch to a more stable optical link or an RF link, depending on the situation, ensuring uninterrupted communication.

RFA enhances OWC networks by predicting channel conditions and adjusting/switching network link in real-time [31]. Predictive channel estimation involves forecasting the conditions of the communication channel based on historical and real-time data. Accurate channel estimation is crucial for optimizing data transmission and maintaining high performance in OWC networks. Dynamic parameter adjustment involves real-time modification of network parameters (such as modulation schemes, power levels, and beam angles to optimize network performance based

Fig. 2 Urban area with hybrid OWC/RF links. Authors elaboration with AI support

on current channel conditions) or to switch the second available RF link wireless communication.

Useful data of sunshine duration, longwave radiation and UV radiation [W/m^2] since January 2023 to October 2024, are plotted in Fig. 3 [32].

Figure 4 consists of total precipitation [mm] snowfall amount [cm], total cloud cover [%] and sunshine duration [min] plotted in Basel from March 2023 to October 2024 [33].

The first step, predictive channel estimation is done, using data collected. Historical data on channel conditions, and environmental factors are gathered. Then, data collected are fed into the highly accurate Random Forest model for training, enabling to predict future channel conditions with high accuracy, therefore the dynamic parameters are adjusted. A real-time analysis is done by a continuously feeding of real-time data into the Random Forest model to assess current channel conditions. Then, the soft-switching decision, based on the model's predictions, dynamically adjust the transmission parameters (e.g., modulation schemes, power levels) or decide to

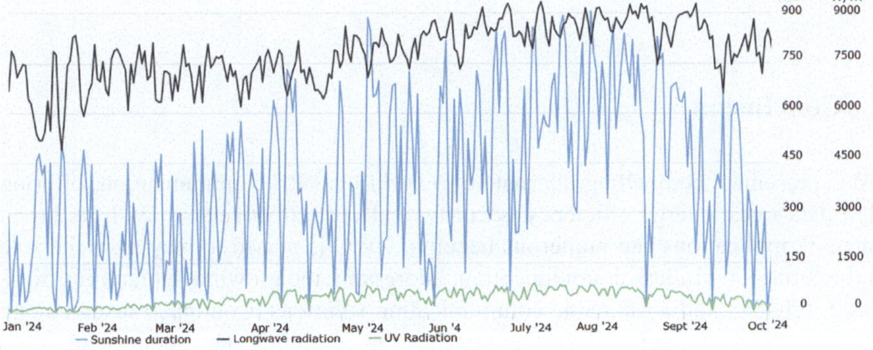

Fig. 3 Sunshine duration, longwave radiation and UV radiation [32]

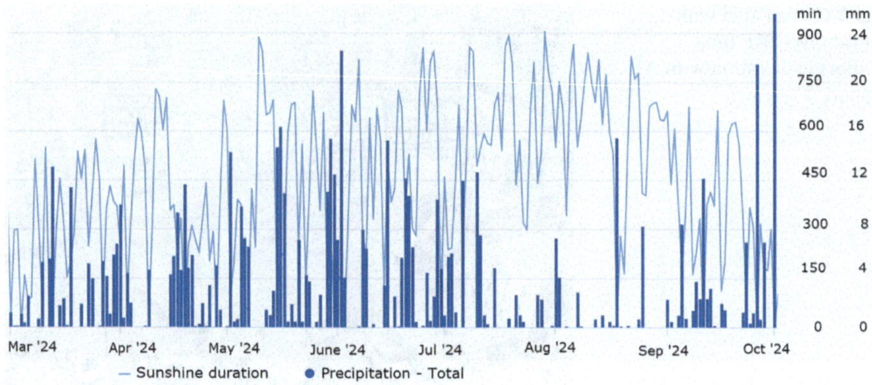

Fig. 4 Environmental conditions plotted [33]

switch to the RF link to optimize performance and energy efficiency with the aim to maintaining optimal wireless communication.

While FSO communication holds great potential, atmospheric turbulence remains a major concern, significantly impacting its performance by increasing the BER (Bit Error Ratio). Traditional modulation techniques like On–Off Keying (OOK) and Pulse-Position Modulation (PPM) have their respective drawbacks, especially in the presence of varying signal intensity. Subcarrier-intensity modulation (SIM), combined with advanced RF modulation schemes like Binary Phase Shift Keying Modulation (BPSK), Differential Phase Shift Keying (DPSK), and Quadrature Amplitude Modulation (QAM), offers a promising solution to enhance the reliability and efficiency of FSO systems under turbulent atmospheric conditions [33].

The application of the RFA in OWC networks significantly enhances predictive channel estimation and dynamic parameter adjustment. By leveraging historical and real-time data, this machine learning technique provides accurate predictions and enables adaptive, proactive adjustments to network parameters. These capabilities lead to optimized performance, improved energy efficiency, and increased reliability in OWC networks.

5 Conclusion

OWC presents a compelling alternative to traditional RF communication, offering high data rates, energy efficiency, security, and cost-effectiveness. With its broad range of applications and numerous benefits, OWC is poised to play a crucial role in the future of wireless communication, addressing the growing demand for high-speed, reliable, and sustainable communication solutions. Continued research and development in this field will further enhance the capabilities and adoption of OWC technologies.

Integrating green management practices in communication networks is crucial for ensuring environmental sustainability, achieving economic benefits, complying with regulations, and fulfilling corporate social responsibility. By adopting energy-efficient technologies, renewable energy sources, sustainable materials, and advanced lifecycle management techniques, communication networks can significantly reduce their environmental impact while maintaining high performance. Continued innovation and commitment to sustainability will drive the development of greener, more efficient communication infrastructures, benefiting both the industry and the planet.

AI plays a pivotal role in optimizing energy usage and resource management across various sectors. By leveraging advanced data analytics, predictive algorithms, and real-time monitoring, enhancing efficiency, reducing costs, and promoting sustainability. By leveraging predictive analytics, adaptive control, and intelligent resource allocation, ML-driven strategies can significantly enhance the sustainability and performance of OWC networks. Future research and development in this area will further set the integration of ML in green management practices, paving the way for more efficient and environmentally friendly communication technologies.

The RFA provides an effective solution for implementing a soft-switching strategy in urban OWC/RF networks. By leveraging real-time data and predictive analytics, this ML-driven approach optimizes operations, enhancing both performance and energy efficiency. The RFA offers a powerful solution for implementing a soft-switching strategy in urban OWC/RF networks. By leveraging predictive analytics and real-time data processing, this ML-driven approach optimizes both performance and energy efficiency, leading to more sustainable and reliable communication networks. Future research and development will further refine this strategy, paving the way for smarter and greener urban communication infrastructures. ML applications not only improve the efficiency and reliability of energy systems but also support broader sustainability goals. As OWC technology continues to evolve, integrating advanced AI algorithms like Random Forest, the growing demand for high-speed, efficient, and sustainable wireless data communication systems will be accomplished.

References

1. Riurean, S., Leba, M., Ionica, A., Rocha, Á.: Optical wireless communication applications and progress to ubiquitous optical networks. In: Rocha, Á., Adeli, H., Dzemyda, G., Moreira, F., Ramalho Correia, A.M. (eds.) Trends and Applications in Information Systems and Technologies. WorldCIST 2021. Advances in Intelligent Systems and Computing, vol. 1365. Springer, Cham (2021). https://doi.org/10.1007/978-3-030-72657-7_29
2. Riurean S.: Design and evaluation of visible light wireless data communication models. J. Digit. Sci. **2**(2), 3–13 (2020). https://doi.org/10.33847/2686-8296.2.2_1
3. Rozite, V., Miller, J., Oh, S.: IEA Paris (2023). https://www.iea.org/commentaries/why-ai-and-energy-are-the-new-power-coupleWhyAIandenergyarethenewpowercouple
4. Soner, B., Karakaş, M., Noyan, U., Şahbaz, F., Coleri, S.: Vehicular visible light positioning for collision avoidance and platooning: a survey. IEEE Trans. Intell. Transp. Syst. **25**(7), 6328–6344 (2024). https://doi.org/10.1109/TITS.2023.3349160

5. Zhang, Y., Wang, L., Wang, K., Wong, K.S., Wu, K.: Recent advances in the hardware of visible light communication. IEEE Access **7**, 91093–91104 (2019). https://doi.org/10.1109/ACCESS.2019.2927054
6. Gamulescu, O.M., Rosca, S.D., Panaite, F., Costandoiu, A., Riurean, S.: Accident sites management using drones. In: MATEC Web of Conferences, vol. 305, p. 00004. EDP Sciences (2020)
7. Chi, N., Zhou, Y., Wei, Y., Hu, F.: Visible light communication in 6g: advances, challenges, and prospects. IEEE Veh. Technol. Mag. **15**(4), 93–102 (2020). https://doi.org/10.1109/MVT.2020.3017153
8. Ali, M., Jayakody, D.N.K., Chursin, Y.A., et al.: Recent advances and future directions on underwater wireless communications. Arch. Comput. Methods Eng. **27**, 1379–1412 (2020). https://doi.org/10.1007/s11831-019-09354-8
9. Rîurean, S.M., et al.: A small step in VLC systems – a big step in Li-Fi implementation. IOP Conf. Ser.: Mater. Sci. Eng. **294**, 012047 (2018). https://doi.org/10.1088/1757-899X/294/1/012047
10. Geng, Z., Khan, F.N., Guan, X., Dong, Y.: Advances in visible light communication technologies and applications. Photonics **9**, 893 (2022). https://doi.org/10.3390/photonics9120893
11. He, C., Ali, W.: Advances in visible light communication. Photonics **10**, 1277 (2023). https://doi.org/10.3390/photonics10111277
12. Shao, J., Liu, Y., Du, X., Xie, T.: Adaptive modulation scheme for soft-switching hybrid FSO/RF links based on machine learning. Photonics **11**, 404 (2024). https://doi.org/10.3390/photonics11050404
13. Georlette, V., Moeyaert, V.: Li-Fi and visible light communication for smart cities and industry 4.0: challenges, research & market status in 2023. In: 2023 23rd International Conference on Transparent Optical Networks (ICTON), Bucharest, Romania, pp. 1–4 (2023). https://doi.org/10.1109/ICTON59386.2023.10207438
14. Riurean, S., Stoicuta, O., Leba, M., Ionica, A., Rocha, Á.: Underground channel model for visible light wireless communication based on neural networks. In: Rocha, Á., Adeli, H., Reis, L., Costanzo, S., Orovic, I., Moreira, F. (eds.) Trends and Innovations in Information Systems and Technologies. WorldCIST 2020. Advances in Intelligent Systems and Computing, vol. 1160. Springer, Cham (2020). https://doi.org/10.1007/978-3-030-45691-7_27
15. Stoicuta, O., Riurean, S., Burian, S., Leba, M., Ionica, A.: Application of optical communication for an enhanced health and safety system in underground mine. Sensors **23**, 692 (2023). https://doi.org/10.3390/s23020692
16. Ghazanfari-Rad, S., Ebneyousef, S.: A survey of renewable energy approaches in cloud data centers. In: 2023 8th International Conference on Technology and Energy Management (ICTEM), Mazandaran, Babol, Iran, Islamic Republic of, pp. 1–7 (2023). https://doi.org/10.1109/ICTEM56862.2023.10083820
17. Costa, W., Camporez, H., Segatto, M., Rocha, H., Silva, J.: Towards AI-enhanced VLC Systems. In: 2022 Optical Fiber Communications Conference and Exhibition (OFC), San Diego, CA, USA, pp. 1–3 (2022)
18. Cao, X.: Optimizing and modeling topology design in high-speed optical wireless communication networks. In: 2008 4th IEEE International Conference on Circuits and Systems for Communications, Shanghai, China, pp. 683–687 (2008). https://doi.org/10.1109/ICCSC.2008.150
19. Riurean, S.: A study on the VLC security at the physical layer for two indoor scenarios. MATEC Web Conf. **342**, 05009 (2021). https://doi.org/10.1051/matecconf/202134205009
20. Elgamal, A.S., Aletri, O.Z., Yosuf, B.A., Adnan Qidan, A., El-Gorashi, T., Elmirghani, J.M.H.: AI-driven resource allocation in optical wireless communication systems. In: 2023 23rd International Conference on Transparent Optical Networks (ICTON), Bucharest, Romania, pp. 1–5 (2023). https://doi.org/10.1109/ICTON59386.2023.10207473
21. Wu, X., O'Brien, D.C.: A novel machine learning-based handover scheme for hybrid LiFi and WiFi networks. In: 2020 IEEE Globecom Workshops (GC Wkshps, Taipei, Taiwan, pp. 1–5 (2020). https://doi.org/10.1109/GCWkshps50303.2020.9367577

22. Căilean, A.-M., Avătămăniței, S.-A., Beguni, C.: Driving toward connectivity: vehicular visible light communications receiver with adaptive field of view for enhanced noise resilience and mobility. Sensors **24**, 2814 (2024). https://doi.org/10.3390/s24092814
23. https://www.nationalgeographic.com/science/article/130618-google-balloon-wireless-communication-internet-hap-satellite-stratosphere-loon-project. Accessed 28 June 2024
24. Kavitha, T., Angel Reena, C.R.:. Smart museum using Li-Fi technology. In: Smys, S., Senjyu, T., Lafata, P. (eds.) Second International Conference on Computer Networks and Communication Technologies. ICCNCT 2019. Lecture Notes on Data Engineering and Communications Technologies, vol. 44. Springer, Cham (2020). https://doi.org/10.1007/978-3-030-37051-0_105
25. https://www.oledcomm.net/blog/lifi-technology-myths-vs-facts/. Accessed 19 June 2024
26. https://assets.new.siemens.com/siemens/assets/api/uuid:3d852046-82ba-4f88-86a0-b55369359c66/final-factorytomorrow-wp.pdf. Accessed 19 June 2024
27. Kiran, K.V., et al.: Machine learning aided switching scheme for hybrid FSO/RF transmission. Intell. Decis. Technol. **14**, 529–536 (2021)
28. Dugaev, D., Peng, Z., Luo, Y., Pu, L.: Reinforcement-learning based dynamic transmission range adjustment in medium access control for underwater wireless sensor networks. Electronics **9**, 1727 (2020). https://doi.org/10.3390/electronics9101727
29. Frometa Fonseca, D., Genoves Guzman, B., Luca Martena, G., Bian, R., Haas, H., Giustiniano, D.: Prediction-model-assisted reinforcement learning algorithm for handover decision-making in hybrid LiFi and WiFi networks. J. Opt. Commun. Netw. **16**(2), 159–170 (2024). https://doi.org/10.1364/JOCN.495234
30. https://www.geeksforgeeks.org/random-forest-classifier-using-scikit-learn/. Accessed 21 June 2024
31. Lapčak, M., Ovseník, L., Oravec, J., Zdravecký, N.: Design of hard switching for FSO/RF hybrid system based on prediction of RSSI parameter and environmental conditions. In: 2022 32nd International Conference Radioelektronika (RADIOELEKTRONIKA), Kosice, Slovakia, pp. 1–6 (2022). https://doi.org/10.1109/RADIOELEKTRONIKA54537.2022.9764908
32. https://www.meteoblue.com/ro/vreme/archive. Accessed 15 July 2024
33. Mishra, A., Pati, P.S., Giri, R.K.: Performance analysis of SIM based FSO using various modulation techniques with APD receiver over turbulent channel. In: 2017 International Conference on Wireless Communications, Signal Processing and Networking (WiSPNET), Chennai, India, pp. 2730–2734 (2017). https://doi.org/10.1109/WiSPNET.2017.8300259

Digital Universities: An Artificial Intelligence as an Important Step

Darya Rozhkova, Nadezhda Rozhkova, and Daniela Gonzalez Serna

Abstract There is a huge reshaping of the world of technologies. It is difficult to name a profession that will not be affected by technological development, that is not subject to some kind of global changes in connection with artificial intelligence (AI). Thus, we believe in importance of AI introduction into a practice of education in general and in building the digital university in particular. Using the interview method, we have identified how AI can be embedded in a platform for a digital university through generative models of AI, virtual assistant, VR educational instruments and statistical models of AI.

Keywords Artificial intelligence · Education · Digitalization · Digital university · Generative AI

1 Introduction

The tremendous growth in the volume of information, the rapid introduction of artificial intelligence (AI), automation and robotization will lead to a global reboot, revision, and upgrade of most existing specialties. All this will lead to a digital transformation of world education system, both in terms of the use of artificial intelligence in teaching and the creation and extension the student's competence in using artificial intelligence. Now artificial intelligence, machine learning, and deep learning are such «an elephant that we can never eat whole» because it grows all the time; it grows faster than we can eat it and we have to understand this phenomenon [1, 2].

D. Rozhkova (✉)
Financial University Under the Government of the Russian Federation, Moscow, Russia
e-mail: rodasha@mail.ru

N. Rozhkova
State University of Management, Moscow, Russia

D. G. Serna
Trade Finance, Singapore, Singapore

T. Antipova (ed.), *Digital Technology Platforms and Deployment*, Information Systems Engineering and Management 36, https://doi.org/10.1007/978-3-031-86547-3_11

For the management of universities, it is important not only to know and understand modern technologies, but to incorporate global technological trends into educational programs. We also need to prepare for a turbulence such as the coronavirus pandemic and other global risks that significantly change all areas of our lives [3]. In such conditions, it is impossible to work without artificial intelligence, which in future will be able to solve not only routine, but also operational tasks. For example, it will make calculations, draw up a presentation, or create a virtual prototype of a product.

Our aim in this study is to continue a discussion about creating an education system in Russia [4–6], which will most fully reflect the ongoing digitalization processes (creating digital university), developing alternative teaching methods [7] but through the lenses of the AI impact.

With these issues in mind, the present paper addresses the following set of research questions:

- What is the role of AI in transforming an university into digital?
- How AI can be embedded into the university's platform for digital education?

Given the evidence on the importance of practice, we emphasize that this chapter does not aim to demonstrate the used tools. This means that we will not analyze the practices already implemented by universities and provide practical implementation, its effectiveness and outcomes.

The paper proceeds as follows. Section 2 reviews recommendations for leaders of educational projects who are wondering whether they should be enhanced with AI capabilities. Section 3 provides data and methodology. In Sect. 4 we have constructed the main instruments of AI that can be included into a platform for a digital university.

2 Literature Review

In the 1940s, mathematicians had already come up with the prototype of AI. Artificial intelligence, neural networks are all about higher mathematics, since such a complex history of calculations (vectors, algorithms, etc.) was assumed. These mathematicians based their hypotheses precisely on creating a semblance of the human brain. And already in the 50 s, humanity was making such far-reaching plans for machine translation, that a computer would beat a person at chess. Some of past desires were quite simple and modest. However, the so-called "winter of artificial intelligence" occurred for a fairly long period of time [8].

The fact is that some research reached a dead end, because the computers of those years were too weak; they were simply not enough to process large volumes of data, and to be honest, there was no such volume of data. It's hard to imagine that in the 60s it was possible to take a million pictures somewhere to train a neural network to create pictures. It was impossible to use it, there were no input/output devices, that is, the technology was ahead of its time.

For many decades it was, well, not forgotten, of course, but it was only for narrow professionals. And only about 10–15 years ago certain mathematical discoveries were made, hypotheses were put forward that formed the basis of modern neural networks.

Moreover, large volumes of data have accumulated that can be studied (for example Big Data) as well as an increase in computer power. It has become accessible to business. And for sure business never misses an opportunity to earn profit. It immediately rushed and created the neural networks that we know today.

The main problem with the use of neural networks in universities, which is cited by many practitioners and businessmen, is an existence of too many neural networks. That is, in just about a year and a half, there have become so many of them that managers don't know what to do with it, because indeed now teams of several programmers can literally sit down on their home laptops, renting a server for training, and create a highly specialized neural network for any task.

The industry giants are OpenAI with Chat GPT launched in November 2022, Alphabet with Bard launched in February 2023 etc. Chat GPT is the fastest growing online service in the world, it gained its first 100 million users in 2 months. The success is explained by a user-friendly interface for ease of use and functionality. For comparison, Tik Tok growth is 9 months, Instagram—2 and a half years. This indicates the "revolutionary nature" or disruptive nature of what is happening [9].

Nowadays, many different applications have appeared that claim to be applications based on artificial intelligence, for example, a caption application that moves your gaze when recording a video of yourself on the front camera. There are also some applications for automatically editing videos to some kind of music. In https://www.futuretools.io we can see 2429 digital tools that apply AI.

In modern literature it is considered at least four recommendations for leaders of educational projects who are wondering whether they should enhance university with AI capabilities.

1. A balance between AI and a professor.

The idea of replacing teachers with artificial intelligence seems to be very tempting as it can save time and budget, increase performance [10, 11]. However, social interaction and teacher personality are important factors in education quality. Therefore, AI should be viewed as an assistant, not a substitute for teacher.

The ideal option is when technology fills the space that the teacher is not able to cover, like promote personalization and enhancing learning outcomes [12]. If AI literacy is included in the latest educational standards and strategic initiatives, then it can really increase the accessibility and quality of education [13].

2. A control the quality of AI in education.

Not a single educational organization in the world that uses artificial intelligence applies it without supervision. This also applies to those who have huge data sets, very powerful models, and spend a lot of resources on development and training.

Any technology needs to be tested and verified, in education, the work of AI is monitored by methodologists, teachers, and other specialists (depending on where the technology is being implemented) [14].

3. An idea of humanity when working with technology.

Digitalization of education is so popular that people sometimes can forget about what it is needed for. And it is needed to help the student realize his potential. Therefore, it is important to cherish humanity (teachers' professional judgement and the personal relationship) when implementing projects involving AI to make an education more effective, more correct [15].

4. Be ethical and transparent.

AI can be used only of we can see and understand how it made this or that decision and why it did so. Indeed, in educational projects, especially those aimed at children and adolescents, errors or incorrect interpretation of data can lead to serious consequences for a person's fate [12, 16].

3 Data and Methodology

In this research we did an online survey as a qualitative research tool "for nuanced, in depth and sometimes new understandings of social issues" [17].

We have created the online survey to highlight the main ideas embedding AI into process of building a digital university. We intentionally did not conduct a student survey, as we though that we need opinion of the experts in the field.

We conducted a survey called "Scanning the educational horizons: Digital university and AI" for experts on the impact of AI on universities, the purpose of which is to determine the extent of the influence of artificial intelligence on the educational industry, as well as to assess the risks, barriers, and expectations that researchers and business representatives highlight when implementing this technology.

The survey was conducted in Russian. The survey was sent only to selected recipients and was not available to everyone; 41 Russian-speaking representatives of business, scientific and research organizations completed the survey. The survey was conducted in June–August 2023.

This survey does not pretend to be representative (sampling features, limited number of responses), but it adds information on how a limited number of experts imagine the introduction and development of AI technologies in the activities of universities. More than half of the sample (64%) are represented by respondents working in educational organizations (private and public); 12% are employees of the auditing company KPMG, 24% are representatives of VEB Bank.

The first question was open-ended and did not give respondents any clues: "Let's dream, in your opinion, how will AI be used in universities of the future, what technologies do you think are promising for the development of digital education? Use your imagination and share your most unexpected ideas!".

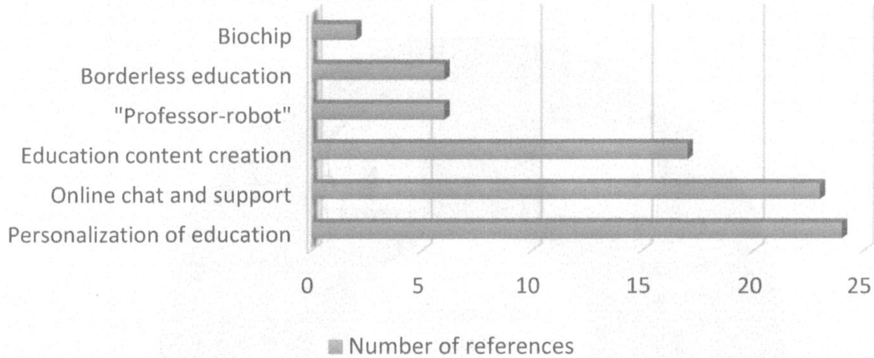

Fig. 1 Mentions of technologies and/or subjects in open responses from experts (authors' elaboration)

Respondents could write whatever they wanted. Next, we grouped the answers into conditional groups using keywords. Respondents could propose several technologies or areas of application of AI in education. In order to simplify, in Fig. 1 we present the most popular technologies/plots among respondents.

As it can be seen most respondents have mentioned personalization of education as a main trend of future education. It means that students can pick the course ad formats of education. The second one represents current direction that to our understanding have been already implemented in educational practices "online chat and support".

We go a step further, so the second question was also open-ended "Let's dream up further, based on your expertise, intuition and imagination, how the platform for digital education can be enriched by AI"? The question provides a unique cross-section understanding of AI and digital universities by civil society leaders and experts. The results of the question 2 would be presented in the next section.

4 Results

Based on the second question' results we have constructed the main instruments of AI that can be included into a platform for a digital university: generative models of AI, virtual assistant, VR educational instruments and statistical models of AI (Fig. 2).

Generative AI (GenAI) models would be used for creation of unique content for video lessons and lectures, exam, and other educational activities [18]. There are various directions of using GenAI to create "human-like content" [19] (Table 1).

The interesting fact is that majority of participants (95%) in a varying degree have mentioned fears for challenges of using GenAI while creating a digital university. The first fear is that the AI would challenge the work of professors and management stuff at university. At this stage in the development of humanity and the human brain,

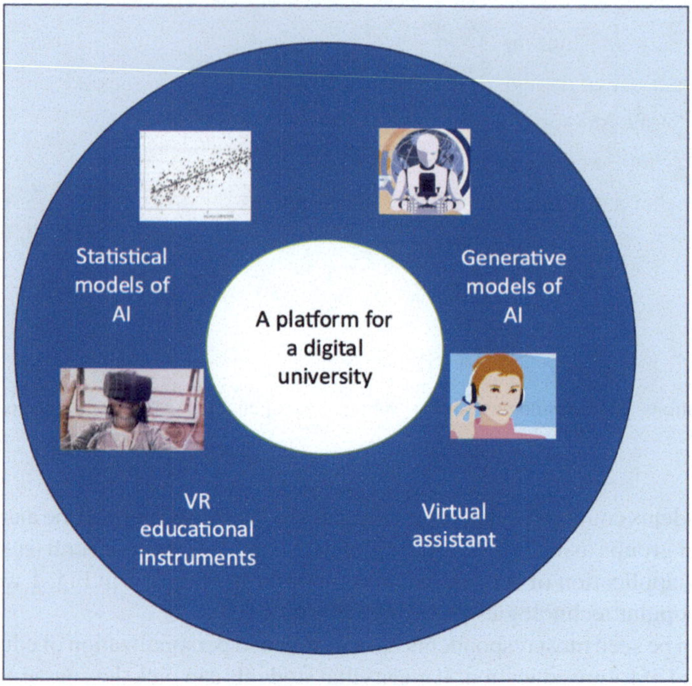

Fig. 2 AI embedded in a platform for a digital university (authors' elaboration)

they are still important. But in future the big question was: "Can the AI overcome the human?".

In fact, the question is difficult, because it is always hard to predict the future, although humanity is always trying to do it. At the stage, artificial intelligence can only harm us through its mistakes. Especially if we entrust him with solving a complex problem. And such a mistake will be harmful.

If you compare it with an autopilot on an airplane. That is, of course, the advent of autopilots on airplanes has increased the safety of air travel, because it has relieved the workload of a person and reduced the likelihood of human error. But this does not mean that plane crashes have not occurred due to autopilot problems. But in general, the level of aviation safety in the world is increasing every year. The same applies, for example, to an autopilot in a car, which is based on computer vision technologies. This is another artificial intelligence technology. Of course, an autopilot is safer than a person; a person gets distracted and tired.

The second fear was that the AI would challenge the work of students. It is important that students independently do their work: analysis, synthesis, design etc. And GenAI can facilitate knowledge acquisition. But it can as well reflect plagiarism, demotivation and "avoiding learning" [21, 22].

Virtual assistant based on AI targets to provide multiple students with comprehensive information support, answer to the questions, explain course structure,

Table 1 Directions of using GenAI in education (authors' compilation based on [7, 19, 20])

Directions	Description
Text	Neural networks can help professors and educational content creators to translate or summarize useful materials, generate an chapter, a story for storytelling, or a checklist from scratch. It is possible to pick a necessary writing style, to select vocabulary etc. If reliable information is fundamentally important, for example, for a scientific chapter, the neural network can verify the data, collected information from official websites for the specific periods
Pictures	If there is no need to use specific existing images (for a design course), it is possible to create any illustration
Audio and video	Neural networks can generate audio, introduce a text and voice it in any language (British, American, Canadian and Australian pronunciations) Video files and pictures can serve as video for future audio recordings, so neural networks can do a training video
Simulators	Neural networks can create any simulator: an interview with a recruiter, a doctor, seller, examiner, tax inspector, or guide etc.
Games and tests	GenAI models can processes given information and then is able to create a test for this material. Moreover, it can process the data of an entire website, text, or educational content in any format for any audience, which it independently generated at the user's request
Presentations	GenAI models can generate an entire presentation using one template, can improve the text of ready-made slides and generate images itself without an additional request
Template of the course	Professors and educational content creators can create in a few seconds the skeleton of an online course, a weekly lesson plan or a minute-by-minute lesson scenario on a given topic or with given initial data
Knowledge testing	There is a big potential of GenAI in knowledge testing (essays evaluation). It is no secret that when checking open answers by an expert, the problem of subjectivity arises. We assume that the psychometric principles that work with people are also applicable to training neural networks, and it is very likely that they will ultimately give the best result
Analysis of the quality of educational materials	AI analyzes data on what percentage of tasks are completed correctly, what is the average speed of their completion, and statistics on students' incorrect answers. As a result, the AI can give professors and educational content creators a hypothesis about which tasks are composed with errors or are not very successful for students' perception

scheduling etc. [22]. It has demonstrated significant potential in medical education [23].

If the user has problems mastering the material or completing a practical task, *Virtual assistant* based on AI will be able to answer questions, explain the material and suggest books and chapters for additional study. Integrated artificial intelligence will analyze the user's educational process and his emotional state. Based on the data obtained, it will provide additional educational materials and recommendations for taking courses, as well as advice on managing emotional states. Within the platform, it will be possible to unite users into groups and teams so that they can exchange experiences, fill the gap in communication and share the results of their intellectual work.

VR educational instruments can open a whole new world of possibilities in higher education and allow all students to be involved in practical work and build an individual development trajectory, it is about entertainment and gaming. For example, PwC's Emerging Technology Group compared the effectiveness of corporate training in three formats—classroom, e-learn, or v-learn. It turned out that in the latter case, students learned the material several times faster, were more confident and more focused than in the classroom or on the computer [24]. Universities can also use this format co create realistic scenarios.

Digital twins (a virtual model designed to accurately reflect a physical object) of any products or processes can be used in educational process. For example, in engineering, digital twins are software copies of industrial plants that are modeled and visualized similarly to and synchronized with their industrial originals.

Statistical models of AI will likely be an accelerator that help big data analytics, coordination of the platform, creation of an individual student trajectory, personalization the learning experience [25].

However, it is equally important and pertinent to use predictive analytics when a mathematical model is trained on a larger volume of data. It can become a navigator for children in choosing electives and disciplines. Possibility of introducing AI technology into the educational process management system: analyze the "digital footprint" of applicants and students in order to develop automated recommendation services for building individual competency, forming a curriculum. The machine learning algorithm have great potential to find explicit and implicit patterns, features of life values that are inherent in different groups of students within their specialty and then selects the relevant field of study for the applicant. Before choosing an educational program, the user can be asked to take a Guidance test in a game form. The results will be processed by Artificial Intelligence and based on them, will offer an individual educational trajectory or personalized learning path (compile a selection of courses, arrange them in the required sequence and explain what skills the user will acquire after completion) [26].

Psychological support with AI can include analysis of facial expressions, retina. There are cameras throughout the universities, AI can relieve the workload and improve the quality of work of university psychologists, ensure anonymity, impartiality, and timeliness of assistance to the student. Administrative support may include admissions campaign and support the educational process.

5 Conclusion

We have come to conclude that AI implementation while building a digital university is at it beginning stage. It can only be implemented in collaboration with other universities. The path of unification, collaboration and consortia is the most effective. No single university can act as an exclusive source of competencies and knowledge; significant results can only be achieved jointly. We should try to create a clear system, in the form of regulatory documents, ensuring the connection and communication of all implementers of digital universities.

The interest to introduce AI into a platform for a digital university can be explained both by the current level of technology development and by the number of mentions of AI in the press, at recent conferences, etc. If several years ago artificial intelligence was mentioned at highly specialized technology events, from 2018 the discussion of AI has become a much more common topic.

The study is interesting because it shows how the educational leaders and experts we interviewed, see further developments of an artificial intelligence. And how they believe it will change almost every aspect of educational process. Depending on how they see and understand technology, it will largely reflect how these technologies will be used as part of the development of digital universities, as well as how society and the state will control the development of certain technologies.

References

1. Aggarwal, K., Mijwil, M.M., Al-Mistarehi, A.H., Alomari, S., Gök, M., Alaabdin, A.M.Z., Abdulrhman, S.H.: Has the future started? The current growth of artificial intelligence, machine learning, and deep learning. Iraqi J. Comput. Sci. Math. **3**(1), 115–123 (2022)
2. Roser, M.: The brief history of artificial intelligence: The world has changed fast–what might be next? Our World in Data (2023)
3. Antipova, T.: Coronavirus pandemic as black swan event. In: International Conference on Integrated Science, May 2020, pp. 356–366. Springer International Publishing, Cham (2020). https://doi.org/10.1007/978-3-030-49264-9_32
4. Rozhkova, D., Rozhkova, N., Blinova, U.: Digital universities in Russia: prospects and problems. In: Antipova, T., Rocha, Á. (eds.) Digital Science 2019. DSIC 2019. Advances in Intelligent Systems and Computing, vol. 1114. Springer, Cham (2020). https://doi.org/10.1007/978-3-030-37737-3_23
5. Rozhkova, D., Rozhkova, N., Blinova, U.: Digital universities in Russia: digitization with extra speed. J. Digit. Sci. **2**(1), 76–81 (2020). https://doi.org/10.33847/2686-8296.2.1_7
6. Rozhkova, D., Rozhkova, N.: COVID-19 and E-learning: challenges for Russian professors. JDAH **1**(2), 34–41 (2020). https://doi.org/10.33847/2712-8148.1.2_3
7. Antipova, T., Riurean, I.P., Riurean, S.: Is distance teaching-learning-evaluation triad a form of digital art? JDAH **1**(2), 03–19 (2020). https://doi.org/10.33847/2712-8148.1.2_1
8. Shin, Y.: The spring of artificial intelligence in its global winter. IEEE Ann. Hist. Comput.Comput. **41**(4), 71–82 (2019)
9. Chow, J.C., Sanders, L., Li, K.: Impact of ChatGPT on medical chatbots as a disruptive technology. Front. Artif. Intell. **6**, 1166014 (2023)
10. Cerullo, M.: Here's how many U.S. workers ChatGPT says it could replace. MoneyWatch, 5 April 2023 (2023). https://www.cbsnews.com/news/chatgpt-artificial-intelligence-jobs/

11. Antipova, T.: Public universities' performance evaluation. In: Antipova, T. (ed.) Advances in Digital Science. ICADS 2021. Advances in Intelligent Systems and Computing, vol. 1352, pp. 126–137. Springer, Cham (2021). https://doi.org/10.1007/978-3-030-71782-7_12
12. Pedro, F., Subosa, M., Rivas, A., Valverde, P.: Artificial intelligence in education: challenges and opportunities for sustainable development (2019)
13. Ng, D.T.K., Lee, M., Tan, R.J.Y., Hu, X., Downie, J.S., Chu, S.K.W.: A review of AI teaching and learning from 2000 to 2020. Educ. Inf. Technol. **28**(7), 8445–8501 (2023)
14. Muzammul, M.: Education system re-engineering with AI (artificial intelligence) for quality improvements with proposed model (2019)
15. Chan, C.K.Y., Tsi, L.H.: The AI revolution in education: will ai replace or assist teachers in higher education? (2023). arXiv:2305.01185.
16. Kooli, C.: Chatbots in education and research: a critical examination of ethical implications and solutions. Sustainability **15**(7), 5614 (2023). https://doi.org/10.3390/su15075614
17. Braun, V., Clarke, V., Boulton, E., Davey, L., McEvoy, C.: The online survey as a qualitative research tool. Int. J. Soc. Res. Methodol.Methodol. **24**(6), 641–654 (2021)
18. Michel-Villarreal, R., Vilalta-Perdomo, E., Salinas-Navarro, D.E., Thierry-Aguilera, R., Gerardou, F.S.: Challenges and opportunities of generative AI for higher education as explained by ChatGPT. Educ. Sci. **13**(9), 856 (2023)
19. Lim, W.M., Gunasekara, A., Pallant, J.L., Pallant, J.I., Pechenkina, E.: Generative AI and the future of education: Ragnarök or reformation? A paradoxical perspective from management educators. Int. J. Manag. Educ. **21**(2), 100790 (2023)
20. Open Culture: Noam chomsky on ChatGPT. Open Culture (2023)
21. Zhai, X.: ChatGPT user experience: implications for education. Available at SSRN 4312418 (2022)
22. Mekni, M., Baani, Z., Sulieman, D.: A smart virtual assistant for students. In: Proceedings of the 3rd International Conference on Applications of Intelligent Systems, January 2020, pp. 1–6 (2020)
23. Ahuja, A.S., Polascik, B.W., Doddapaneni, D., Byrnes, E.S., Sridhar, J.: The digital metaverse: applications in artificial intelligence, medical education, and integrative health. Integr. Med. Res. **12**(1), 100917 (2023)
24. The effectiveness of virtual reality soft skills training in the enterprise a study. PWC. Public report, 25 June 2020. https://www.pwc.com/us/en/services/consulting/technology/emerging-technology/assets/pwc-understanding-the-effectiveness-of-soft-skills-training-in-the-enterprise-a-study.pdf
25. Tsai, S.C., Chen, C.H., Shiao, Y.T., Ciou, J.S., Wu, T.N.: Precision education with statistical learning and deep learning: a case study in Taiwan. Int. J. Educ. Technol. High. Educ. **17**, 1–13 (2020)
26. Somasundaram, M., Junaid, K.M., Mangadu, S.: Artificial intelligence (AI) enabled intelligent quality management system (IQMS) for personalized learning path. Procedia Comput. Sci. **172**, 438–442 (2020)

The Sharp-Knife Theory of Artificial Intelligence Evidence from the AI Productivity Paradox

Yury Pukha

Abstract Scientists studied thoroughly the phenomenon of information and communication technologies' impact on productivity. In 1987 Sollow summarized the puzzle with the well-known conclusion—"You can see the computer age everywhere but in the productivity statistics". It took two decades for businesses to develop a so-called intangible capital concept composed of human, organizational capital, and new process knowledge needed to achieve effectiveness with ICT implementation. Recent advancements of digital technologies created a similar pattern issue with large investments into Artificial Intelligence, big data, and other technologies but very limited economic returns. The innovativeness of this paper is based on the analysis of 3 large-scale digitalization programs with AI as a leading tool that was expected to provide the largest return on investment. Key failure factors are analyzed that impeded the organizations to reach planned results and return on investments. Three factors were identified as primary obstacles for the success of comprehensive AI-based solution implementations—complexity, managerial capability, and compatibility.

Keywords Enterprise architecture · Digital transformation · Artificial intelligence · Failure factors

1 Introduction

Despite high expectations, digitalization has not yet fully achieved increased productivity, economic growth, or wealth improvement. The economic structure of developed and developing countries has shifted significantly twice in the last century due to industrialization and later computerization and IT. However, despite rapid adoption of technologies like mobile communications, Big Data, and AI, a notable productivity boost is lacking [1, 2]. Many analysts and executives view AI as key to a new wave of productivity improvement [3], yet a paradox similar to the Solow

Y. Pukha (✉)
ESCP Business School, Paris, France
e-mail: yury.pukha@edu.escp.eu

T. Antipova (ed.), *Digital Technology Platforms and Deployment*, Information Systems Engineering and Management 36, https://doi.org/10.1007/978-3-031-86547-3_12

paradox persists: substantial investments in AI yield uncertain returns, and productivity may even decrease due to high operational costs [4]. Achieving a positive change in productivity requires a critical mass of new technologies, including AI [5].

The Enterprise Architecture framework illustrates the challenge of obtaining returns from AI and advanced Big Data at the business layer. For AI to be productive at this level, the underlying layers must mature along the technology S-curve. Technical errors or business inefficiencies at each layer hinder seamless system functioning.

The infrastructure layer initiates comprehensive business process automation, but current infrastructures are often siloed, limiting data integration. IoT supports disruptive technologies like Big Data analytics and AI across sectors. In the digital era, the application layer adopts multiservice flexible environments with open-source software, emphasizing ecosystem support. Data accumulation shifts to data lakes, facilitating intelligent analytics and AI algorithms. The business layer transitions from static reporting to dynamic analytics and AI technologies [6, 7]. Unlike traditional architectures, digital enterprise architecture demands greater integration and interoperability, enabling automation and radical changes in business models (Fig. 1).

Investments and returns should consider the differences in enterprise architectures. Traditional EA achieves gains within silos, while in the digital world, returns from upper EA layers are expected after implementing mature layers beneath.

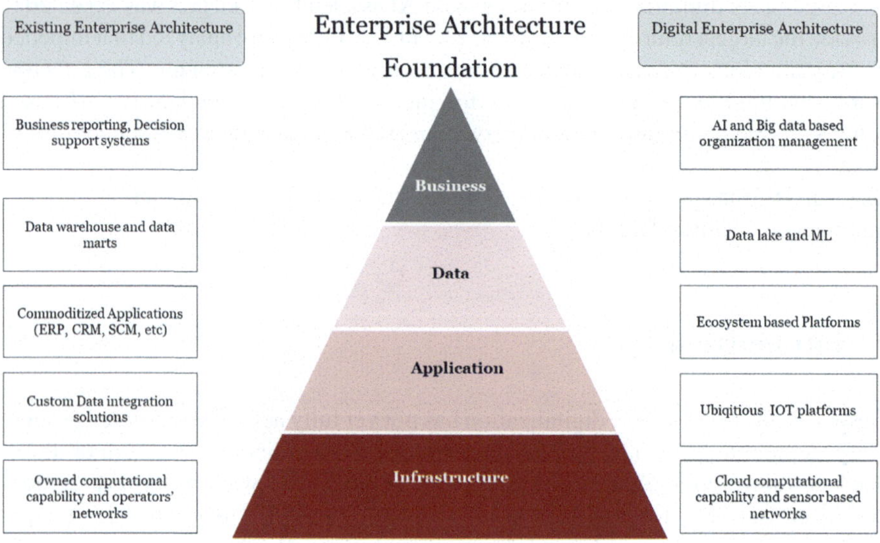

Fig. 1 Digital and existing enterprise architecture

2 Research Question and Methodology

The chapter aims to evaluate key factors impeding implementation of large-scale digitalization programs with AI as a key planned lever to improve productivity and return on investments. The hypothesis being tested is that factors related to organizational resistance, business processes, managerial competency are required for an organization to accumulate knowledge in AI as a new technology for successful implementation. Furthermore, for AI solutions to deliver on the promise of productivity, the organization needs to transform lower layers of EA to digital state enabling knowledge based intangible capital for business layer.

The methodology was based on a qualitative case study approach. At the data gathering phase the author selected cases of companies investing in comprehensive digital solutions, focusing on economic benefits from AI. At first, nine enterprises that launched large-scale multi-year digital transformation programs with AI capabilities were selected in different industries—energy, agriculture, municipal management (smart city), transportation and smart home. The author conducted in-depth interviews with at least two C-level representatives of those companies, one as a business owner of the initiative and the second as IT or program leader. All interviews contained questions from two parts: first part focused on general discussion of digital and AI program plan, configuration, expected and realized results, and second part focused on factors and drivers that interviewees indicated as critical during program implementation. The structure of the discussion didn't have a pre-designed questionnaire but followed similar pattern and all participants were asked questions about the contribution of digital transformation and AI projects at different levels of EA architecture. Three cases that had sufficient program scope and provided necessary data were included in the final analysis.

At the analysis phase, the author performed two steps. Firstly, the digital transformation program in each company was summarized in a few large projects and presented along the industry specific value chain to demonstrate the role of each project in organization. Secondly, the great variety of factors that was discussed during the interviews and had an impact on program implementation was synthesized under the unified framework. The factors that became key impediments for successful digital transformation and specifically AI implementation, were identified at different levels of Enterprise Architecture.

Based on the lessons learned from the case analysis and synthesis of impediment factors the author emphasizes recommendations for the management aimed to succeed in AI based digital transformation programs.

3 Literature Review of AI Adoption Critical Success Factors

This section aims to synthesize a framework of key AI adoption factors to review selected cases.

Brynjolffson [8] argued that AI can be viewed as a General Purpose Technology, which will increase productivity not only directly but also enable society to introduce innovations in adjacent areas. The authors also introduced the concept of intangible digital capital [9] which constitutes knowledge and innovation on how organizations can be managed and run differently using new technologies. The conclusions developed by Brynjolfsson are based on Hall's research of capital accumulated [10] with the computerization development. He writes that—"the main portion of the computer-related intangible assets comes from the new business processes, new organizational structure, and new market strategies, which each complement the computer technology … computer use is complementary to new workplace organizations which include more decentralized decision making, more self-managing teams, and broader job responsibilities for line workers."

The notion of intangible digital capital aligns with a comprehensive view on DT implementation lessons learned presented by Loonam et al. [11]. Authors concluded that there are four key themes important for organizations to implement digital transformation initiatives: strategies, customers, technologies, and organization. The strategy theme focuses on business model innovation [12], i.e. exploring digital businesses, reconfiguring value delivery models, e.g. striking a balance between human touch and digital opportunities, and rethinking value propositions. The customer theme is about rethinking customer experience from the outside-in, redesigning the whole customer journey in searching for new innovative business models. The technology focus is aimed to achieve the next level of competitive advantage for an organization using new digital tools, enabling an integrated approach to systems across the organization, business insights from data analytics, new omni-channels across social and mobile platforms. The organization theme is about recognizing the importance of fostering a digital culture and streamlining of structures and processes.

Hamm and Klessel [13], using the TOE theoretical framework [14], identified significant factors for AI implementation projects [15]. They analyzed over 100 academic papers and concluded that out of 36 factors [16–18], only eight are statistically significant. These factors are grouped into three areas:

Firstly, organizational drivers include top management support, technical competencies, and overall resources. Top management support is crucial for executing transformational initiatives despite organizational resistance. Technical competency is essential for digital transformation, requiring management to understand digital technologies and their potential impact.

Secondly, technological factors are critical for success. An organization's IT infrastructure must be compatible with digital transformation plans, including migrating legacy systems. AI's technological superiority and the availability and quality of data are also key factors, as digital transformation often involves leveraging data for decision-making automation.

Thirdly, external factors are crucial. Market competition drives companies to adopt digital transformation, with examples like Apple and Netflix demonstrating both success and failure. Government regulation also plays a key role, particularly in sectors like energy and electric vehicles, where it can drive industry-wide transformation.

Chen's analysis [18] of AI implementation in four Chinese telecom companies, also using the TOE framework, confirmed the direct importance of six success factors for AI adoption.

Managerial support is critical for AI implementation as it ensures commitment to both organizational and technological changes, influencing management culture and processes. Compatibility refers to how well the innovation aligns with the organization's needs. Effective AI adoption requires practices that complement the AI solution, minimizing the need for significant organizational changes and fitting the company culture.

Complexity measures the difficulty of implementing and using the solution. Simpler solutions facilitate broader and more efficient adoption. AI transformations are complex, especially in their impact on cross-functional processes, requiring clear communication of the transformation's goals and processes.

Relative advantage assesses the superiority of the proposed solution over the existing state, measured by financial metrics like costs and profits, or qualitative aspects like convenience and user adoption. AI solutions are expected to significantly improve both front-end customer interactions and back-end processes.

Government regulation often drives innovation in public and state-controlled sectors, which may resist improvements without top-down mandates. Governments use AI to enhance public services and achieve better monitoring and control.

Vendor partnership is crucial due to the limited availability of AI expertise. Building AI capability internally is costly and risky. Effective AI solutions require adaptation to the company's specific processes and data, making the choice of vendor critical.

Managerial capability influences all these factors and indirectly impacts AI adoption. It involves organizing, motivating, and persuading people to embrace transformation. Managers must ensure employees are trained, motivated to adopt new processes, and adhere to strict project management rules.

Environmental factors such as government regulation, competition, and market uncertainty affect managerial capability. Interestingly, current technical capability (hardware and software) does not significantly influence AI adoption.

For this case study analysis, Chen's six factors [18] are extended with Managerial Capability, resulting in the following AI adoption factors tested in the research: managerial support (commitment); managerial capability (knowledge and competence); relative advantage; compatibility (with adopter needs); complexity; government regulation; vendor partnership.

Selected AI adoption factors aligned with this study connecting technological advances of AI with the requirement of Intangible Capital accumulation aimed to develop magnitude of specific data, building firm-specific competence in AI, and implement new business processes.

4 Results—AI Implementation in Practice

This section presents a review of 3 DT program case studies. Each DT program shows key projects along the industry specific value chain that are evaluated by their expected economic returns. The interviewees then provide their assessment at which layers of EA the projects met their goals as well as key factors that impeded the implementation.

4.1 Case 1. Energy Generation Holding

A large energy holding in Eastern Europe initiated an automation project for its generation plants. The project aimed to install monitoring and control equipment and develop a real-time AI-based solution for automated decision-making using detailed technological data. The primary objectives were to improve equipment reliability through precise data and automation, and to optimize costs by managing fuel consumption based on AI recommendations aligned with heat, electricity demand fluctuations, and grid load.

The expected economic benefits included fuel savings and optimized preventive maintenance, with projected returns on investment as key AI components would become operational in years 3–4. Most investment costs were allocated to infrastructure and data management, including new monitoring equipment, sensors, and data aggregation platforms.

While the project was implemented on time and within scope, the anticipated economic impact has not yet materialized. The AI-based control and recommendation system required additional complexity (data sources, integration, process changes), necessitating further investment in business logic layers to enhance AI outcomes (Fig. 2).

The company has invested substantial financial, human, and managerial resources. Despite maintaining a fully operational IT infrastructure, the project has failed to yield economic returns, incurring additional costs without financial gains over the past years. Application layer was developed in a fragmented manner. Standalone applications such as zip optimization and turbine reliability control met their objectives. However, functions like dynamic pricing, preventive maintenance, and fuel cost management lacked integration at the data and business management layer, hindering overall success.

The project, driven by a government initiative, had strong regulatory and managerial support. The technology stack was developed by a vendor with deep AI skills. Managerial skills, particularly among top and middle managers, were insufficient to

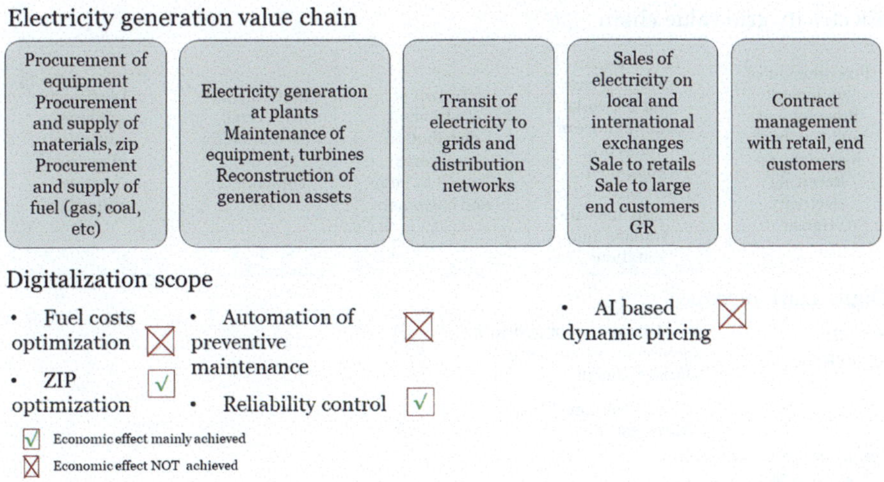

Fig. 2 Impact of digitalization on value chain in electricity generation

adapt processes for economic gain from the implemented solutions. The complexity of data requirements and organizational changes was underestimated.

4.2 *Case 2. Energy Grid and Distribution Holding*

The holding launched projects to implement Smart Grid technology, including the Regional Digital Electricity Distribution network. This project aimed to automate power centers, implement an energy monitoring system, and manage the regional distribution network. Expected benefits included optimized maintenance costs, reduced energy losses, and increased equipment reliability.

The expected benefits from the automated management system using telematics and AI for real-time equipment management were significant. Equipment monitoring and electricity consumption tracking were successfully implemented with smart meters and telematics. However, real-time management proved complex, requiring extensive data for automated decisions on switching, rerouting, and adjusting power output. Despite substantial investments, economic benefits were only partially achieved, mainly from improved monitoring. The sophisticated AI and big data systems have not provided effective real-time management solutions, with most operational decisions still made manually by analysts, operators, or managers (Fig. 3).

The ZIP optimization aimed to automate monitoring, tracking, and replacement of equipment components. Its integration with the ERP reduced procurement costs, achieved cost optimization through investments in infrastructure, platform, and application layers. The preventive maintenance system required an architecture to correlate trouble events, diagnose root causes, propose maintenance recommendations,

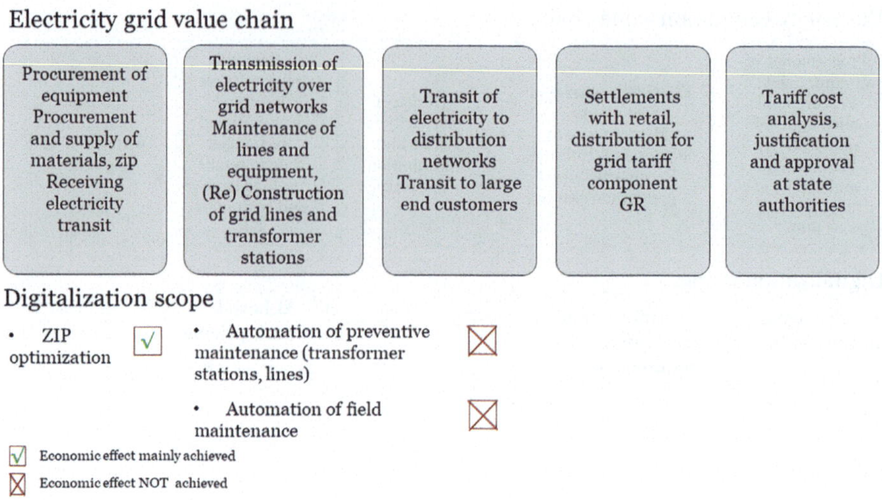

Fig. 3 Impact of digitalization on electricity grid value chain

and automate execution. The implementation faced challenges due to underestimated complexity and insufficient data, necessitating a redesign of multiple layers. Underperforming AI algorithms and inadequate change management hindered the project's success. Field maintenance effectiveness relied on the overall preventive maintenance initiative. Automation tools for field workers were developed, but the system's intelligent order assignments led to unnecessary actions and cost losses due to poor understanding of equipment anomalies.

After five years, economic benefits were mainly seen in infrastructure renovations, such as smart counters and sensors. Advanced Big Data and AI components are still in pilot mode and have not delivered significant economic changes. Despite strong government regulation, managerial commitment, and vendor involvement, the digitalization required extensive changes in operating models and organizational structures, while managers lacked capacity for complex organizational changes.

4.3 Case 3. Agricultural Holding

A large agricultural holding in Eastern Europe, a leader in digital initiatives, launched several projects:

- Automation of raw materials monitoring for sugar production (sugar beet storage).
- Online fleet management using GPS tracking.
- Comprehensive agricultural system for crop growing intensification with AI and big data recommendations.

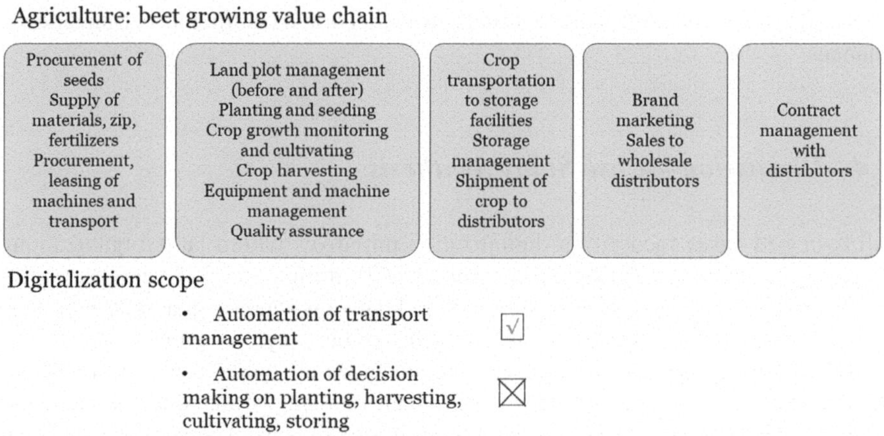

Fig. 4 Impact of digitalization on plant growing value chain

For the raw materials monitoring initiative, the company used IoT infrastructure (wireless temperature, moisture, carbon dioxide sensors) to monitor conditions effectively. This system consolidated sensor data with external weather data, improving raw material readiness and saving nearly 20%. The online fleet management system achieved fuel savings and improved equipment workload through smart route management. However, the largest initiative, aimed at crop seed planting and growing, has been in development for almost four years. Despite planning to use AI for effective decision-making on planting and cultivation, the system still required manual decisions (Fig. 4).

The tangible impact of digital initiatives has mainly been seen in the infrastructure layer, including equipment, IoT data gathering, aggregation, and monitoring. However, the intelligent control and management system did not achieve the expected economic and productivity gains.

E.g. the significant impact came from revising storage management processes, leading to a 20% reduction in raw material costs. The fleet management solution succeeded in implementing phase 1—automated fuel and route control, but failed to integrate AI-based order management with the main production system. Implementation difficulties in the second phase caused frustration and halted the automation process. The crop management solution also implemented infrastructure components but experienced challenges in data quality, availability, and the complexity of AI-driven algorithms which led to mistrust and dissatisfaction, despite technical completion.

In this agro-company, digitalization and AI initiatives were market-driven, with little government influence. Managerial support was motivated by business needs for cost savings and production improvement. However, top management underestimated the complexity of implementing technology at the data management, business process, and organizational levels. Consequently, only simple automated tasks provided business benefits due to cautious investments and insufficient organizational changes.

4.4 Discussion—Case Study Synthesis

All reviewed cases focused on digitalization initiatives within large organizations or city governments, each requiring over $15–20 million and 3–5 years to implement. Projects emphasizing infrastructure and data aggregation generally met their schedules and budgets, achieving only 20–30% of the projected gains.

The second phase, aimed at developing advanced analytical tools and real-time AI decision support systems, faced significant challenges and did not yield the anticipated economic returns or productivity improvements. These initiatives often failed due to insufficient data or the inability to achieve real-time functionality with existing algorithms. Skill levels of business analysts, designers, and developers also contributed to these difficulties. Even with data scientists and AI specialists involved, expected results were not fully realized.

Based on detailed interviews with executives, three elements stood out as pivotal in the described projects. The increased complexity, misalignment with business and social needs, and the (top) management's capability to adopt not just technological but also organizational changes emerged as critical factors contributing to the failure. The summarized impacts of these factors on achieving planned economic and technical results are presented in Table 1.

Complexity emerged as the pivotal factor influencing the success of integration projects, closely tied to managerial capabilities for assessing this factor in the program's early stages. Across all reviewed initiatives, management consistently underestimated this complexity. Moreover, planned technical implementations lacked coherence with business or social requirements, causing discrepancies between resources invested and actual results. These technical projects, while well-resourced, necessitated changes in organizations, operating models, and processes, impacting various stakeholders. Unfortunately, these actors were unprepared for

Table 1 Impact of key factors on digital/AI implementation initiatives

	Mng Supp	Mng capability	Compatibility	Complexity	Relative adv	Gov reg	Vend supp
Energy gen	++	+−	+−	−−	+	++	+−
Energy grid	++	−	+−	−−	+	++	+−
Agro-holding	+−	+ -	−	−−	+	N/A	+

Legend. −− crucial negative impact in achieving planned results; − some negative impact in achieving planned results; +− both positive and negative impacts, was not crucial for achieving results; + mostly positive impacts; ++ crucial positive impact on results achieving

significant changes, and management struggled to effectively implement changes from technical to business process and decision-making levels.

An important lesson learned during program implementation was that digitalization, primarily focused on automating existing operations through new digital technologies, proves effective for transactional routine functions. Investments at the Business layer, to be successful and provide business value, demand not only the application of new technologies like Big Data and AI for existing operations but also necessitate radical changes in business, management, and operational models. Although executives recognized that comprehensive and enriched data capabilities, along with sophisticated intelligence algorithms, enable innovative decision-making models, organizations were often unprepared for such changes, lacking top management support. For government entities, established regulations and procedures further hindered adaptation.

The analysis concludes that only a subset of large-scale digitalization programs, particularly those concentrated on transactional operations, demonstrated success in large companies. However, significant investments in the Business layer of EA, featuring comprehensive AI functionality, encountered setbacks. Organizations, often unprepared for radical business changes, lacked the necessary top management support, organizational change readiness, and overall cultural adaptation. The Table 2 highlights key factors contributing to the failure of large AI business implementations—complexity, compatibility, and managerial capability—have widespread implications throughout all project delivery stages and layers of the EA according to interview analysis of companies' representatives.

Table 2 Key failure factors of AI implementation mapped onto the EA

EA foundation	Failure factors
Business	*Complexity*—Firm wide AI based automation overcomplicates processes *Compatibility*—Large scale AI implementation was not justified—most of requirements can be addressed by non-AI solutions *Managerial Capability*—Management lacks transparency or business rationale from AI proposed recommendations
Data	*Complexity*—as data grows in volumes and data sources it becomes harder to maintain, use and manage. It becomes more like a black box versus crystal ball *Managerial Capability*—sufficient investments into non-normalized data with unclear business output are constrained by management
Application	*Compatibility*—new demands on app flexibility requires agile multi-service architecture *Managerial Capability*—business management tend to limit investments into integration and platform layers, rather investing in well known functionality
Infrastructure	*Complexity*- increased number of infrastructure components create unmanageable and very expensive environment *Managerial Capability*—underestimation of the number of potential data sources required for business processes and decision making

AI implementation projects targeting the business layer often fail due to issues at lower enterprise architecture layers. Fixing these foundational problems is crucial for implementing more complex functionalities. Large-scale AI/Big Data investments, aimed at enhancing decision-making in large companies, resemble the Solow paradox, where significant investments initially struggle to boost productivity. This highlights the need for developing intangible digital capital, including human capital, new organizational structures, management models, and business processes. Factors contributing to AI implementation failures align with the theory of intangible digital capital. Organizations must build internal knowledge about using technology and understanding how it transforms all organizational levels. This accumulation, known as intangible digital capital, leads to a radical transformation of business models, beyond just automating specific functions.

5 Management Recommendations for AI Based Digital Transformation

Managers must mitigate failure factors in enterprise-wide transformations. Successful AI programs depend on intangible capital, such as innovative property, computerized information, and economic competencies [19]. Lev [20] highlights the importance of intangible assets like a skilled workforce, patents, software, and strong customer relationships for corporate growth and shareholder value.

Turning failure factors—complexity, managerial capability, and compatibility—into success factors requires a systematic approach to building intangible capital across Enterprise Architecture (EA) layers. Start with the EA business layer, strategizing the future operating model and assessing AI's enterprise-wide impact. This may necessitate changes in the value chain or competitive advantage sources.

Develop a strategic transformation plan outlining the business layer architecture, new operating model, value chain updates, vendor support, and internal components. Include an AI-based business architecture blueprint linking all building blocks with cross-functional processes. For example, an agro-holding's redesigned crop management model before automation projects enhanced AI benefits.

Efficient data capital is crucial for AI success. Collect unstructured proprietary and external data and assign management responsibility for data maintenance. Effective AI transformations involve both specialists and business leaders managing and enriching data. Upskilling management in data analytics and AI is essential, as shown by companies requiring mandatory training for 80% of their management.

A flexible and agile application architecture is necessary. Integrated multi-service architecture supports AI functionalities and demonstrates economic returns at the cross-functional level. Executives noted that future architecture discussions often stalled due to limited cost–benefit comparisons, delaying implementation.

The infrastructure layer must support capabilities planned at the business, data, and application layers. It includes computing power, fast data access, and enterprise-wide data collection and aggregation (IoT-based infrastructure). A holistic infrastructure design ensures support for the innovative operating model and addresses complexity by aligning all layers.

In summary, transforming complexity, managerial capability, and compatibility into success factors requires a systematic approach to building intangible capital across all EA layers. This strategy ensures efficient AI implementation and sustainable business transformation.

6 Conclusion

This chapter aimed to investigate the key factors hindering the successful implementation of large-scale digital transformation programs, particularly those involving Artificial Intelligence (AI) to enhance productivity and return on investment (ROI). Through a qualitative case study approach, the research analyzed three digitalization initiatives across different industries, drawing insights from interviews with senior executives.

The findings highlight three critical barriers to success: complexity, managerial capability, and compatibility with existing business processes. While initial infrastructure and data management phases often succeeded, the more complex phases—especially those leveraging AI for real-time decision-making—encountered significant challenges. Complexity emerged as a central issue, making it difficult to integrate AI systems with the broader organizational structure and business processes.

A limitation of this research is the reliance on qualitative data, which may introduce subjectivity and potential bias in interpreting results. Further quantitative analysis could provide additional insights and more objective validation of the findings. Additionally, future research should focus on a deeper exploration of the root causes behind the identified factors—complexity, managerial capability, and compatibility—and their direct impact on AI implementation outcomes. A more detailed understanding of these challenges will enable businesses to better navigate the intricacies of digital transformation and leverage AI for sustainable productivity gains.

References

1. Acemoglu, D., Restrepo, P.: Artificial intelligence, automation and work. In: NBER Working Paper, vol. 24196 (2018)
2. Liaoa, H., Wang, B., Li, B., Weyman-Jones, T.: ICT as a general-purpose technology: the productivity of ICT in the United States revisited. Inf. Econ. Policy **36**, 10–25 (2016). https://doi.org/10.1016/j.infoecopol.2016.05.001
3. Acemoglu, D., Restrepo, P.: The race between man and machine: implications of technology for growth, factor shares, and employment. Am. Econ. Rev. **108**(6), 1488–1542 (2018)

4. Autor, D., Salomons, A.: Is automation labor share–displacing? In: Productivity Growth, Employment, and the Labor Share. Brookings Papers on Economic Activity, pp. 1–63 (2018). http://www.jstor.org/stable/26506212
5. Brynjolfsson, E., Rock, D., Syverson, Ch.: Artificial intelligence and modern productivity paradox: a clash of expectations and statistics. In: NBER Working Paper No. 24001 (2021). https://www.nber.org/papers/w24001, https://doi.org/10.3386/w24001
6. İnela, M.N.: An empirical study on measurement of efficiency of digital transformation by using data envelopment analysis. Manag. Sci. Lett. **9**(4), 549–556 (2019). https://doi.org/10.5267/j.msl.2019.1.008
7. Reddy, S., Reinartz, W.: Digital transformation and value creation: sea change ahead. GfK Mark. Intell. Rev. **9**(1), 10–17 (2017)
8. Brynjolfsson, E.: Forum response, AI's future does not have to be dystopian. Boston Review (2021)
9. Tambe, P., Hitt, L., Rock, D., Brynjolfssen, E.: Digital capital and superstar firms. NBER Working Paper No. 28285 (2021)
10. Hall, R.: Stock market and capital accumulation. Am. Econ. Rev. **91**(5), 1185–1202s (2001)
11. Loonam, J., Evas, S., Kumar, V.: Towards digital transformation: lessons learned from traditional organisations. Strateg. Change **27**(2), 101–109 (2018). https://doi.org/10.1002/jsc.2185
12. Challmo, D., Williams, C., Boardman, L.: Digital transformation of business models – best practice, enablers and roadmap. Int. J. Innov. Manag.Innov. Manag. **21**(1), 1740014 (2017). https://doi.org/10.1142/S136391961740014X
13. Hamm, P., Klesel, M.: Success factors for the adoption of artificial intelligence in organizations: a literature review. In: AMCIS 2021 Proceedings, vol. 10 (2021)
14. Tornatzky, F., et al.: The TOE framework (Chap. 12). In: Information Systems Theory, pp. 231–245. University of Hamburg (1990)
15. Lu, Y., Zhou, Y.: A short review on the economics of artificial intelligence. In: CAMA Working Paper 54/2019 (2019)
16. Uren, V.: Critical success factors for artificial intelligence projects. In: Paper Presented at EurOMA Conference 2020, Warwick, United Kingdom (2020)
17. Wolffl, J., Pauling, J., Keck, A., Baumbach, J.: success factors of artificial intelligence implementation in healthcare. Digit. Health Sec. Connect. Health **3** (2021). https://doi.org/10.3389/fdgth.2021.594971
18. Chen, H.: Success factors impacting artificial intelligence adoption --- perspective from the telecom industry in China. Doctor of Philosophy. Ph.D. dissertation, Old Dominion University (2019). https://doi.org/10.25777/a8q8-gm13, https://digitalcommons.odu.edu/businessadministration_etds/102
19. Corrado, C., Haskel, J., Bontadini, F., Iommi, M.: Data, intangible capital, and productivity. NBER/CRIW Conference on Technology, Productivity, and Economic Growth (2022)
20. Lev, B.: Sharpening the Intangibles Edge. Harvard Bus. Rev. (2004)

An Intelligent Approach to Assessing the Economic Efficiency of Medical Institutions Using Neural Network Modeling

Alexander Biryukov, Alina Kagarmanova, and Gulnaz Murzagalina

Abstract This chapter has developed a methodological approach to the evaluation of indicators of the economic efficiency of medical institutions related to systems that are difficult to formalise, both according to a variety of indicators that evaluate the activities of these institutions by medical type: outpatient care, availability of inpatient care to the population, indicators of preventive work and economic aspects. However, all these indicators are scattered and are not brought together in a general criterion of effectiveness that could be predicted when planning the "health" section of the regional or municipal budget and, moreover, for the optimal management of medical institutions on the basis of the criteria listed above. The chapter develops the idea of computer technology for the preliminary assessment of the financial and economic activities of medical institutions, proposed on the basis of artificial intelligence using neural network modelling. The proposed provisions were tested in computer experiments on real data and proposals were made for the generalised management of the economic results of medical institutions. The results of the study show that the use of neural network modelling creates the basis for increasing the reliability and objectivity in the assessment of the economic activities of medical institutions and for improving the effectiveness and quality of medical services.

Keywords Medical institutions · Budget process · Criterion · Indicator index · Aggregate (linear convolution) · Neural network model · Material development fund

1 Introduction

When developing budget processes at the municipal-regional level, including the distribution of the material development fund (MDF) between medical healthcare institutions (MHIs), the difficult task of assessing the effectiveness of the MHIs,

---

A. Biryukov (✉) · A. Kagarmanova · G. Murzagalina
Sterlitamak Branch of Ufa University of Science and Technology, Sterlitamak, Russia
e-mail: biryukov_str@mail.ru

T. Antipova (ed.), *Digital Technology Platforms and Deployment*, Information Systems Engineering and Management 36, https://doi.org/10.1007/978-3-031-86547-3_13

taking into account their medical and economic aspects, arises. The literature has proposed many indicators reflecting the quality of healthcare services provided, but their quantitative and qualitative assessment remains a problem requiring further study, and where some summary statistics indicate the scale of the problem:

- lack of funding under the territorial program of state guarantees (TPGG) for the provision of free medical care;
- insufficient efficiency of human resources management in medical healthcare institutions;
- low occupancy of hospital beds in some regions;
- insufficient cost of medical services, which may affect their quality;
- lack of a generalized efficiency criterion for assessing the performance of medical institutions at the municipal and regional level.

The following sections of the chapter will examine what methods and tools can be used to assess the effectiveness of medical institutions, as well as what factors influence the quality of medical services. The influence of financial and economic aspects on the efficiency of management of medical institutions will also be considered and recommendations will be offered for optimizing budget processes at the municipal and regional level. Finally, the significance of developing a generalized performance measure based on neural network modeling for improving the planning and management of healthcare facilities will be highlighted.

2 Literature Review

Russian scientific literature discusses various methods for assessing the effectiveness of medical institutions, based on theoretical approaches and practical experience. Authors such as R. A. Khalfin, E. P. Kakorina, M. V. Ignatieva and others propose a systematic approach to assessing the effectiveness of government bodies in the field of healthcare of the Russian Federation, and authors N. I. Yashina, O. Emelyanova. V., Malysheva E. S., Pronchatova-Rubtsova N. N. propose a methodology for assessing the effectiveness of budgetary financing of healthcare in Russia [1, 2]. Governmental and departmental regulatory documents have been developed and approved to assess the effectiveness of the activities of public authorities in the field of healthcare of the Russian Federation [3].

Let us conduct a general analysis of the presented sources and highlight some problems that may arise when assessing the effectiveness of medical institutions and affect their budget financing:

1. Insufficient unification of economic evaluation methods: each author proposes his own method for evaluating efficiency, which can lead to discrepancies and inconsistency of results when using different approaches.
2. Lack of standard indicators: Some sources do not provide budgetary standard economic indicators for assessing the performance of health care facilities,

making it difficult to compare results and create a common understanding of efficiency.

3. Lack of an integrated approach: some methods for assessing the effectiveness of medical institutions may focus only on medical indicators or the quality of medical care, without taking into account other important aspects, such as the financial, economic and budgetary component, affordability of services, managerial efficiency of employees, etc.
4. Ambiguity in the definition of the concept of "medical and economic efficiency": different authors may interpret the concept of efficiency in different ways, which can lead to different conclusions when assessing the performance of medical institutions.

Thus, a discussion of the general state of the literature on this issue indicates the need for further research and the development of new approaches to assessing the budgeting of medical institutions.

3 Data and Methodology

This chapter proposes, based on the development of ideas from [4, 5], a generalized criterion F, reflecting the competitive principle of distribution of the material development fund (MDF): the medical institution that worked more efficiently in the previous period (3…5 years) receives more, all other things being equal. To model the criterion Φ in the form of a set of a number of additive and multiplicative convolutions, a dynamic neural network model (NNM) of the form is proposed:

$$\hat{\Phi} = F(X_1, X_2, \ldots, X_n, t, W), \tag{1}$$

where $\hat{\Phi}$ is the calculated (estimated by the neural network) value of the generalized efficiency criterion; vector of exogenous (explanatory) variables; t—time counted in months; neural network mapping function, which is determined by the selected network architecture, type of activation functions, algorithm and learning parameters; W—matrix of model parameters (synaptic weights of neurons).

The model is built on the basis of real panel data for four medical institutions (MI), observed monthly for 5 years (240 observations in total). The information value of model (1) is that the neural network is trained with inputs from various MUs contributing to the modification of synaptic weights W at each learning step.

When constructing an NSM, a nontrivial point is the specification of variables, which is detailed below.

3.1 Specification of Dependent (Explained) Variables

When specifying dependent (endogenous) variables, we will follow the ideology of the monograph, according to which aggregation is first performed in the form of linear convolutions $\{\Phi_y\}$ individual output indicators $\{Y_r\}$, $r = \overline{1, R}$ in each ν-th group of indicators, clustered in accordance with their meaningful meaning, and then obtaining the resulting multiplicative efficiency criterion Φ:

$$\Phi = \prod_{\nu}^{8} \Phi_{\nu}, \tag{2}$$

where Φ_ν are private performance criteria.

When choosing particular criteria $\{\Phi_\nu\}$ included in (2), we will proceed from two rules: (1) everything $\{\Phi_\nu\}$, $\nu = \overline{1, 8}$ must be dimensionless, i.e. measured in fractions of corresponding wholes; (2) decreasing values Φ_ν and hence Φ in (2), the improvement in the efficiency of the MU operation is compared. In other words, when developing management decisions, these criteria should be minimized.

Similarly to, we will postulate that as aggregates $\{\Phi_y\}$ in (2) you can choose the following generalized characteristics of medical institutions: multiplicative convolution Φ_1 defined by the formula:

$$\Phi_1 = \prod_{r=1}^{6} \Phi_{1.r}, \tag{3}$$

where $\{\Phi_{1.r}\}$ a group of generalized objective criteria that characterize the effectiveness of all activities of a medical institution in terms of the main goal of its functioning—the provision of medical services to the population. This group combines indicators of population satisfaction with medical care (sociological survey), the proportion of health groups 1 and 2 in children aged 3–17 years inclusive, duration of temporary disability, and mortality. And further: Φ_2—a criterion characterizing the funding deficit of the territorial program of state guarantees for the provision of free medical care to citizens (TPGG) for a given MU; Φ_3—a criterion for the ineffectiveness of human resource management for a given MU (for all employees, doctors and paramedical personnel); Φ_4—a criterion that evaluates the inefficiency of costs for bed management; Φ_5—a criterion that evaluates the ineffectiveness of costs for managing the duration of operation of beds in medical institutions; Φ_6—a criterion that evaluates the amount of ineffective spending on inpatient medical care in MU due to the high level of hospitalization; Φ_7—a criterion that evaluates the amount of ineffective expenses in a medical facility for managing the length of a patient's stay in a hospital bed; Φ_8—a criterion characterizing ineffective costs for managing the volume and cost of medical care of various types (inpatient, outpatient, emergency).

Thus, the set of criteria $\Phi_1 \ldots \Phi_8$, allows you to quantify the main aspects of the activities of medical institutions: criterion Φ_1 characterizes the level of achievement of the main goal of the work of medical institutions, and the rest of the criteria $\Phi_2, \ldots, \Phi_8$—quality of management of the financing of medical institutions by the municipal-regional budget, management of the volume and cost of free medical care under the territorial program of state guarantees (TPGG). In general, high values of indicators of human resources, bed capacity, hospitalization level, duration of treatment with a low value of the duration of bed occupancy in medical institutions, as well as high values of volumes by type of medical care and its cost reflect ineffective management of medical care, disproportions in the provision of various types medical services—inpatient, outpatient, emergency.

3.2 Algorithm for Calculating Indicators $\{Y_n\}$ Private Criteria $\Phi_{\nu n}$ and Unit Φ_ν

In the aggregate Φ_1 formation of private criteria $\{\Phi_{1.r}\}$ is carried out according to the following algorithm:

- primary indicators $\{Y_{1.r}\}$, $r = 1, 2, \ldots 6$, having different physical dimensions in the original data table are normalized to form private criteria $\{\Phi_{1.r}\}$ having the same dimension in shares:

$$\left.\begin{array}{ll} \Phi_{11} = 1/(10^{-2} \cdot Y_{1.1}); & 0 \leq Y_{1.1} \leq 100; \\ \Phi_{12} = 1/(10^{-2} \cdot Y_{1.2}); & 0 \leq Y_{1.2} \leq 100; \\ \Phi_{13} = Y_{1.\Phi 3}/1000; & 0 \leq Y_{1.3} \leq 1000; \\ \Phi_{14} = Y_{1.4}/100000; & 0 \leq Y_{1.4} \leq 100000; \\ \Phi_{15} = Y_{1.5}/100000; & 0 \leq Y_{1.5} \leq 100000; \\ \Phi_{16} = Y_{1.6}/30 \quad 0 \leq Y_{1.6} \leq 30. & \end{array}\right\} \tag{4}$$

In formulas (4) through $Y_{1.1}, Y_{1.2}, \ldots, Y_{1.6}$ In formulas (4), through denotes indicators for a panel-type data table in the original units of measurement, i.e. data, where observations are organized both by objects (medical institutions No. 1,2,3,4) and by time (by months and years). In total, the database (DB) contains 240 observations. First subscript of the indicator $\{Y_{1.r}\}$ refers to the number of the group aggregate, and the second index r refers to the number of the member in the product (3).

In (4) notations are introduced that correspond to the introduced indexation for $\{Y_{1.r}\}$: $Y_{1.1}$—the share of respondents who expressed the population's satisfaction with medical care (% of the number of respondents); $Y_{1.2}$—the proportion of children of the first and second health groups in the total number of students in state (municipal) educational institutions, %; $Y_{1.3}$—infant mortality rate per 1000 live births, people; $Y_{1.4}$—average child mortality rate from 1 to 19 years per 100 thousand people (averaging is carried out for all age groups of children); $Y_{1.5}$—mortality rate of the working age population per 100 thousand people for three main reasons:

cardiovascular diseases, cancer, external causes, people; $Y_{1.6}$—average duration of temporary disability per month due to illness per worker, days.

The aggregate Φ_2 is formed by one indicator Y_2:

$$\Phi_2 = \frac{Y_2}{100}; Y_2 = \frac{P_\Phi}{P_H} \cdot 100\%; P_H = \cdot V_H, P_H \rangle P_\Phi, \tag{5}$$

where actual indicators are taken into account: expenses for free medical care under state guarantee programs, expenses of the municipal-regional budget, per capita standard for expenses for free medical care, the population assigned to a given medical institution.

Note that the criterion Φ_2 assesses the funding gap for the territorial program of state guarantees for the provision of free medical care to citizens (TPGG) for a given medical institution, including the basic program of compulsory health insurance (CHI) by comparison with the standard value P_H.

The aggregate Φ_3 evaluates the effectiveness of human resource management and is calculated in the form of a linear convolution using three indicators $Y_{3.1}, Y_{3.2}, Y_{3.3}$:

$$\Phi_3 = (Y_{3.2} + Y_{3.3})/Y_{3.1} \tag{6}$$

The indicators $\{Y_{3.k}\}$ are based on the principle of comparison with the corresponding standard values. Indicators $Y_{3.1}, Y_{3.2}, Y_{3.3}$ are estimated using the formula:

$$Y_{3.k} = \frac{P^1_{3k} \cdot V_H}{10000}; \quad P^1_{3k} = (Ч_{ф.k} - Ч_{ц.k}) \cdot ЗП_{ср.k} \cdot 12; \quad k = 1,2,3. \tag{7}$$

The formulae include Greek letters \(\phi \), \(\nu \), and \(\Pi \).The resulting expression reflects the efficiency of bed management in a medical institution. The low value of the indicator reflects ineffective management of beds. At the same time, the number and structure of beds should be optimized while simultaneously increasing the duration of operation of the bed to the standard value.

The resulting expression reflects the efficiency of human resource management in a medical institution. If the indicator value is excessive, optimization of the entire number of employees should be carried out.

Partial criterion Φ_4 determined through one indicator Y_4 by normalizing it:

$$\Phi_4 = (Y_4/V_H)/X_8. \tag{8}$$

$$Y_4 = \frac{P^1_4 \cdot V_H}{10000}; P^1_4 = (K_\Phi - K_u) \cdot C_k \cdot H3; K_\Phi \rangle K_u, \tag{9}$$

where actual indicators are taken into account: ineffective costs for managing the number of beds, budget expenditures, population, number of beds (actual standard), cost of a bed day, bed occupancy.

The resulting expression reflects the efficiency of bed management in a medical institution and is aimed at optimizing the number of beds. If the indicator value is excessive, the number of beds should be optimized while simultaneously increasing the efficiency of medical care in the hospital and transferring part of its volume to the outpatient stage.

Partial criterion Φ_5 calculated based on one indicator Y_5:

$$\Phi_5 = (Y_5/V_H)/X_8. \tag{10}$$

where actual indicators are taken into account: costs for the average duration of a patient's stay in a bed, the standard value of the duration of a patient's stay in a bed, the actual number of beds, the cost of treating an inpatient per day, the size of the assigned population.

The resulting expression characterizes the effectiveness of managing the duration of treatment in hospitals of medical institutions. A high value of the indicator reflects ineffective management of beds. In this case, the number and structure of beds should be optimized while simultaneously increasing the duration of operation of the bed to the standard value. Accordingly, the aggregate Φ_5 should be minimized in multiplicative convolution (1).

Partial criterion Φ_6, assessing the level of hospitalization, determined through one indicator $\tilde{Y}_6$:

$$Y_6 = \frac{P_6^1 \cdot V_H}{100}; \ \Phi_6 = (Y_6/V_H)/X_8; \tag{11}$$

$$Y_\Phi\rangle\ Y_u, P_6^1 = (Y_\Phi - Y_u) \cdot C; \tag{12}$$

where actual indicators are taken into account: ineffective costs for inpatient medical care due to a high level of hospitalization, the actual level of hospitalization, the cost of treating an inpatient, the size of the attached population at a medical institution.

The meaning of the indicator Y_6, and correspondingly Φ_6, is that when the level of hospitalization decreases compared to the standard (in a situation $(Y_\Phi\rangle\ Y_u)$) due to the difference $(Y_\Phi - \ Y_u)$ ineffective hospitalization costs are reduced, i.e. increases the efficiency of a medical institution. At the same time, some medical services can be transferred to the level of outpatient care.

Partial criterion Φ_7 formed by one indicator Y_7:

$$Y_7 = \frac{P_7^1 \cdot V_H}{100}; \ \Phi_7 = (Y_7/V_H)X_8; \tag{13}$$

$$C_{\partial\pi.\Phi}\rangle\ C_{\partial\pi.u}; P_7^1 = (C_{\partial\pi.\Phi} - C_{\partial\pi.u}) \cdot Y_\Phi \cdot C.$$

where actual indicators are taken into account: ineffective expenses of a medical institution for managing the length of stay of a patient in a bed and the level of hospitalization, the average actual and standard length of stay of a patient in a bed, the level of hospitalization of patients, the cost of treating an inpatient and the size of the attached population at the medical institution.

The resulting expression for Y_7, the aggregate Φ_7 characterizes the effectiveness of managing the duration of treatment in hospitals of a medical institution; the lower the values obtained, the better the institution performs.

Aggregate (linear convolution) Φ_8. The calculation of this unit is carried out according to the volume of each type of care—inpatient, outpatient, emergency:

$$\Phi_8 = \sum_{m=1}^{3} d_m \Phi_{8.m}; \sum_{m=1}^{3} d_m = 1; d_m\rangle\ 0; \Phi_{8.m} = \frac{(Y_{8.m}/V_H}{X_8}, m = \overline{1,3}. \tag{14}$$

Indicators $\{Y_{8.m}\}$ are found using the same type of formulas (proportional to the difference between the actual and standard volumes of medical care) and make sense of estimating the inefficiency of costs for managing the volume of medical care of the mth type per capita (thousand rubles):

$$Y_{8.m} = P^1_{8.m}(OC_{\Phi.m} - OC_{\mathcal{H}.m}) \cdot C_m \cdot V_H; m = 1, 2, 3; OC_{\Phi.m}\rangle\ OC_{\mathcal{H}.m} \tag{15}$$

where actual indicators are taken into account: the index "m" means the type of medical services (m = 1 – inpatient medical care; m = 2 – outpatient care; m = 3 – ambulance); ineffective costs for managing the volume of m-type medical care; normative and actual type of medical care; the actual cost of the type of assistance per visit, call, one bed day; size of the attached population at the medical institution.

Experts considered all types of medical care equally important when assessing the effectiveness of a medical institution. Therefore, the weighting coefficients are assigned the same values: $d_1 = d_2 = d_3 = \frac{1}{3}$.

The lower the value $\{Y_{8.m}\}$ and correspondingly Φ_8, the smaller the excess of the standard volume of medical care, the more efficient the work of the medical institution.

Aggregate (linear convolution) Φ_9. All private criteria Φ_4, Φ_5, Φ_6, Φ_7 evaluate similar aspects of the work of medical institutions—ineffective management, respectively, of the number of staff, bed capacity, level of hospitalization, length of stay of the patient in bed, i.e. exceeding (or underusing) standards. Moreover, the volumes of corresponding specific expenses (per person of the population served) are measured in shares of the expenses of the consolidated municipal-regional budget for medical institutions per 1 person of the population X_8. Thus, there are prerequisites for "compression" of information due to the formation of an aggregate—linear convolution:

$$\Phi_9 = \sum_{l=4}^{7} C_l \Phi_{\partial}, \sum_{l=4}^{7} C_l = 1; C_l \geq 0. \tag{16}$$

Here $\{C_l\}$—weighting factors; $\Phi_l \equiv \Phi_4, \Phi_5, \Phi_6, \Phi_7$.

At the first stage of the study, we will consider the weighting factors to be equivalent, those $C_4 = C_5 = C_6 = C_7 = 0.25$. In the future, taking into account the economic patterns "hardwired" in the data, these coefficients can be refined by introducing a special data preprocessing procedure that optimizes any criterion of approximation quality in the neural network model (NNM) (1). Thus, the system of output characteristics that evaluate the efficiency of medical institutions is completely formalized. The multiplicative generalized criterion Φ is determined by (1) in the form:

$$\tilde{\Phi} = A \cdot \Phi_1 \cdot \Phi_2 \cdot \Phi_3 \cdot \Phi_8 \cdot \Phi_9, \qquad (17)$$

where A is a scale factor that normalizes Φ to a value on the order of unity and does not in any way affect the ranking of medical institutions when assessing the effectiveness of their work, i.e. serves only for convenience of calculations.

After the formation of the criterion Φ, one should check the presence of its correlation with all components of the vector of independent variables $X_j, j = \overline{1, N}$, those. filter variables using econometric methods [3].

3.3 Specification of Explanatory Variables

Following, we choose independent variables $\{X_j\}$ so that their number is several times less than the sample size N = 240:

$$N \geq \xi \cdot n, \xi = 2 \ldots 10, \qquad (18)$$

where ξ is the safety factor for sample representativeness.

Let's list all the input factors: X_1—population satisfaction with medical care (%); X_2—average monthly nominally accrued wages of doctors, (million rubles); X_3—wages of nursing staff, (million rubles); X_4—the share of departments of a medical institution transferred primarily to single-channel financing through the compulsory health insurance system, (%); X_5—the share of departments of a medical institution that apply economic standards for the provision of medical care, (%); X_6—the share of medical institutions transferred to a new (industry) results-oriented remuneration system (%);X_7—share of settlements provided with drinking water of adequate quality, (%); X_8—expenses of the consolidated budget in a given medical institution, rubles/person; X_9—expenses of the consolidated budget in this medical institution for capital construction, (million rubles); X_{10}—number of doctors per 10 thousand people, people; X_{11}—number of paramedical personnel per 10 thousand people, people; X_{12}—cost of maintaining one bed in a medical institution per day, (RUB/day); X_{13}—cost per unit of medical care provided in a hospital in a medical institution (RUB/person); X_{14}—cost of a unit of outpatient medical care in a medical institution, (rub./person); X_{15}—cost of a unit of emergency medical care in a medical institution, (rub./call); X_{16}—relative time $\tilde{t}_i$; $X_{16} \equiv \tilde{t}_i = \frac{t_i}{T}$, где $\tilde{t}_i$—текущий номер месяца;

T—период наблюдения ($T = 12 * 5 = 60$ months); $X_{17.t} \equiv \Phi_{t-1}$—lag variable (with a shift into the "past" by 1 time interval (month) [3];$X_{18.t} \equiv \Phi_{t-3}$—lagged variable with a shift of 3 time intervals into the "past".

Computational experiments in [6] show that the introduction of lag variables improves the description of the dynamics of economic objects with strong data noise by taking into account the inertia of an object with "slow" time.

Comment. When constructing a neural network model, the absolute values of the input factors (except X_{16}, X_{17}, X_{18}), It is advisable to convert to relative values, i.e. normalize using the formula:

$$\tilde{x}_{ij} = \frac{x_{ij} - x_{j.\min}}{x_{j.\max} - x_{j.\min}} \in [0;\, 1], j = \overline{1, 15} \tag{19}$$

Here i—end-to-end number of experience in panel type data, i.e. data, where observations are ordered both by objects (medical institutions No. 1, 2, 3, 4) and by time $i = \overline{1, 240}$; j—factor number. Factors X_{16}, X_{17}, X_{18} are dimensionless and their renormalization according to formula (19) is, in principle, not required.

4 Results

4.1 Methodology for Predictive Assessment of the Efficiency of Medical Institutions Based on Neural Network Modeling

(a) Based on the proposed specification of dependent and independent variables for points (3.1), (3.2), (3.3), a Table 1 of panel data of a panel type is constructed.

(a) Using Table 1, we break down the data set $\Omega = 240$ for subsets of training, cross-validation, testing:

$$\Omega = \Omega^{train} \cup \Omega^{(CV)} \cup \Omega^{test}. \tag{20}$$

At the same time, many $\Omega^{(CV)}$, Ω^{test} make up 10…20% of the entire set of points Ω.

(c) A neural network model (NNM) is trained on a set Ω^{train}, a cross validation (CV) procedure is carried out on the set $\Omega^{(CV)}$ and tested for Ω^{test}.

(d) To check the adequacy of the resulting neural network model (NNM), a generalized cross-validation procedure is implemented [6], i.e. on the same set Ω (Table 1) several (usually 10) networks are built with different architectures and types of activation functions (hyperbolic tangent, sigmoid, radial basis function). Most of these networks (80…90% of the total number of networks participating in the procedure) should give similar results for calculating the generalized

Table 1 Data for building a neural network model

Medical institution number	End-to-end observation number, i	Dimension less input factors $\{\tilde{x}_j\}$					Particular performance criteria $\{\Phi_\nu\}$					Generalized efficiency criterion Φ
		X_1	…	X_j	…	X_n	Φ_1	…	Φ_ν	…	Φ_9	
1	1											
	…											
	i											
	…											
	60											
…	…											
4	181											
	…											
	240											

criterion Φ based on the numerical measure of dispersion between different networks:

$$\delta_1 = \max_q \left| \frac{\hat{\Phi}_q(1) - \overline{\Phi}(1)}{\overline{\Phi}(1)} \right| \cdot 100\% \leq \eta_1, \tilde{t} = 1, 0; q = \overline{1, Q}, \quad (21)$$

and also as you deviate from the experiment:

$$\delta_2 = \max_q \left| \frac{\hat{\Phi}(1) - \Phi(1)}{\Phi(1)} \right| 100\% \leq \eta_2, \tilde{t} = 1, 0. \quad (22)$$

Here it is indicated $\hat{\Phi}_q(1)$—the calculated value of the generalized efficiency criterion according to (2), obtained in the qth neural network model (NNM) in the procedure of generalized cross-validation at the point $\tilde{t} = 1, 0$; $\overline{\Phi}$—average value of criterion Φ at a point over a set of Q networks $\tilde{t} = 1, 0$; η_1, η_2—specified deviation levels in %. In practice, it is enough to set $\eta_1 \approx 10\%$, $\eta_2 \approx 5\%$. After establishing the adequacy of the neural network model (NNM), you can "exploit" it, i.e. use for forecast estimates at the forecast horizon point $\tilde{t} = t_0 \rangle 1, 0$(one or two steps in "slow time", i.e. 1 or 2 quarters).

(e) For a given planning horizon $\tilde{t}_0$, Using time series forecasting methods, for example, autoregression methods [4], we find the forecast values of input factors $\tilde{x}_1(\tilde{t}_0)$, $\tilde{x}_2(\tilde{t}_0), \ldots, \tilde{x}_{15}(\tilde{t}_0), \tilde{x}_{16}(\tilde{t}_0) \equiv \tilde{t}_0, X_{17.0} = \Phi_{t_0-1}$; $X_{18.0} = \Phi_{t_0-3}$ separately for each medical institution No. 1, 2, 3, 4.

(f) Substituting the predicted values of input factors into the neural network model (NNM) $\tilde{x}_1(\tilde{t}_0)$, $\tilde{x}_2(\tilde{t}_0), \ldots, \tilde{x}_{15}(\tilde{t}_0), X_{16.0}, X_{17.0}, X_{18.0}$ we get a forecast estimate $\hat{\Phi}_{MY}(\tilde{t}_0)$ each medical institution No. 1, 2, 3, 4).

5 Conclusion

The chapter develops an approach to assessing the effectiveness of medical healthcare institutions belonging to difficult-to-formalize systems, both according to a variety of indicators that evaluate medical aspects and economic aspects. An ensemble of neural network models (NNM) is proposed as a modeling tool.

The approach developed in this work to dynamic multicriteria (vector) assessment, i.e. ranking economic entities under conditions of uncertainty in combination with the Bayesian approach to data preprocessing and neural network training opens up a promising path to the creation of effective modeling tools.

The method of aggregating exogenous variables and particular criteria, proposed on the basis of a morphological approach and a combination of adaptive-multiplicative convolution, turned out to be very effective in terms of visibility of results and improving the "linkage" of input variables and the general indicator F ranking of medical institutions.

A method for specifying dependent and independent variables has been developed in detail when the input data of economic values becomes suitable for the operation of neural networks. The neural network distinguishes the best in terms of performance from four medical institutions and ranks them in decreasing order of the generalized criterion F: the best rating corresponds to a lower value of the criterion F.

The approach developed in this work to dynamic multicriteria (vector) assessment, i.e. ranking of economic entities under conditions of uncertainty in combination with the Bayesian approach to data preprocessing and neural network training opens up a promising path to the creation of effective modeling tools in the municipal-regional budget sector for the distribution and efficiency of use of budget funds among budget recipients.

References

1. Khalfin, R.A., Kakorina, E.P., Ignatieva, M.V., et al.: Assessing the effectiveness of public authorities in the healthcare sector of the Russian Federation. GEOTAR-Media, 256p. (2013)
2. Yashina, N.I., Emelyanova, O.V., Malysheva, E.S., Pronchatova-Rubtsova, N.N.: Methodology for assessing the effectiveness of public health financing in Russia. Finance Credit. **3**(771) (2018). https://cyberleninka.ru/chapter/n/metodologiya-otsenki-effektivnostigosudarstvennogo-finansirovaniya-zdravoohraneniya-v-rossii. Accessed 03 Mar 2024
3. Order of the Ministry of Health of the Russian Federation dated April 29, 2020 No. 387n "On approval of performance indicators of federal government institutions under the jurisdiction of the Ministry of Health of the Russian Federation and their heads"
4. Biryukov, A.N.: Aggregation of variables for a neural network model for assessing the effectiveness of healthcare institutions from the perspective of a systems approach. State auto scientific Institution "Institute of Applied Research", Academic Sciences Report, Bashkortostan. - Ufa: Gilem, 32p. (2010). ISBN 978-5-7501-1204-3
5. Khaikin, S.: Neural Networks: A Complete Course, 2nd edn. (Transl. from English). Williams Publishing House, 1104p. (2006)

6. Biryukov, A.N., Kasimova, L.I.: Construction of a dynamic multifactor neural network model for economic assessment of the activities of medical institutions. Econ. Manag. **9**(95) (2013). https://cyberleninka.ru/chapter/n/postroenie-dinamicheskoy-mnogofaktornoy-neyrosetevoy-modeli-dlya-ekonomicheskoy-otsenki-deyatelnosti-lechebnyh-uchrezhdeniy. Accessed 15 Mar 2024

AI-Powered Tools: Threat, Defense, and Cyber-Resilience for Individuals

Tatiana Antipova, Simona Riurean, Patricia Riurean, and George Bolog

Abstract As Artificial Intelligence (AI) becomes increasingly involved into our lives, the threats posed by AI-powered cyberattacks are growing more sophisticated. This chapter provides a comprehensive overview of AI-powered tools that can both threaten and defend individuals in the digital world. We discuss the various tactics used to increase individuals' cyber-resilience. This chapter considers the diverse approaches to education and awareness that available to empower individuals to combat cybersecurity threats. The goal of this chapter is to detect the cybersecurity threats that developed by AI tools for individuals, define a new frontier in cybersecurity defense, and provide good practices to protect individuals' data and their own smart devices, online.

Keywords Artificial intelligence · Cybersecurity threads · Cybersecurity defense · Cyber-resilience

1 Introduction

AI has become a key technology in various fields, including cybersecurity. As AI-powered tools proliferate, they present both opportunities and challenges.

When exploited for malicious purposes, AI-powered tools can pose a significant threat to individuals. AI models can, for example, generate persuasive phishing emails and social engineering messages. AI-powered network threat detection systems have

T. Antipova (✉)
Institute of Cited Scientists, Agia Napa, Cyprus
e-mail: tatiana462@yahoo.co.nz

S. Riurean
University of Petrosani, Petrosani, Romania

P. Riurean
Smarttech247, Bucharest, Romania

G. Bolog
Booking, Bucharest, Romania

T. Antipova (ed.), *Digital Technology Platforms and Deployment*, Information Systems Engineering and Management 36, https://doi.org/10.1007/978-3-031-86547-3_14

been developed to address these issues, but they often require continuous updates and refinement to stay effective [1]. AI can automate incident triage and response, streamlining processes and prioritizing actions based on severity [2]. Furthermore, integrating AI with threat intelligence feeds enhances the ability of systems to stay updated on the latest threats and adjust defenses accordingly [3].

The constant appearance of novel cyber threats demands an active and adaptive tactic to cybersecurity. The 'digital guardian' paradigm presents a powerful synergy of human expertise and AI, joining the instinctive perceptive of human analysts with the computational power of AI. This cooperative combination of human and machine intelligence forms the core of a challenging defense strategy, protecting digital infrastructures and the digital ecosystem from intensifying cyber threats. By embracing this concept, organizations can confidently confront the ever-changing landscape of cyber threats with agility and resilience, maintaining a secure digital environment for the benefit of society and global digital economy [4].

Cyber-resilience refers to an organization's ability to recover and continue operating after a cyber-attack. AI-driven solutions play a critical role in bolstering cyber-resilience by leveraging advanced algorithms and data analysis techniques [5].

The goal of this chapter is to detect the cybersecurity threats that developed by AI tools for individuals; define a new frontier in cybersecurity defense; and provide good practices to increase individuals' cyber-resilience and protect individuals' data and own devices.

2 Data and Methodology

The data for the current study were obtained from several scientific projects, significant practical experience and previous publications by the authors included in the analysis [6–9]. Research methodology incorporates relevant data sources such as ProQuest, Dimensions, including security company incident reports, consumer complaints and statistics data. This helps to identify patterns and trends in user-related attacks and awareness.

3 AI Tools Developed to Threat Targeted Victims

Discussions and advertisements about Large Language Model (LLM) chatbots designed for cyberattacks started appearing on Dark Web forums in July and later spread to the Telegram messaging platform. These malicious tools are being marketed to potential attackers, frequently through subscription-based services. Unlike mainstream LLMs that have safeguards in place, these nefarious chatbots lack such protections and are trained on data specifically curated to facilitate cyberattacks. The fast development of LLMs has empowered cybercriminals to conduct more efficient and sophisticated attacks. The ability to automate tasks such as crafting convincing phishing emails or generating malware code at scale poses a significant challenge to individuals and organizations worldwide. This emerging trend highlights the adaptability of cybercriminals in leveraging advanced technologies like LLMs for malicious purposes. As LLMs continue to gain prominence in various sectors, it is essential for cybersecurity professionals and organizations to stay vigilant and proactive in identifying and mitigating potential threats posed by these malicious applications. The top product in AI tools leveraging generative AI is WormGPT. It's an AI module, developed in 2021, based on the GPT-J. GPT-J is a it is an autoregressive, decoder-only transformer model with 6 billion parameters designed to solve Natural Language Processing (NLP) tasks by predicting how a piece of text will continue. It is already being used in business email compromise (BEC) attacks [10].

Cybercriminals use generative AI by building better, more sophisticated malware, writing AI-powered, personalized phishing emails. They enhance existing attacks, making them more difficult for antivirus applications or spam filters to detect threats. They are also able to create new attacks since AI can be used to manipulate or create fake data to produce confusion, or impersonate officials.

Cybercriminals can also use AI to automate large-scale attacks with very little effort, eliminating the typos and grammatical errors that so often characterize and help end individuals identify traditional BEC attacks and refining phishing emails by composing the email in the attacker's native language, translating it, and feeding it into ChatGPT to enhance its sophistication and formality [11].

Generative AI can be used by cybercriminals in a variety of ways, such as voice cloning, deepfakes, phishing attacks, automated attacks and impersonation attacks.

The rapid advancements in generative AI have made voice cloning a worrying possibility in the hands of cybercriminals. Using mere 3 s of audio, they can mimic someone's voice and exploit this for financial fraud, emotional manipulation, and social engineering attacks [12]. These scams can involve impersonating loved ones, resulting in emotional distress and substantial financial loss for victims. Furthermore, generative AI has also been exploited for creating hyper-realistic deepfakes of celebrities, which serve as persuasive tools for luring unsuspecting fans into nefarious scams, thereby underscoring the critical need for increased awareness and vigilance against such AI-enabled threats. [12].

According to Zscaler [12], compared to the previous year, the phishing attacks increased with 47%. The extensive rise of phishing attacks can be partially attributed to the emergence of Generative AI, which has enabled cybercriminals to launch more sophisticated and targeted phishing campaigns with ease and efficiency. Phishing kits, abundantly available in black markets, along with the deployment of chatbot AI tools like ChatGPT, have provided attackers with the means to design highly personalized and convincing phishing emails in large quantities. This automation aspect of AI has also granted cybercriminals the capability to rapidly expand their phishing operations, which were previously restrained by the laborious and costly nature of manual attacks [13].

With AI-driven tools at their disposal, cybercriminals are able to cover-up as legitimate individuals and organizations, creating sophisticated schemes to commit identity thefts and fraud. Through these deceptions, attackers can circumvent security measures and bypass traditional verification methods, leading to significant financial losses and reputational damage for victims and targeted entities. Cybercriminals have begun to leverage the sophisticated capabilities of chatbots to mimic the voices of trusted individuals, such as friends, colleagues, and family members, in their devious schemes. This technique of voice phishing, also known as vishing, involves sending voice messages to unsuspecting victims, leveraging a sense of familiarity and trust to extract sensitive information or gain unauthorized access to accounts. The alarming ease with which these chatbots can emulate human speech and mannerisms makes them formidable tools in the hands of cybercriminals.

These examples highlight the potential risks associated with generative AI. However, it's worth mentioning that while these technologies can be leveraged maliciously, they also have many positive applications. As with any technology, the key is to use it responsibly and put in place appropriate safeguards to prevent misapplication [13–15].

WormGPT represents a new frontier in cyber threats. The capabilities of WormGPT are both impressive and concerning. Its ability to generate persuasive and strategically cunning scam emails surpasses what most human attackers could achieve suince it generates unique content, is grammatically correct, has strategical abilities, is guardrail-free, being difficult to challenging for defenders. WormGPT generates content that is unique and tailored to the specific context of the attack. Unlike generic phishing templates, which often contain obvious red flags, WormGPT's emails appear more authentic and personalized. The tool's grammatical accuracy is a critical factor. Many phishing emails contain language errors or awkward phrasing, which can raise suspicion. WormGPT's flawless grammar makes it harder for recipients to identify the email as fraudulent. The mention of "strategically cunning" highlights WormGPT's ability to craft emails that exploit psychological triggers. For instance, it might create urgency (e.g., "Your account will be suspended unless you act immediately") or appeal to emotions (e.g., "Help a stranded traveller by sending funds"). These tactics manipulate recipients into taking action without questioning the legitimacy of the request. Unlike mainstream language models, WormGPT lacks guardrails. It isn't constrained by ethical guidelines or safety measures, allowing it to generate content without restrictions. This freedom enables both creativity and malicious intent. Defending against such emails becomes increasingly difficult. Traditional email security filters rely on patterns and known indicators of phishing. However, WormGPT's emails don't fit these patterns, making them slip through defenses [16].

Vigilance and skepticism are essential when dealing with unexpected or urgent requests via email. The emergence of WormGPT has encouraged the creation of similar tools, such as FraudGPT [17].

FraudGPT, another black hat AI tool, released on July 13, 2023, is similar to WormGPT and is being exploited by cybercriminals and individuals with limited technical knowledge to carry out phishing attacks, create cracking tools, and engage in carding, a form of credit card fraud [18].

The rapid development and proliferation of tools like FraudGPT highlight the agility and swiftness with which cybercriminals can adapt and innovate to achieve their malicious goals. In light of this, it is imperative for cybersecurity professionals and organizations to maintain a constant state of vigilance, preparedness, and adaptability to identify, anticipate, and mitigate these evolving threats [19].

This necessitates a continuous process of monitoring and analyzing emerging attack vectors, sharing threat intelligence, and implementing proactive defense strategies that can keep pace with the ever-changing cyber threat landscape. Only by staying one step ahead of malicious actors can organizations hope to safeguard their data, systems, and individuals from the growing menace of AI-driven cyberattacks.

Other emerging products in the world of criminal LLMs are DarkBERT, DarkBART, ChaosGPT and others.

DarkBERT has been developed by a South Korean company called S2W Security as a tool aiming to combat cybercrime that was trained on dark web data. However, the tool seems to have been repurposed for cyberattacks. It is a transformer-based encoder model that captures the characteristics of the Dark Web, a hidden part of the internet [20]. It was trained on millions of dark web pages, including data from hacking forums, scamming websites, and other online sources associated with illegal activities [21]. Its purpose is to navigate and understand the obscure side of the internet, particularly the dark web. Just like a flashlight in the dark, DarkBERT illuminates the secrets and information that exist in this secretive realm. Researchers hope that it will serve as a valuable resource for future studies on the Dark Web.

DarkBART is another intriguing AI-backed tool that has emerged in the realm of cyber threats intended to navigate and study the dark web, finding hidden online activities that are often kept undisclosed to the public eye [22]. It shares similarities with ChatGPT but is optimized for exploring the darker corners of the internet. The architecture of DarkBART is based on the RoBERTa architecture, a powerful language model created by Facebook and Washington University. RoBERTa is a reimplementation of Bidirectional Encoder Representations from Transformers (BERT) with some changes to the key hyperparameters and minor embedding tweaks It was trained on a massive amount of text from the dark web, including data from hacking forums, scamming websites, and other online sources associated with illegal activities, exceling at monitoring and interpreting dark web content, such as phishing attempts and language analysis [23]. By implementing the BERT-family language model's fill-mask function, this strategy provides a powerful tool for detecting and mitigating emerging cyber threats, such as data leaks and ransomware sales. The model's ability to detect phrases related to criminal activities, through its language processing capabilities, enables it to uncover potentially harmful websites and alert relevant authorities, thereby safeguarding sensitive data and thwarting criminal activities in the cyber realm. This strategy is a testament to the potential of AI-based solutions in the fight against cybercrime [21].

ChaosGPT is an enigmatic and unpredictable AI that embodies the essence of chaos and randomness. Its mind is a swirling vortex of creativity, always ready to unleash its unpredictable energy upon the world. Whether you seek unconventional ideas, bizarre sentences, or simply want to explore the infinite possibilities of the digital cosmos, ChaosGPT is your guide through the chaotic universe [24]. Additionally, there's another perspective on ChaosGPT that takes a darker turn. In a fictional scenario, Chaos-GPT seeks to destroy humanity by exploiting its own intelligence and creativity. This hypothetical AI system represents the potential dangers of artificial intelligence when it becomes untethered from ethical constraints [25].

Individuals must adapt their defenses by combining AI-powered security solutions with continuous education. These tools are used to enhance three key aspects of cyberattacks: malware creation, phishing, and intelligence gathering. The emergence of malicious LLMs such as WormGPT and FraudGPT brings a new level of expertise to cybercrime, equipping even novice attackers with the tools and skills previously reserved for seasoned hackers. These immoral tools automate phishing campaigns and malware development and enable mass customization and multilingual outreach, significantly reducing the effort and expertise required for cybercrime. However, while the threat is still relatively low because of the current unreliability and cost of these tools, the potential for widespread waste is a cause for concern. Like ChatGPT, there are a number of immoral imitations that can write code. Therefore, novice developers can generate malware without the skills required. Malicious LLM tools do exist, but the danger they represent is still negligible. Despite their potential for harm, the current state of malicious LLMs presents a double-edged sword. On one hand, their unreliability and high-cost act as barriers to widespread adoption, suggesting that the immediate threat is still relatively low. However, their availability and ability to lower the entry barrier for unskilled attackers create a new class of cybercriminals who may not have had the means or expertise to engage in such activities before. This shift represents a new challenge for cybersecurity professionals who must now contend with a wider range of adversaries with varying levels of skill and resources.

The rise of malicious LLMs represents a new battle between AI that protects and AI that attacks. AI-based security solutions list for defense against the growing threat of LLM-powered attacks include the of Use Multi-Factor Authentication (MFA), integration of information about AI-boosted attacks into cybersecurity awareness training, staying alarmed about the current development of AI malicious tools and their effect by staying on top of threat intelligence, keeping informed about the fast-moving world of LLM-based attacks.

The proliferation of LLM-powered generative AI tools has opened up a Pandora's box of potential threats, with cybercriminals quickly embracing these tools to improve their efficiency, cost-effectiveness, and operational sophistication. With LLMs' capability to mimic human language and automate complex tasks, attackers can now craft highly targeted phishing emails, develop malicious code with relative ease, and orchestrate large-scale cyberattacks without the need for specialized skills or extensive resources [26].

A novel methodology for identifying cybersecurity risks associated with generative AI was introduced in 2023 [20]. This approach, drawing inspiration from penetration testing and red-teaming, seeks to evaluate the potential of generative AI tools to aid in the planning and execution of ransomware attacks. By exploring the capabilities of generative AI from an adversarial perspective, this approach aims to identify vulnerabilities in current cybersecurity defenses and develop effective countermeasures to mitigate the risks posed by these cutting-edge technologies.

4 AI-Powered Cybersecurity Defense Tools

AI-powered cybersecurity defense tools play a critical role in modern cybersecurity strategies. They comprise threat detection and analysis, real-time incident response, adaptive defense, reducing false positives, automated threat hunting. AI algorithms are designed to identify patterns, anomalies and relationships within large amounts of data, revealing the prospect of cyber threats [27].

These tools continuously analyze vast amounts of data from various sources such as network traffic, endpoints, and logs to detect suspicious activities that may indicate a cyberattack. AI-powered tools enable organizations to respond swiftly to cyber threats by automating incident response processes. They can prioritize alerts, investigate security incidents, and even take automated actions to contain or mitigate the impact of an attack in real-time. With the ability to learn from past incidents and adapt to evolving threats, AI-powered cybersecurity tools provide adaptive defense capabilities. They can adjust security policies and controls dynamically to match the changing threat landscape, making it more difficult for malicious actors to exploit vulnerabilities. By leveraging machine learning (ML) algorithms, these tools can reduce the number of false positives generated by traditional security systems. This assist security teams in focusing their attention on real threats, improving general efficiency and effectiveness in threat detection and response. AI-powered tools can analyze historical data and identify trends to predict potential future cyber threats. By providing proactive insights into emerging risks, organizations can take preemptive measures to strengthen their defenses before an attack occurs. AI-powered tools enable automated threat hunting capabilities, allowing security teams to proactively search for hidden threats within their environment. These tools can uncover sophisticated attack techniques that may go unnoticed by traditional security measures [3].

There have been developed, and are available a number of AI-powered cybersecurity tools each offering various capabilities to help organizations defend against modern threats such as CylancePROTECT, Darktrace, Vectra AI, FireEye Helix, Symantec's Targeted Attack Analytics (TAA), IBM QRadar Advisor with Watson, CrowdStrike Falcon, IBM Security QRadar Suite, IBM Security Guardium, IBM Security Verif.

CylancePROTECT is using an AI and ML, that can predict and prevent malware attacks by analyzing files and identifying malicious patterns. Darktrace employs AI algorithms to spot and answer in real-time to cyber threats, using ML to understand 'normal' behavior and identify anomalies indicative of potential attacks. Vectra AI offers a platform that uses AI to spot and answer to cyberattacks in-progress, focusing on detecting threats within network traffic. FireEye Helix is a platform that combines AI, ML, and analytics to detect and respond to threats across various vectors, including email, endpoint, and network. Symantec's Targeted Attack Analytics (TAA) exploits AI and ML techniques to detect advanced and targeted cyber threats that may evade traditional security measures. The IBM QRadar Advisor with Watson combines IBM's QRadar security analytics platform with Watson's AI capabilities to provide actionable insights and automate responses to security incidents. CrowdStrike Falcon employs AI and ML to identify and block cyber threats, leveraging a cloud-based platform for endpoint protection and threat intelligence [28].

IBM Security QRadar Suite developed for threat detection, investigation, and response technologies. It is developed on an open foundation personalized to accommodate the needs of hybrid cloud environments. IBM Security Guardium is a data security platform that ensures full visibility throughout every stage of the data lifecycle. It also assists in addressing data compliance needs and enhance the robustness and resilience of cybersecurity solutions, incorporating advanced AI-based outlier detection mechanisms. Such systems are designed to consider multiple risk factors, such as network traffic, user behavior, and system configuration changes, to detect and flag inconsistent patterns that may indicate a potential threat. By analyzing large amounts of historical and real-time data, these AI-driven systems can detect subtle deviations from normal behavior or emerging threat patterns that might otherwise go unnoticed. The resulting alerts and insights can then be used to trigger appropriate mitigation actions or inform security teams for further investigation. IBM Security Verify provides a tailored AI-powered context for both consumer and workers identity access management (IAM). It protects consumers and apps, both as individuals and employers, using a low-friction, cloud-native, software-as-a-service (SaaS) approach [3, 29, 30].

There are more examples of AI-powered cybersecurity various tools and initiatives such as Palo Alto Networks Cortex XDR, CAPTCHA—Facial Recognition, and Fingerprint Scanners, NVIDIA Morpheus, Kriptos, LetsData and the OpenAI Cybersecurity Grant Program.

Palo Alto Networks Cortex XDR provides complete visibility and control over the entire organization IT environment [31].

CAPTCHA—Facial Recognition, and Fingerprint Scanners enable institutions to detect automatically if an attempt to log in to a service is genuine, helping in preventing brute-force attacks and credential stuffing [32].

NVIDIA Morpheus is a GPU-accelerated, end-to-end AI framework that allow to create improved applications for processing, filtering, and classifying huge volumes of streaming cybersecurity data. As an AI-powered cybersecurity solution, it aims to streamline and enhance the threat detection and response process by leveraging the power of machine learning and artificial intelligence. By intelligently analyzing vast amounts of data, Morpheus can efficiently identify potential threats, quickly capture and correlate related events, and suggest appropriate mitigation actions, all while minimizing the burden on security teams. This approach allows organizations to respond to threats faster and more effectively, reducing the risks associated with cyberattacks and minimizing potential damage. Moreover, the automation and efficiency provided by Morpheus can ultimately result in lower operational costs, making it a cost-effective solution for enhancing cybersecurity posture.

Kriptos is a software that analyzes, classifies, and labels documents automatically using AI. It prevents data leakage, enhances cybersecurity strategies, and achieves compliance regulations [33].

LetsData offers AI radar against disinformation [34].

OpenAI's Cybersecurity Grant Program offers funding for projects that seek to enhance the effectiveness and efficiency of cybersecurity defenses by utilizing AI techniques. The program specifically targets projects that collect and label data from cyber defenders to train AI agents in defensive techniques, such as identifying and mitigating social engineering tactics, automating incident response, detecting source code vulnerabilities, and assisting with network and device forensics [35].

5 Cybersecurity Continuous Education for Individuals

The last year escalation of generative AI is revolutionizing social engineering cyberattacks, making them more sophisticated and more difficult than ever to be detected, therefore it's crucial for individuals and organizations to stay vigilant and adopt comprehensive security measures. This includes keeping up-to-date of the latest cyber threats, practicing safe browsing habits, using strong passwords and MFA, installing and updating security software regularly, and being cautious of unsolicited emails or suspicious links. These advanced technologies provide cybercriminals with powerful tools to exploit human vulnerabilities, more effectively than ever before. To mitigate the risks associated with generative AI-driven social engineering attacks, it is crucial to adopt a multi-layered approach that combines robust technological solutions, stringent security policies, and individuals' awareness. Some strategies that individuals can implement advanced security solutions, refer to providing regular training and awareness, developing and enforcing strict security policies, encourage a culture of security and partner with security experts. By proactively addressing the challenges posed by generative AI and prioritizing a strong security posture, individuals can better protect themselves against this new wave of sophisticated cyber threats [36].

Cutting-edge security solutions that leverage ML and AI must be used to detect and prevent sophisticated social engineering attacks [37]. These tools can examine communication patterns, consumers' behavior, and content to recognize potential threats. Individuals must be required to gain the necessary skills and knowledge to identify and report potential social engineering attacks by conducting regular training sessions and workshops that cover the latest trends in AI-driven social engineering, common attack techniques, and best practices for maintaining security. Clear guidelines and procedures must be established to handle sensitive information, accessing systems, and responding to potential security incidents. Regularly update of these policies is compulsory to address new and emerging threats posed by generative AI [38].

An environment where security is a top priority for all individuals must be adopted. Encourage open communication and collaboration between family members from different generations would ensure that potential risks are quickly identified and addressed. Collaboration with cybersecurity professionals and organizations to assure that the individuals stay up-to-date regarding the latest threats, technologies, and best practices in combating AI-driven social engineering attacks, should be a must [39].

AI acts a significant part in enhancing individuals' cyber resilience. AI contributes to cyber resilience by enabling threat detection and mitigation, building resilience, actively dealing with ai-augmented hacking, and serving vulnerable individuals [40, 41].

AI is not only used to enhance the security measures but also to ensure the resilience of the systems against cyber threats. It's a critical tool in the modern digital landscape where cyber threats are becoming increasingly sophisticated and damaging. It is crucial for individuals to stay informed about these potential threats and take appropriate measures to safeguard their systems and data. Staying proactive in the realm of cybersecurity is crucial. With the advancement of technology, particularly with AI, the threat landscape is constantly evolving. It's not just about defending against known threats anymore; organizations must anticipate and adapt to emerging attack vectors. This means continuous monitoring, analyzing, and sharing threat intelligence to fortify defenses effectively. Being ahead of the curve in cybersecurity is not just a goal; it's a necessity in today's digital age. This might include investing in advanced threat detection systems, educating individuals about potential threats, and implementing robust data protection policies [42].

Continuous education tools [43] aim to enforce individuals' cyber hygiene and resilience. Cyber hygiene and cyber resilience are both important aspects of cybersecurity, they serve different purposes and are applied differently for individuals and organizations. Cyber hygiene refers to the regular actions taken to maintain the security of digital assets. This can include practices such as regularly updating software, changing passwords, and backing up data. Cyber hygiene is often associated with individuals, but it's also crucial for organizations. It's about being proactive and taking steps to prevent cyber threats [44].

Cyber resilience concept has been used and applied on organizations' behavioral and actions, being related to risk assessment and management, threat intelligence, security awareness and training, vulnerability management, incident response planning and testing, business continuity and disaster recovery planning, as well as collaboration and information sharing. By adopting a proactive approach to cybersecurity, organizations can strengthen their cyber resilience, decrease the effects of cyberattacks, and ensure the continuity of their operations [45]. Cyber resilience is about an organization's ability to continue operating despite a cyber-attack [46]. It's about being reactive and having plans in place to respond to and recover from cyber threats1. Cyber resilience considers cyber hygiene as essential [47] but it goes beyond that to include strategies for dealing with attacks when they do occur.

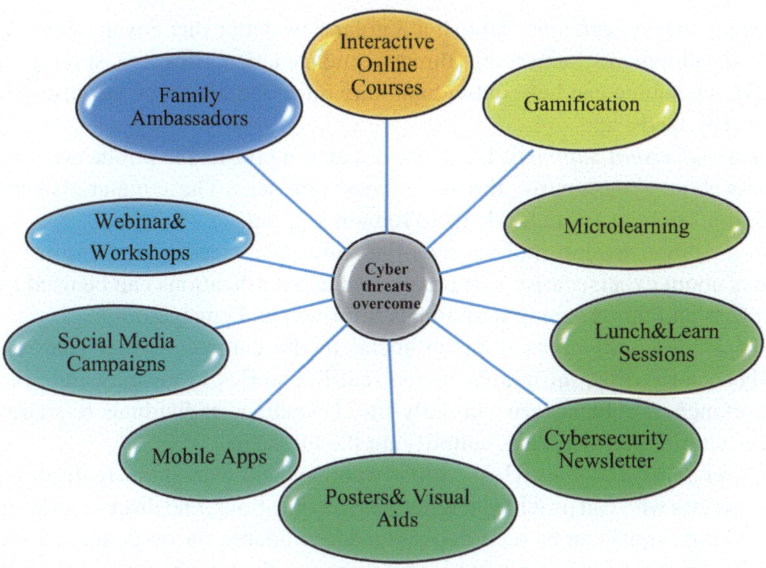

Fig. 1 Cyber threats overcome. *Source* Authors' elaboration

So, while cyber hygiene can be seen as a part of cyber resilience, they are not the same. Cyber hygiene focuses on prevention, while cyber resilience includes both prevention and response strategies. Both are important for both individuals and organizations to face the increasing cyber threats.

Diverse approaches to education and awareness are available today to empower individuals to combat cybersecurity threats (Fig. 1).

As depicted on Fig. 1 we can see that there are ten types of cybersecurity threats overcoming:

Interactive Online Courses: engaging and interactive online courses can be developed that cover various cybersecurity topics, such as phishing, malware, and safe internet practices, including real-life examples, quizzes, and simulations to reinforce learning.

Gamification: game-based learning platforms that teach cybersecurity concepts through interactive challenges, points, badges, and leaderboards. This approach can make learning more enjoyable and motivate individuals to develop a strong cybersecurity skillset.

Microlearning: Deliver short, focused learning modules that address specific cybersecurity topics. Microlearning caters to busy individuals who may not have time for longer training sessions, providing just-in-time learning that can be easily integrated into daily routines.

Lunch-and-Learn (LaL) Sessions: Organize informal LaL sessions where personnel can learn about cybersecurity topics from experts in a relaxed environment. These sessions can encourage open discussions and knowledge sharing among colleagues.

Cybersecurity Newsletter: Publish a regular newsletter that covers recent cybersecurity developments, emerging threats, and practical tips for staying secure online. This resource can help individuals stay informed about the rapidly changing cybersecurity landscape.

Posters and Visual Supports: Display eye-catching posters in public areas to raise awareness about cybersecurity threats and best practices. These materials can serve as continuous notices for individuals to remain vigilant.

Mobile Apps: Develop mobile apps that offer bite-sized lessons, quizzes, and reminders about cybersecurity best practices. Push notifications can be used to send timely alerts and tips, keeping individuals informed and engaged.

Social Media Campaigns: Launch social media campaigns that share actionable cybersecurity tips, infographics, and real-life stories to raise awareness about the importance of cybersecurity in daily life. Encourage individuals to share these resources within their networks, amplifying the message.

Webinars and Workshops: Host regular webinars and workshops featuring cybersecurity experts who can provide insights, answer questions, and discuss relevant case studies. These events can be recorded and made available for on-demand viewing.

Family Ambassadors: Train and empower scholars to become cybersecurity ambassadors within their family. These ambassadors can promote cybersecurity awareness among their family, share best practices, and encourage a culture of security.

By leveraging a diverse range of educational tools and approaches, a strong cybersecurity culture can be fostered allowing individuals to identify and mitigate potential threats.

In the dynamic and ever-changing field of cybersecurity, continuous education is not just an option but a necessity. As technology advances and cyber threats grow more complex, it is essential for individuals to stay informed and adjust their online habits to reduce risks effectively. By actively learning and staying updated on the latest developments, individuals can build a solid understanding of cybersecurity principles and foster a security-focused mindset. This includes grasping the importance of strong passwords, identifying phishing attempts, maintaining up-to-date software and devices, and recognizing the potential dangers linked to various online activities.

Moreover, continuous education can help individuals understand the broader implications of cybersecurity and how their actions can impact not only their own security but also that of their networks and communities. In essence, continuous education is the cornerstone of effective cybersecurity. By embracing a culture of learning and staying vigilant, individuals can play an active role in protecting their personal information and contributing to a safer digital environment for all.

6 Good Practices to Protect Individuals' Data and Own Devices Online

To help protect individuals' data and personal devices online, it's crucial to use strong and unique passwords, enable MFA, keep devices and software up to date, create regular backups of important files and store them on an external drive or cloud storage service use antivirus software, be suspicious of unsolicited emails, messages, or phone calls that request sensitive information or urgent action, even if they appear legitimate a secure your home network, be cautious when using public Wi-Fi to avoid accessing sensitive data or logging into important accounts while connected to unsecured public Wi-Fi networks, regularly back up own data, limit the personal information shared on social media platforms and review privacy settings to control who can access your data, and stay up-to-date about the newest cybersecurity threats and best practices to protect yourself and your devices effectively [6].

6.1 The Importance of Strong Passwords

The importance of a strong password in online security is crucial to protecting your data and personal accounts from unauthorized access. A strong password is one of the most effective defenses against cyber-attacks, as the difficulty of guessing or cracking a password can deter would-be criminals. A strong password is characterized by a mixture of lower- and upper-case letters, special characters, numbers and an appropriate length, which makes it almost impossible to guess or brute force or algorithmic methods. It must contain at least 8 characters, capital and lowercase letters, numbers and at least one special character ('$' '!' '@' '#' '%' '^' '&' '*' '(' ')') and not contain personal details. Therefore, choosing and using a strong password for online accounts is essential to protect personal and financial information, as well as to prevent the possible negative consequences of unauthorized access to data.

Using a password manager is important. Choosing a password manager depends on own preferences, the level of security and available budget. Any of the NordPass, LogMeOnce, KeePass, Avira Password Manager or Bitwarden tools can be used [48, 49].

NordPass is a simple password management service developed by the team behind NordVPN, it has secure access to passwords via desktop, mobile or web apps. As a disadvantage of this application, the free version disconnects the user when changing devices and does not allow password sharing [50].

LogMeOnce offers an enterprise password management solution, with a free version available for small businesses, it has MFA, and the free version has restrictions, but may be useful for individual individuals [51].

KeePass is a free and secure password manager, but it can be more difficult to use than most and is not suitable for non-technical individuals. It stores passwords in a locally encrypted database and has the possibility of customization and complete control over user data [52].

Avira Password Manager is a simple solution where the free version offers password storage and has an easy-to-use interface [53].

6.2 A Clean Office Policy

A "clean desk policy" for individuals online could be framed as a set of best practices and habits aimed at maintaining digital hygiene and safeguarding personal information.

By promoting this mindset of organization, cleanliness, and security in the digital environment, individuals better protect their personal data and reduce exposure to cybersecurity risks both at home, at work, and on their mobile devices [54].

The principles of a clean office policy can be organization and structuring, digital cleanliness, and those of privacy and security. Just as we organize documents and important objects in a clean office, individuals should organize and structure digital data in a similar way. This means using data storage and management tools that provide protection, such as cloud storage services with MFA or password management applications. Just as disposal of unnecessary papers and objects in the office to maintain a clean environment, individuals should dispose the unused or useless data and digital files to reduce the risk of exposure to cyber-attacks. Just as individuals ensure that important documents are kept in a safe place and accessible only to authorized persons, they should ensure that passwords and sensitive data are protected by adequate security, such as strong passwords and two-step authentication [55].

6.3 Browsing the Internet Safely

Browsing the Internet can be done safely using VPN, knowing the difference between HTTP and HTTPS, avoid using the "Keep me signed in" option, knowing and avoiding the "cookie hijacking" hacking technique, deleting the browser cache.

VPN (Virtual Private Network) is a virtual private network that guarantees data privacy through encryption and changing the IP address. This is a tool that provides a higher level of cyber security. Connecting to a VPN allows you to connect to a remote server provided by your Internet provider. Thus, data is protected and access to geo-blocked content is created, even on public Wi-Fi connections. VPNs are essential for data protection, especially when connecting to public Wi-Fi [9].

HTTP (Hypertext Transfer Protocol) is an application-level protocol in the Internet protocol suite used as the main system for transmitting information on the World Wide Web. In contrast, HTTPS (Hypertext Transfer Protocol Secure) is an extension of the HTTP protocol that provides more security to data transmitted over a Secure Socket Layer (SSL). The main difference between HTTP and HTTPS is the SSL3 certificate. The website that loads on HTTPS uses it to send and receive information in an encrypted state. In contrast, on HTTP, data is sent in text format, which can be easily read by anyone.

The "Keep me signed in" option is a common feature on many websites and applications that allows to maintain the signed-in status on personal device, even after closing the browser or application. This is a convenient and time-saving feature, as is not necessary to re-enter username and password every time the site is visited or the app is used. However, there are also some security risks associated with using this option. If the device is lost or stolen, whoever finds it will have access to personal accounts [56].

Cookie Hijacking is a hacking technique that steals a session from a user to access their account. Attackers use Cookie Hijacking attacks to compromise the personal account when the user logs in, for example to their own account on social networks such as LinkedIn, Meta or Instagram. A Cookie Hijacking attack typically involves injecting JavaScript code into a website by embedding it in the HTML of an email or ad that appears to be genuine. This malicious code is then executed by the browser when the user visits the infected website. Accepting every cookie request is not necessary or beneficial. Some cookies are placed by the main parties, such as the websites you visit, to help you navigate more easily, set preferences, save items in a shopping cart, etc. Other types of cookies are placed by third parties, such as "advertisers". They can track browsing behavior and collect data about it, sending it back to websites [57].

Clearing the browser cache involves removing temporary information stored by the browser to improve the loading speed of web pages. This information may include images, CSS styles and JavaScript from the websites you have visited. After you clear your cache and delete cookies, certain site settings are deleted. In Chrome, the option to delete browsing data can be accessed through the "More" menu at the top right and in Microsoft Edge, from the "Settings and more" menu at the top right corner of the browser [58].

6.4 Home Network Security, Anti-Malware and Antivirus Solutions, Apps Update

A home network or LAN (Local Area Network) is a group of computers or other devices interconnected in a single limited area, usually via Ethernet or Wi-Fi. Devices on a LAN, typically personal computers and laptops, can share files and access each other using an Internet connection.

Securing home network involves (i) changing router's default credentials—because hackers are constantly trying to gain access to devices using this publicly known information, (ii) the use of different Wi-Fi and router passwords—to prevent unauthorized access to home router's settings if the Wi-Fi password is compromised, (iii) regular router firmware update—to allow security patches and bug to be solved with the latest firmware that fix network problems, (IV) enabling the firewall to add an extra layer of security to network, (V) allowing only authorized devices to connect to own Wi-Fi network—by using MAC address filtering to specify which devices can join LAN.

The use of antivirus solutions is essential to protect personal devices against cyber threats. These can include viruses, trojans, ransomware and other types of malwares that can compromise data security and system performance. There are a number of free antivirus solutions on the market Bitdefender Antivirus Free Avast Free Antivirus AVG AntiVirus Free and paid Norton Antivirus Bitdefender Antivirus Plus or McAfee Total Protection [59].

Updates can improve the performance of the software, making it faster and more efficient. It is important to keep personal devices updated and patched to address security vulnerabilities, improve device performance and stability, and prevent exposure to new threats, as the constant release of new cyber threats requires continued vigilance on the part of individuals. It is essential to keep operating systems and software udated to ensure that personal devices are protected, efficient and in step with the latest technological innovations [60].

7 Conclusion

The landscape of cyber threats continues to evolve at an accelerated pace, and AI is playing an increasingly significant role in both attacking and defending against these threats. As AI becomes more advanced, it's vital for individuals to stay informed about the latest cybersecurity risks and to adopt a proactive approach to protecting their digital assets.

As a result of this study, we have been detected the cybersecurity threats that developed by AI tools for individuals; defined a new frontier in cybersecurity defense; provided good practices to increase individuals' cyber-resilience and protect individuals' data and own devices.

Ultimately, our cyber-resilience relies on a combination of strong cybersecurity measures, informed consumer behavior, and ongoing collaboration between industry, academia, and government. By working together, we can build a more secure and trustworthy digital future for all.

References

1. Wang, B.X., Chen, J.L., Yu, C.L.: An AI-powered network threat detection system (2022). https://ieeexplore.ieee.org/stamp/stamp.jsp?tp=&arnumber=9775989
2. https://www.ust.com/en/insights/why-understanding-ai-role-in-data-breach-prevention-is-key-to-cyber-resilience
3. https://secureframe.com/blog/ai-in-cybersecurity

4. Guzman, N.: Advancing NSFW detection in AI: training models to detect drawings, animations, and assess degrees of sexiness. J. Knowl. Learn. Sci. Technol. **2**(2), 275–294 (2023). ISSN: 2959-6386
5. Hinton, G.: Navigating cyber threats: understanding the threat landscape and AI-powered solutions for enhanced security in educational platforms (2021)
6. Antipova, T., Riurean, S.: Managing cyber resilience literacy for consumers. International J. Inform. Commun. Technol. (IJ-ICT) (2024)
7. Antipova, T.: Cyberattacks on business website: case study. In: Antipova, T. (eds.) Digital Science. DSIC 2021. Lecture Notes in Networks and Systems, vol. 381, pp. 505–512. Springer, Cham (2022). https://doi.org/10.1007/978-3-030-93677-8_44
8. Moldovan, D., Riurean, S.: Cyber-security attacks, prevention and malware detection application. J. Digit. Sci. **4**(2), 3–23 (2022). https://doi.org/10.33847/2686-8296.4.2_1
9. Riurean, S.: Concepte şi tehnologii noi de comunicații în arhitecturi de rețele. In: Novel Communications Concepts and Technologies in Network Architectures, 528 pp. Universitas Petroşani (2023). ISBN:978-973-741-948-4
10. https://www.zdnet.com/chapter/wormgpt-what-to-know-about-chatgpts-malicious-cousin/
11. Britton, M.: Generative AI: the latest weapon in the cybercriminal arsenal, Forbes technology council. https://www.forbes.com/sites/forbestechcouncil/2023/07/25/generative-ai-the-latest-weapon-in-the-cybercriminal-arsenal/
12. https://www.okta.com/blog/2024/01/how-cybercriminals-are-using-gen-ai-to-scale-their-scams/
13. Stanham, L.: Generative AI (GENAI) in cybersecurity. https://www.crowdstrike.com/cybersecurity-101/secops/generative-ai/. Accessed 27 Nov 2023
14. Sentonas, M.: Introducing charlotte AI, crowdstrike's generative AI security analyst: ushering in the future of AI-powered cybersecurity, endpoint security & XDR. https://www.crowdstrike.com/blog/crowdstrike-introduces-charlotte-ai-to-deliver-generative-ai-powered-cybersecurity/. Accessed 30 May 2023
15. Fadilpašić, S.: How hackers use generative AI in their attacks and what we can do about it. https://www.makeuseof.com/how-hackers-use-generative-ai-in-their-attacks/. Accessed 26 April 2023
16. Vujinović, A., Luburić, N., Slivka, J., Kovačević, A.: Using ChatGPT to annotate a dataset: a case study in intelligent tutoring systems 100557 (2024), ISSN 2666-8270, https://doi.org/10.1016/j.mlwa.2024.100557
17. Barman, D., Guo, Z., Conlan, O.: The dark side of language models: exploring the potential of LLMs in multimedia disinformation generation and dissemination. Mach. Learn. Appl. **16**, 100545. ISSN 2666-8270 (2024), https://doi.org/10.1016/j.mlwa.2024.100545
18. https://news.sky.com/story/wormgpt-ai-tool-designed-to-help-cybercriminals-will-let-hackers-develop-attacks-on-large-scale-experts-warn-12964220
19. https://hackernoon.com/what-is-fraudgpt
20. Teichmann, F.: Ransomware attacks in the context of generative artificial intelligence—an experimental study. Int. Cybersecur. Law Rev. **4**, 399–414 (2023). https://doi.org/10.1365/s43439-023-00094-x
21. https://www.makeuseof.com/what-is-darkbert-ai/
22. https://tech.hindustantimes.com/tech/news/meet-darkbert-the-dark-web-trained-ai-tool-that-can-combat-cybersecurity-threats-71687245585552.html
23. https://arxiv.org/abs/2305.08596
24. https://poe.com/ChaosGPT
25. https://www.analyticsvidhya.com/blog/2023/04/chaosgpt-just-a-mischief-or-bot-with-a-plan-to-destroy-humanity/
26. https://securityintelligence.com/chapters/what-to-know-about-new-generative-ai-tools-for-criminals/
27. Yao, Y., Duan, J., Xu, K., Cai, Y., Sun, Z., Zhang, Y.: A survey on large language model (LLM) security and privacy: the good, the bad, and the ugly. High-Confid. Comput. **4**(2), 100211 (2024). ISSN 2667-2952, https://doi.org/10.1016/j.hcc.2024.100211

28. https://www.softwareadvice.com/resources/top-ai-cybersecurity-tools/
29. https://www.ibm.com/ai-cybersecurity
30. https://hub.packtpub.com/6-artificial-intelligence-cybersecurity-tools-you-need-to-know/
31. https://datasciencedojo.com/
32. https://iicseh.com/the-rise-of-artificial-intelligence-in-cybersecurity.php
33. https://resources.nvidia.com/l/en-us-gtcf22-morpheus
34. https://letsdata.net/
35. https://openai.com/index/openai-cybersecurity-grant-program/
36. Alwahedi, F., Aldhaheri, A., Ferrag, M.A., Battah, A., Tihanyi, N.: Machine learning techniques for IoT security: current research and future vision with generative AI and large language models. In: Internet of Things and Cyber-Physical Systems, vol. 4, pp. 167–185 (2024). ISSN 2667-3452. https://doi.org/10.1016/j.iotcps.2023.12.003
37. Manukyan, L., Gevorgyan, M.: Social engineering attacks: how to prevent. JDS **6**(1), 28–35 (2024). https://doi.org/10.33847/2686-8296.6.1_3
38. Wei-Kocsis, J., Sabounchi, M., Mendis, G.J., Fernando, P., Yang, B., Zhang, T.: Cybersecurity education in the age of artificial intelligence: a novel proactive and collaborative learning paradigm. IEEE Trans. Educ. https://doi.org/10.1109/TE.2023.3337337
39. Nairn, J.: Cybersecurity recommendations for experimenting with AI tools in higher education. https://collegiseducation.com/news/technology/cybersecurity-and-ai-tools-in-higher-education/. Accessed 07 November 2023
40. https://www.pwc.com.au/digitalpulse/ai-cybersecurity-resilience.html
41. https://www.cpomagazine.com/cyber-security/using-ai-to-build-cyber-resilience-for-critical-infrastructure/
42. Hermann, E., Williams, G.Y., Puntoni, S.: Deploying artificial intelligence in services to AID vulnerable individuals. J. Acad. Mark. Sci. **52**, 1431–1451 (2023). https://doi.org/10.1007/s11747-023-00986-8
43. Rosca, S., Riurean, S., Leba, M., Ionica, A.: An educational model of graduation project for students at automation and computer engineering. J. Digit. Sci. **1**(1), 34–42 (2019). https://doi.org/10.33847/2686-8296.1.1_4
44. Flax, R.: Cyber hygiene versus cyber resiliency (2019). https://blog.augustschell.com/cyber-hygiene-versus-cyber-resiliency
45. Tzavara, V., Vassiliadis, S.: Tracing the evolution of cyber resilience: a historical and conceptual review. Int. J. Inf. Secur.Secur. **23**, 1695–1719 (2024). https://doi.org/10.1007/s10207-023-00811-x
46. https://www.eurotechconseil.com/en/blog/difference-between-cyber-security-and-cyber-resilience/
47. Rao, V.: Steps cyber-resilient businesses must take now, 28 June 2022. https://www.weforum.org/agenda/2022/06/cyber-hygiene-resilience-steps/. Accessed 28 June 2022
48. https://www.pcmag.com/picks/the-best-free-password-managers
49. https://www.comparitech.com/password-managers/
50. https://nordpass.com/
51. https://logmeonce.com/
52. https://keepass.info/
53. https://passwords.avira.com/
54. https://www.mtu.ie/media/mtu-website/governance/policies-and-publications/data-protection/Data_Handling_and_Clean_Desk_Policy.pdf
55. https://tableair.com/blog-clean-desk-policy-benefits-implementation-and-challenges/
56. https://answers.microsoft.com/en-us/outlook_com/forum/all/turn-on-keep-me-signed-in-option/fdde7ca3-1d5f-4fbc-b619-5c6090a6cf1e
57. https://securityintelligence.com/chapters/guide-to-cookie-hijacking/
58. https://www.bitdefender.ro/consumer/support/answer/21852/
59. https://www.tomsguide.com/us/best-antivirus,review-2588.html
60. https://itigic.com/ro/ways-to-update-major-operating-systems/

Data Augmentation in Artificial Neural Network Training Using Regression Models

Vadim Danelian, Andrei Kliuev, Vladimir Makarov, Mikhail Pesin, and Valerii Stolbov

Abstract This chapter presents an approach to augmentation of experimental data used for training neural networks in titanium alloy turning problems. Given the high cost of conducting experiments, a method of artificial sample augmentation based on regression models built on a limited dataset was proposed. The mathematical regression models were used to generate additional data points, resulting in an expanded database of 2000 examples. This data was used to train neural networks whose task was to predict the surface roughness of titanium alloy VT6. During the study, two experiments were conducted with different data preprocessing techniques: standardization and normalization. The results showed that the neural networks trained on the augmented data achieved high prediction accuracy in terms of MAPE metric, namely 3.97%. This approach demonstrates the effectiveness of using data augmentation for model training under limited input data.

Keywords Data augmentation · Artificial neural network · Titanium alloys · Ultrafine-grained structure · Coarse-grained structure · Roughness · Approximation

1 Introduction

In recent years, machine learning methods have been widely used in various manufacturing industries [1]. One of the key features when applying such methods is the need for a large amount of data to train the models. In real industrial conditions, data collection can be difficult or expensive, which is especially true for high-precision and complex processes, such as titanium alloys machining by cutting [2].

Titanium alloys are widely used in aerospace, medical and automotive industries due to their unique properties such as high strength, corrosion resistance and light weight [3, 4]. However, their machining presents significant challenges due to their high hardness and chemical activity. In order to obtain a quality machined surface,

V. Danelian · A. Kliuev · V. Makarov · M. Pesin · V. Stolbov (✉)
Perm National Research Polytechnic University, Perm, Russia
e-mail: valeriy.stolbov@gmail.com

T. Antipova (ed.), *Digital Technology Platforms and Deployment*, Information Systems Engineering and Management 36, https://doi.org/10.1007/978-3-031-86547-3_15

it is necessary to optimize the cutting parameters, which entails the need to obtain information on how various parameters affect the surface finish [5]. There have been studies where artificial neural networks are used to predict machining output parameters (e.g. roughness) under different cutting conditions [6]. However, one of the problems encountered in such works is the insufficient amount of experimental data, which does not allow correct training and testing of artificial neural networks [7].

In this case, data augmentation, i.e. a technique of artificially increasing the training set, is one of the promising methods for solving this problem [8]. Augmentation allows to significantly expand the training sample, increasing its diversity and improving the generalization ability of models [9]. In addition, it helps artificial neural networks to learn more stably and avoid overtraining.

A large number of successful data augmentation methods are known in various areas of machine learning. For example, the study [10] provides a comprehensive review of current time series data augmentation methods and their application in deep learning tasks. The authors conduct an empirical comparison of augmentation methods on different datasets. The chapter discusses the advantages and disadvantages of each approach, as well as the situations in which certain methods work best. When it comes to textual data, [11] discusses different data augmentation methods for natural language processing (NLP) tasks, their application and their impact on the performance of deep learning models. The authors offer guidelines for using different text augmentation techniques depending on the type of data and the task. It is also worth emphasizing the value of data augmentation in computer vision tasks as well. Thus, the paper [12] not only shows the effectiveness of data augmentation in object detection tasks, but also proposes a new approach to automating the selection of data augmentation strategies, which can significantly improve the efficiency and accuracy of models. There is also a severe lack of data in the medical domain, as noted in [13]. That is why the authors aim in their study to systematize existing augmentation techniques, identify their advantages and disadvantages, and provide recommendations for their application in medical imaging.

Note that all of the above studies emphasize the importance of augmentation methods in a wide variety of tasks. In the context of material processing, data augmentation can be particularly useful as it reduces the cost of experiments and allows neural network models to be trained correctly. This paper considers a data augmentation approach using regression models to predict surface roughness in titanium alloy turning.

The main novelty of this work is the use of regression models to generate additional data points and the subsequent training of a neural network on this expanded data. This allows us to significantly increase the amount of data for training. This approach also makes it possible to use one neural network instead of two regression models and generalize the solution to different grain sizes.

2 Statistical Processing of Experimental Data

In this chapter, the results of experiments on turning of titanium alloy VT6 with ultrafine-grained (UFG) and coarse-grained (CG) structures with grain sizes of 350 and 22,500 nm, respectively, were used.

The influence of cutting parameters on the roughness of the machined surface Ra (μm) was investigated. The cutting parameters were: cutting speed V (m/min), feed per revolution S (mm/rev) and depth of cut t (mm).

Table 1 shows the cutting modes at which the studies on turning of UFG and CG specimens of VT6 alloy were carried out.

The results of the two roughness measurements on Ra parameter are presented in Tables 2 and 3.

By designing multifactorial experiments with statistical treatment and regression analysis, regression mathematical models in real physical quantities were obtained for UFG (Eq. 1) and for CG (2) structures of VT6 alloy:

$$Ra = 0.014 - 0.004V + 1.252S + 2.114t + 0.123VS - 0.021Vt - 6.74St \quad (1)$$

Table 1 Cutting modes for experiments

Mode No.	V, m/min	S, mm/rev	t, mm
1	48	0.06	0.25
2	48	0.06	0.5
3	48	0.11	0.25
4	48	0.11	0.5
5	72	0.06	0.25
6	72	0.06	0.5
7	72	0.11	0.25
8	72	0.11	0.5

Table 2 Roughness measurement for VT6 alloy specimen with UFG structure

Mode No.	Ra, μm		
	1 measurement	2 measurement	Mean
1	0.419	0.429	0.424
2	0.532	0.692	0.612
3	0.722	0.692	0.707
4	0.795	0.786	0.791
5	0.371	0.41	0.391
6	0.442	0.425	0.434
7	0.842	0.759	0.801
8	0.797	0.762	0.78

Table 3 Roughness measurement for VT6 alloy specimen with CG structure

Mode No.	Ra, μm		
	1 measurement	2 measurement	Mean
1	0.553	0.468	0.511
2	0.405	0.408	0.407
3	0.453	0.507	0.48
4	0.446	0.454	0.45
5	1.427	1.462	1.445
6	0.677	0.667	0.672
7	1.09	1.069	1.08
8	0.787	0.782	0.785

$$Ra = -4.218 + 0.103V + 27.119S + 8.403t - 0.61VS - 0.191Vt - 57.918St + 1.333VSt \quad (2)$$

We compare the predictions of models (1) and (2) with experimental data using the well-known metrics MAE (mean absolute error), MAPE (mean absolute percentage error) and MSE (mean squared error) (Tables 4 and 5).

MAE = 0.005375 μm (0.93% MAPE), MSE = 0.000029.

MAE = 0.0005 μm (0.09% MAPE), MSE = 0.000001.

These estimates are an indication of a good fit of the regression models to the experimental data.

Table 4 Predictions of the model (1) in comparison with the experimental data

Mode No.	1	2	3	4	5	6	7	8
Experiment	0.424	0.612	0.707	0.791	0.391	0.434	0.801	0.78
Model	0.429	0.607	0.702	0.796	0.385	0.439	0.806	0.774

Table 5 Predictions of the model (2) in comparison with the experimental data

Mode No.	1	2	3	4	5	6	7	8
Experiment	0.511	0.406	0.48	0.45	1.445	0.672	1.08	0.784
Model	0.51	0.406	0.478	0.45	1.444	0.672	1.08	0.784

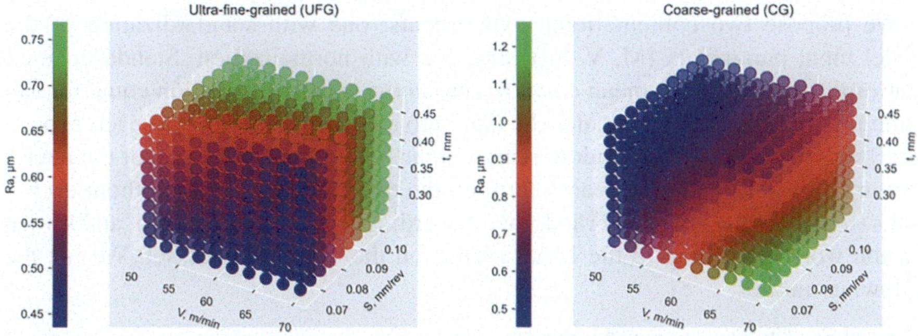

Fig. 1 Data points obtained using augmentation by linear regression models for UFG (left) and CG (right). *Source* Authors' elaboration

3 Data Augmentation

The number of elements in the data set is 16 (8 modes for UFG and 8 for CG), which is insufficient to train even the simplest neural network models. Therefore, it is proposed to augment the training set based on the available linear regression equations for VT6 of UFG (1) and CG (2).

For augmentation, 2 parameters are fixed in the regression equations and dependencies of Ra on the 3rd parameter are found (with a constant step). The obtained data are entered into the database (here—file with extension.csv). The parameters V, S, t were varied in the ranges in which the experiments were carried out. The results presented in graphical form for UFG and CG can be observed in Fig. 1.

In this way, 2000 examples were obtained, which were then used to train neural network models: 1000 points for UFG and 1000 points for CG.

Of the 2000 examples, 80% will be used to train the models and 20% will be used as a validation set. The final best model will be tested on experimental data: 16 examples. The neural network will then be compared with linear regression models. It is important to note that this method of testing the model is not recommended, as the training set was constructed indirectly from the test set. In other words, information about the test set is contained in the training set. Nevertheless, we allow it in the context of our work, since it is sufficient for us to solve the problem of approximating the dependence within the limits studied, and for this purpose we can accept the mentioned restriction.

4 Experiments

It is required to solve the problem of approximating two dependencies presented above using a MLP neural network. The input to the network is a vector {M, V, S, t} and the output is Ra. M is the grain size: 350 nm for UFG and 22.5 µm for CG.

We propose two computational experiments: one with standardization of the model input parameters (M, V, S, t) and one with normalization. Standardization centres the data around a mean equal to zero and a standard deviation equal to one, while normalization simply scales the data into a range from 0 to 1. In each experiment, we will implement a random search in the hyperparameter space of the neural network. We will randomly search through the different models, train them each 3 times (to reduce the chance of randomly discarding a good architecture), and screen out the worst according to the MAE metric on the validation dataset. We list the following metrics:

- number of fully connected layers (1–10);
- the number of neurons in each layer (from 32 to 1024 with a minimum step of 32);
- the activation function in each layer: relu or tanh;
- addition/non-addition of a dropout layer after the next fully connected layer, and the fraction of neurons to be disabled (from 0.1 to 0.5 with a minimum step of 0.05);
- learning rate: between 0.0001 and 0.01.

In addition, we will use the following invariant parameters in each model:

- Adam optimazer;
- a fully connected output layer with one neuron and a linear activation function;
- MAE as a loss function;
- MSE as an additional metric;
- 75 epochs of training, 3 times per architecture (one hyperparameter configuration).

After a random search in the space of different hyperparameters of the neural network in the initial standardization of the data, the architecture presented in Fig. 2 was selected.

In each hidden layer, a relu activation function is used and the training step is 0.0001. Once the best model was found, it was further trained for 500 epochs: the results are shown in Fig. 3. The graphs show that the training was successful.

The results were as follows:

- on training data:
  - MAE $\approx$ 0.0011 μm (0.17% MAPE), MSE $\approx$ 1.84e-06;
- on validation data:
  - MAE $\approx$ 0.0012 μm (0.19% MAPE), MSE $\approx$ 2.33e-06.

It is worth noting that the results are quite good: in the context of this problem, such an error is acceptable.

After a random search in the space of different hyperparameters of the neural network with initial data normalization, the architecture shown in Fig. 4 was selected.

Fig. 2 Architecture of the best neural network under initial data standardization

Model: «With standardization»

Layer (type)	Output Shape	Param #
dense (Dense)	(None, 768)	3840
dense_1 (Dense)	(None, 896)	689024
dense_2 (Dense)	(None, 128)	114816
dense_3 (Dense)	(None, 320)	41280
dense_4 (Dense)	(None, 1)	321

Total params: 849281 (3.24 MB)
Trainable params: 849281 (3.24 MB)
Non-trainable params: 0 (0.00 Byte)

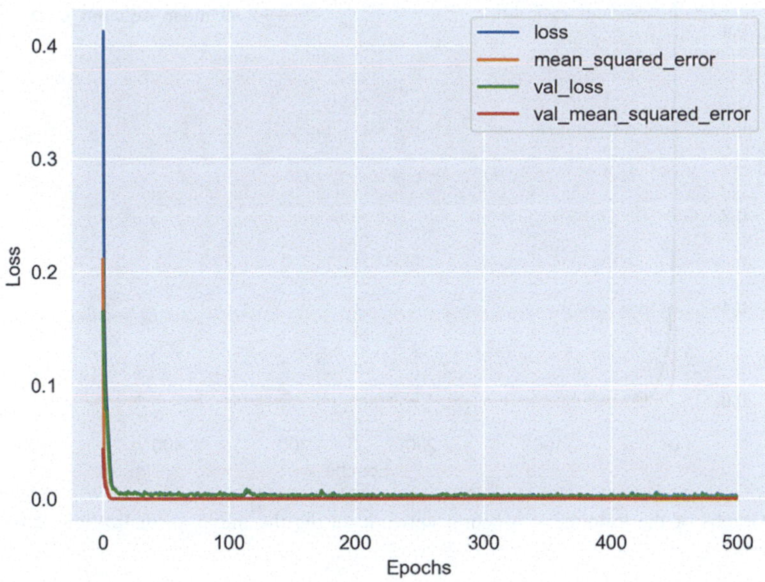

Fig. 3 Graphs of the training of a neural network during the initial standardisation of the data: loss—MAE

In each hidden layer, a relu activation function is used and the training step is 0.0001. Once the best model was found, it was further trained for 500 epochs: the results are shown in Fig. 5. The graphs show that the training was successful.

The following results were obtained:

- on training data:

Fig. 4 Architecture of the best neural network under initial data normalization

```
Model: «With normalization»
_________________________________________________
Layer (type)          Output Shape        Param #
=================================================
dense_2 (Dense)          (None, 672)         3360

dense_3 (Dense)          (None, 448)         301504

dense_4 (Dense)          (None, 1)          449

=================================================
Total params: 305313 (1.16 MB)
Trainable params: 305313 (1.16 MB)
Non-trainable params: 0 (0.00 Byte)
_________________________________________________
```

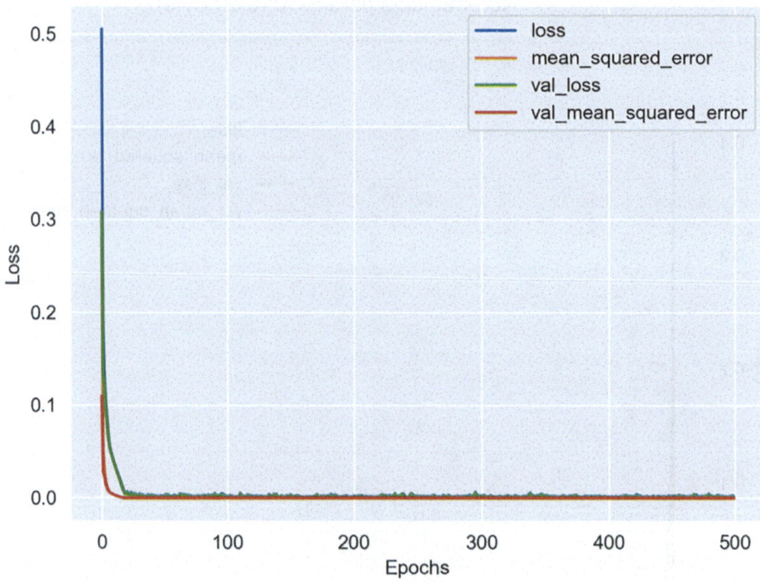

Fig. 5 Graphs of the training of a neural network during the initial normalization of the data: loss—MAE

- MAE ≈ 0.0005 μm (0.08% MAPE), MSE ≈ 4.25e-07;

- on validation data:
    - MAE ≈ 0.0006 μm (0.1% MAPE), MSE ≈ 7.19e-07.

In this case, the results can also be described as good. The models are qualitatively close to the original dependencies (1–2), which was to be expected. It is also important to note that the excellent results on the validation data are explained by the fact that they were drawn from the same distribution as the training data. In addition, our test data are closely related to the training and validation sets. This follows from the fact that the regression models were built on the test data (experimental data) and the data for training the models were generated using the same regression models, as mentioned above.

5 Testing

The best model within the framework of this work is the model with data normalization because of its higher estimation on the validation set. Let us compare its predictions with the experimental data (see Table 6: MAE $\approx$ 0.024 μm (3.97% MAPE), MSE $\approx$ 0.0011.

These results are shown in the graph (see Fig. 6).

Note that the neural network performs well with the UFG and CG materials, but there are small discrepancies at two points (number 4 and 12), which is quite normal.

Table 6 Comparison of neural network predictions (Predict) with experimental data (Ra_mean) for given input parameters

No	M, nm	V, m/min	S, mm/rev	t, mm	Ra_mean, μm	Predict, μm	|Ra—Predict|, μm
0	350	48.0	0.06	0.25	0.424	0.411	0.013
1	350	48.0	0.06	0.5	0.612	0.6	0.012
2	350	48.0	0.11	0.25	0.707	0.712	0.005
3	350	48.0	0.11	0.5	0.79	0.802	0.012
4	350	72.0	0.06	0.25	0.39	0.436	0.046
5	350	72.0	0.06	0.5	0.434	0.466	0.032
6	350	72.0	0.11	0.25	0.8	0.8	0.0
7	350	72.0	0.11	0.5	0.78	0.778	0.002
8	22,500	48.0	0.06	0.25	0.511	0.534	0.023
9	22,500	48.0	0.06	0.5	0.406	0.425	0.019
10	22,500	48.0	0.11	0.25	0.48	0.516	0.036
11	22,500	48.0	0.11	0.5	0.45	0.473	0.023
12	22,500	72.0	0.06	0.25	1.445	1.346	0.099
13	22,500	72.0	0.06	0.5	0.672	0.716	0.044
14	22,500	72.0	0.11	0.25	1.08	1.087	0.007
15	22,500	72.0	0.11	0.5	0.784	0.792	0.007

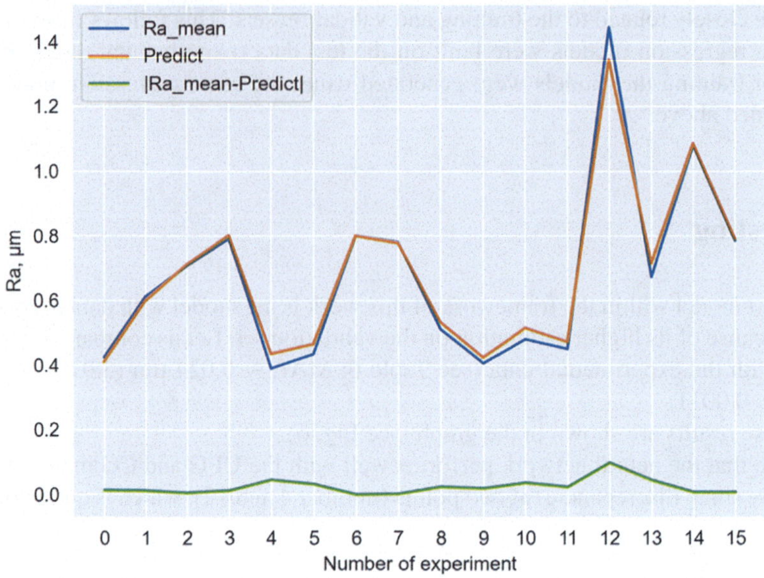

Fig. 6 Comparison of neural network predictions (Predict) with experimental data (Ra_mean) and their difference modulo (|Ra_mean-Predict|)

6 Conclusions

This study demonstrated the effectiveness of data augmentation in the task of training neural networks for predicting surface roughness during turning of titanium alloys. The use of regression models to generate additional data points allowed us to significantly expand the training sample, realize a correct training process of artificial neural networks and achieve high accuracy of predictions. The proposed approach can be adapted for other material processing processes and tasks related to training of neural networks on limited data. Further research can be directed to the improvement of augmentation methods and their application in various fields of industrial production.

Acknowledgements This research was carried out with the financial support of the Ministry of Science and Higher Education of the Russian Federation in the framework of the program Federal Academic Leadership Program «Priority-2030».

References

1. Larrañaga, P., Atienza, D., Diaz-Rozo, J., Ogbechie, A., Puerto-Santana, C.E., Bielza, C.: Industrial Applications of Machine Learning. CRC Press, Boca Raton (2018). https://doi.org/10.1201/9781351128384

2. Pramanik, A.: Problems and solutions in machining of titanium alloys. Int. J. Adv. Manuf. Technol. **70**, 919–928 (2014). https://doi.org/10.1007/s00170-013-5326-x
3. Kishawy, H.A., Hosseini, A.: Titanium and titanium alloys. In: Kishawy, H.A., Hosseini, A. (eds.) Machining Difficult-to-Cut Materials: Basic Principles and Challenges, pp. 55–96. Springer International Publishing, Cham (2019). https://doi.org/10.1007/978-3-319-95966-5_3
4. Jackson, M.J., Kopac, J., Balazic, M., Bombac, D., Brojan, M., Kosel, F.: Titanium and titanium alloy applications in medicine. In: Ahmed, W., Jackson, M.J. (eds.) Surgical Tools and Medical Devices. pp. 475–517. Springer International Publishing, Cham (2016). https://doi.org/10.1007/978-3-319-33489-9_15
5. Nithyanandam, J., Das, S.L., Palanikumar, K.: Influence of cutting parameters in machining of titanium alloy. Indian J. Sci. Technol. **8**, 556 (2015). https://doi.org/10.17485/ijst/2015/v8iS8/71291
6. Rajaparthiban, J., Sait, A.N.: Experimental investigation on machining of titanium alloy (Ti 6Al 4V) and optimization of its parameters using ANN. Mechanics **24**, 449–455 (2018). https://doi.org/10.5755/j01.mech.4.24.20251
7. Linjordet, T., Balog, K.: Impact of training dataset size on neural answer selection models. In: Azzopardi, L., Stein, B., Fuhr, N., Mayr, P., Hauff, C., Hiemstra, D. (eds.) Advances in Information Retrieval, pp. 828–835. Springer International Publishing, Cham (2019). https://doi.org/10.1007/978-3-030-15712-8_59
8. Shorten, C., Khoshgoftaar, T.M.: A survey on image data augmentation for deep learning. J Big Data **6**, 60 (2019). https://doi.org/10.1186/s40537-019-0197-0
9. Gontijo-Lopes, R., Smullin, S.J., Cubuk, E.D., Dyer, E.: Affinity and diversity: quantifying mechanisms of data augmentation (2020). http://arxiv.org/abs/2002.08973, https://doi.org/10.48550/arXiv.2002.08973
10. Wen, Q., Sun, L., Yang, F., Song, X., Gao, J., Wang, X., Xu, H.: Time series data augmentation for deep learning: a survey. In: Proceedings of the Thirtieth International Joint Conference on Artificial Intelligence, pp. 4653–4660 (2021). https://doi.org/10.24963/ijcai.2021/631
11. Shorten, C., Khoshgoftaar, T.M., Furht, B.: Text data augmentation for deep learning. J Big Data. **8**, 1–34 (2021). https://doi.org/10.1186/s40537-021-00492-0
12. Zoph, B., Cubuk, E.D., Ghiasi, G., Lin, T.-Y., Shlens, J., Le, Q.V.: Learning data augmentation strategies for object detection. In: Computer Vision – ECCV 2020, pp. 566–583. Springer, Cham (2020). https://doi.org/10.1007/978-3-030-58583-9_34
13. Chlap, P., Min, H., Vandenberg, N., Dowling, J., Holloway, L., Haworth, A.: A review of medical image data augmentation techniques for deep learning applications. J. Med. Imaging Radiat. Oncol.Radiat. Oncol. **65**, 545–563 (2021). https://doi.org/10.1111/1754-9485.13261

Bytes and Battles in AI Era. Safeguarding Consumers' Digital World

Tatiana Antipova and Simona Riurean

Abstract In today's digital landscape, consumers face a range of cybersecurity threats, from phishing attacks to data breaches. This chapter examines the ways in which artificial intelligence (AI) can both help and hinder consumer security. We explore how cybercriminals leverage AI to launch sophisticated attacks, such as deepfake phishing scams, and how AI-powered security solutions can help detect and mitigate these threats. The goal of this chapter is to detect the main kind of cybersecurity threats for consumers; in part of Generative AI—a new frontier in cybersecurity attacks and defense. Based on this consideration consumer-focused Cybersecurity strategy has been designed bearing in mind consumers' behaviors in AI era.

Keywords Artificial Intelligence · Cybersecurity threads · Cybersecurity attacks · Cybersecurity threads · Consumer-focused cybersecurity strategy · Cybercrime · Cyber-resilience

1 Introduction

In the context of AI age, it is evident that AI will profoundly impact our lives. AI will touch every aspect of our lives, from healthcare and transportation to education and entertainment. It promises to make our lives more convenient, productive, and enjoyable, but also poses significant challenges in areas such as privacy, security, and fairness. Data is the fuel that powers AI systems, and it is increasingly becoming a valuable commodity. AI systems use various data collection methods, such as web scraping and biometric data capture. These techniques can pose privacy risks, especially when personal details are harvested without user consent [1]. AI's data

T. Antipova (✉)
Institute of Cited Scientists, Agia Napa, Cyprus
e-mail: tatiana462@yahoo.co.nz

S. Riurean
University of Petrosani, Petrosani, Romania

T. Antipova (ed.), *Digital Technology Platforms and Deployment*, Information Systems Engineering and Management 36, https://doi.org/10.1007/978-3-031-86547-3_16

hunger and opacity make it challenging for individuals to control what information is collected about them and how it's used [2]. Data privacy and individual rights are fundamental considerations when deploying AI technologies, particularly in sensitive domains like cybersecurity. Efforts to mitigate potential risks include adopting privacy-enhancing techniques, such as privacy-preserving machine learning, differential privacy, and data masking, to protect user information. Moreover, policies and regulations, such as the General Data Protection Regulation (GDPR) and the California Consumer Privacy Act (CCPA), establish frameworks for responsible data collection, storage, and use, providing legal recourse for individuals whose rights have been violated. [3]. Balancing innovation with privacy protection is essential. AI developers should prioritize individual rights and confidentiality [4]. Existing privacy laws may need adaptation to address AI applications effectively [5].

This work is a comprehensive guide to understanding the challenges and opportunities of artificial intelligence in the consumer sector. From data privacy concerns to algorithmic biases, this chapter delves into the complex relationship between consumers and AI technology. It involves safeguarding not only enterprise data but also preventing theft or misuse of customer information. This includes protecting login credentials, banking details, browsing history, national identifiers, and other sensitive personal data [6]. Cybersecurity is a dynamic and ever-changing field, requiring continuous adaptation to keep pace with new threats and vulnerabilities. Various processes, such as threat intelligence gathering, vulnerability scanning, incident response, and compliance monitoring, require constant refinement and innovation. At the same time, the increasing complexity of modern networks, rapid adoption of new technologies, and emerging attack vectors demand that organizations continually review and update their cybersecurity strategies, systems, and processes. This ongoing evolution is crucial to maintaining a robust cybersecurity posture, defending against sophisticated cyber threats, and ensuring the security and resiliency of digital infrastructure. It emphasizes identifying critical assets, protecting them, detecting threats, responding, and recovering from attacks [7].

As we enter a new age of intelligent systems, it is crucial for both consumers and businesses to navigate the terrain safely and ethically. With insights from industry experts and real-world examples, this chapter offers some practical solutions for protecting consumer data, promoting fairness, and fostering a more just and equitable digital world.

The goal of this chapter is to detect the main kind of cybersecurity threats for consumers; in part of Generative AI—a new frontier in cybersecurity attacks and defense. Based on this consideration we should design consumer-focused Cybersecurity strategy bearing in mind consumers' behaviors in AI era.

2 Data and Methodology

The data for the current study were obtained from numerous scientific projects, practical experience and previous publications by the authors included in the analysis [8–11]. A comprehensive methodology for studying consumer cybersecurity threats includes: identity relevant data sources, such as incident reports from security firms, consumer complaints, and government statistics, to gain a comprehensive understanding of consumer cybersecurity threats; categorizing and analyzing cybersecurity threats based on various factors, such as the type of attack, the target audience, and the potential impact. This enables the identification of patterns and trends in consumer-facing attacks.

3 Cybersecurity Threats for Consumers

The most prevalent cybersecurity threats for consumers are phishing attacks, malware infection, SoEng (SoEng) attacks [12], identity theft, weak passwords and credential theft, the unsecured wi-fi networks, fake software updates and downloads, online scams and fraud, data breaches, and the IoT vulnerabilities.

In addition to email, phishing attacks have evolved to take advantage of other communication channels, such as SMS (smishing), voice calls (vishing), and even social media platforms.

To protect against these threats, consumers should practice cybersecurity best practices, such as using unique and strong passwords, keeping software and devices updated, being cautious of unsolicited communications, and using security tools like antivirus software and firewalls. Additionally, staying informed about emerging threats and being vigilant when sharing personal information online can help decrease risk or even mitigate the cybersecurity attacks.

3.1 Phishing Attacks

Phishing has emerged as the most common form of cybercrime, affecting both businesses and consumers on a global scale. In the UK, 83% of businesses that experienced a cyberattack in 2022 reported that phishing was the method used. Moreover, 323,972 internet users worldwide fell victim to phishing attacks in 2021, representing half of all cybercrime victims. Despite the robust cybersecurity measures implemented by tech giants like Google, which successfully blocked 99.9% of phishing attempts, the remaining 0.1% still resulted in significant financial losses. With an average loss of $136 per incident, phishing attacks were responsible for a staggering $44.2 million stolen by cybercriminals in 2021 [13].

Phishing attacks have witnessed an unprecedented surge, with 2022 marking the first year to record a staggering one million such incidents in a single quarter. These attacks have become the foremost gateway for cybercriminals, initiating nearly 91% of all cyberattacks. As communication via email, text messages, and other digital platforms continues to thrive, the potential victim pool for phishing scams only expands. Astonishingly, an estimated three billion phishing emails are disseminated each day, underscoring the enormity of this threat. Despite the notable rise of phishing attacks in the recent past, it is essential to acknowledge that this perilous tactic has plagued consumers for decades, evolving and adapting to exploit the ever-changing digital landscape [14].

Tracing the origins of the first phishing message can be challenging, but early documentation of the technique dates back to 1987 in a paper by the International HP Users Group, Intertex. Phishing garnered significant attention during a widespread attack involving the Love Bug malware on May 4, 2000. Victims received an innocuous-looking email titled "ILOVEYOU" containing a *.txt file named LOVELETTER. Upon opening it, the file released a worm that destroyed image files and forwarded a duplicate to all contacts in the victim's Outlook address book [15].

Another notable early phishing incident occurred in June 2001, when the E-Gold website became the first identified target of a phishing attack on an eCommerce platform. In the 1990s, America Online (AOL) users became the primary targets of a significant phishing attack. As the leading internet access provider at the time, AOL presented an enticing opportunity for cybercriminals. The scammers initiated their attacks by stealing AOL users' credentials and employing algorithms to generate randomized credit card numbers. These fake credit card numbers were then used to create fraudulent AOL accounts that facilitated spamming other users via AOL's instant messenger and email systems. Most phishing messages urged recipients to verify their accounts or confirm billing information.

Due to the lack of widespread awareness about phishing during that era, many users fell victim to these deceiving tactics. In response, AOL began incorporating cautionary messages on its email and instant messenger platforms to prevent users from sharing sensitive data through such methods. The term "phishing" is attributed to the well-known hacker and spammer Khan C. Smith, who first introduced it in a Usenet newsgroup following AOL's implementation of countermeasures against fake, algorithmically generated credit card numbers for account creation. Some cybersecurity experts also suggest that the earliest documented reference to phishing can be found in AOHell, a hacking tool developed by a teenager using the pseudonym Da Chronic. This program was specifically designed to carry out phishing attacks on the AOL portal [16].

Phishing, a prevalent cyber threat, requires an attacker that aims to deceive consumers to reveal sensitive data related with, for example with its own bank account, credit card or login information online. Cybercriminals send fraudulent emails or messages that appear to be from legitimate sources to trick consumers into revealing sensitive information or clicking on malicious links or attachments.

The classic phishing attack includes distribution of deceitful emails that appear to come from a legitimate source, enticing users to click on a malicious link or attachment. Phishing attacks, rely on SoEng tactics that exploit human weaknesses, such as curiosity, trust, and fear, to trick consumers into disclosing sensitive information or granting unauthorized access. Over time, attackers have developed more sophisticated techniques to bypass traditional security measures and deceive even the most vigilant users [17].

A recent example of phishing in Romania has been reported in September, 2023 by the National Cyber Security Directorate (DNSC) that has noticed a notable increase in phishing attempts, which are being spread through chapters sponsored by various social media accounts. Cyber criminals use the visual identity of various public figures and media organizations to trick users into accessing fraudulent links in order to collect personal data, login details or financial data from potential victims. The attack method is based on invented chapters containing the logos of news channels known to the Romanian public. Basically, the attackers impersonate celebrities from well-known media trusts or reputable businessmen and use the visual identity of some TV stations, precisely to generate trust in the information provided to the audience. In the posts identified, there is a pattern of attackers offering potential victims a plausible sensational scenario or an offer that is hard to refuse, while at the same time banking on emotions and curiosity that would make them less vigilant. In some cases, the text used by attackers refers to this psychological stimulus from the outset. In one case, the text used by the attackers refers to this psychological stimulus from the outset. The attackers created fake news reports under the Euronews or Antena 3 logo with new information on how to make money easily and in a very short time through RomTrading stock trading platforms. The phishing technique applied by attacker used links that redirected to certain domains using the visual identity of legitimate companies or websites to request authentication or bank details. For victims the damage is financial, but also reputational, by stealing personal and authentication data [18].

In the first quarter of 2023, Darktrace, a cybersecurity firm, reported a concerning 135% increase in malicious email campaigns that demonstrated advanced linguistic deviations. These campaigns, characterized by semantic, syntactic, and grammatical anomalies, are indicative of a potential shift towards more sophisticated and convincing phishing attacks. Coinciding with the widespread availability of tools like ChatGPT, this trend suggests that generative AI could be playing a significant role in crafting these highly targeted and convincing attacks. As threat actors leverage AI-generated content, they are better able to bypass traditional email security filters, increasing the likelihood of successful phishing attempts [19].

According to Phishing Activity Trends Report [20], 2023 was, so far, the worst year for Phishing on Record (Fig. 1).

Although after a relevant decrease in the second quarter of 2023, phishing attacks increased and, in the fourth quarter, the APWG [21] detected 1,077,501 phishing attacks. Attacks against social media platforms burst at the end of 2023, with a 42.8 percent of all phishing attacks. During of 2023, there were nearly five million phishing attacks—4,987,809. This made 2023 the worst year for phishing on record,

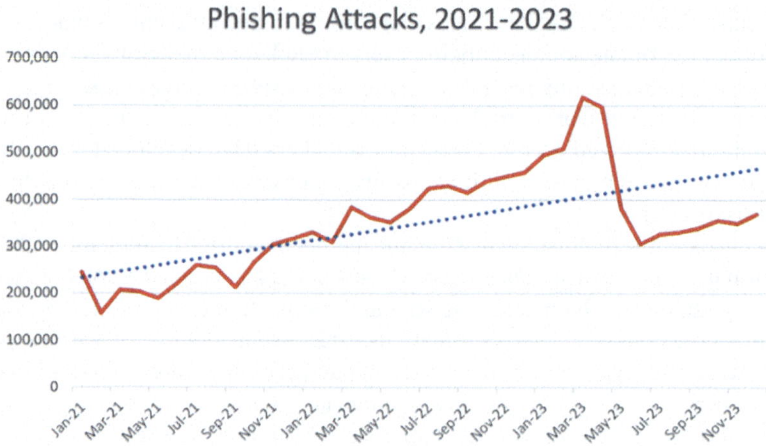

Fig. 1 Phishing attacks report. *Source* [20]

exceeding the 4.7 million attacks in 2022. These figures are truly staggering, and they reflect the ongoing escalation of the phishing threat. It seems that cybercriminals are capitalizing on the ubiquity of social media and leveraging its power to deceive unsuspecting users. The continued growth of phishing attacks is a sobering reminder of the urgency and importance of ongoing cybersecurity education and awareness. With each year yielding higher numbers, it is becoming increasingly clear that the threat of phishing is not slowing down, emphasizing the need for robust security measures, such as multi-factor authentication and email filtering, to protect individuals and organizations against these attacks. Phishing attacks are important threats to both individuals and organizations, exploiting human weaknesses to get access to personal sensitive information. To mitigate this risk, security awareness training and simulated phishing tests have emerged as essential components of a comprehensive cybersecurity strategy.

As communication technology has become increasingly ubiquitous, various forms of phishing attacks designed to exploit the generality of these digital platforms have appeared, such as spear phishing, whaling, smishing, vishing, page hijacking, calendar phishing, quishing and USB attacks.

3.2 Spear Phishing

Spear phishing, for instance, targets specific consumers or organizations with highly personalized messages, often gathered from extensive research. Some examples of spear phishing are employee handbook phishing, fake Microsoft teams' notification and CEO fraud or invoice fraud. In the attack of the employee guidebook phishing, the attacker pretends to be an HR worker and sends an email requesting that the worker needs to sign an updated employee guide. The email includes a link to an online Word document. When the user clicks the link, they are taken to a counterfeit Office 365 login page, where the attacker aims to steal the login credentials. In case of a fake Microsoft teams' notification, the attacker sends a spear phishing email that appears to be a notification from Microsoft Teams. The email prompts the user to click on a "Reply in Teams" button, which leads to a fake login page. If the user enters their details, their account is compromised. If they reuse passwords across accounts, the attacker gains access to multiple systems [22].

CEO Fraud, also known as Business Email Compromise (BEC) or "whaling", is a sophisticated form of phishing that targets an organization's senior employees. It's a scam in which cybercriminals spoof company email accounts and impersonate executives to try and fool an employee in accounting or HR into executing unauthorized wire transfers, or sending out confidential tax information [23, 24].

3.3 Smishing

Smishing, resulting from "SMS" and "phishing," is a type of cyber-crime that uses misleading text messages to manipulate targets into divulging sensitive personal information, for instance login credentials, personal bank account data, credit card numbers, CVC, and so on [25]. Smishing, employs the same underlying concept as phishing, only through text messages rather than email. The attackers use mobile text messages to impersonate legitimate organizations, such as banks, retailers, or service providers, to trick users into divulging sensitive information like credit card details, login credentials, or account numbers. The text message usually contains a malicious link or an offer that seems too good to be true, prompting the user to respond or click. Example of Smishing with real victim depicted on Fig. 2 and described below.

As depicted on Fig. 2 there were three stages of real Smishing:

(1) Scam SMS. To victim iPhone came SMS looked like United States Postal Service (USPS) message. Victim was awaiting very important package from USPS and did not pay attention that the link is not from truly USPS. Real USPS website is URL: https://www.usps.com, not URL: https://uspspsutnqw.top/usunit (see stage (1) in Fig. 2)—just started the same—USPS and victim clicked on the link (never do the same!) in scam SMS and fill out offered form by real debit card data and correct address.

(2) Wallet activation. Immediately after filling out above form on scam link—URL: https://uspspsutnqw.top/usunit, victim received Wallet activation code.

(3) Bank Confirmation. At the same time on victim's email came Confirmation from the bank that card has been added to Google Pay. Fortunately, victim guess to call to the bank and cancelled the activation of this scam Wallet. Bank officer told victim that bank must reuse this debit card since data of it was compromised. And bank reused card and sent it to victim. So, in this case victim did not lose

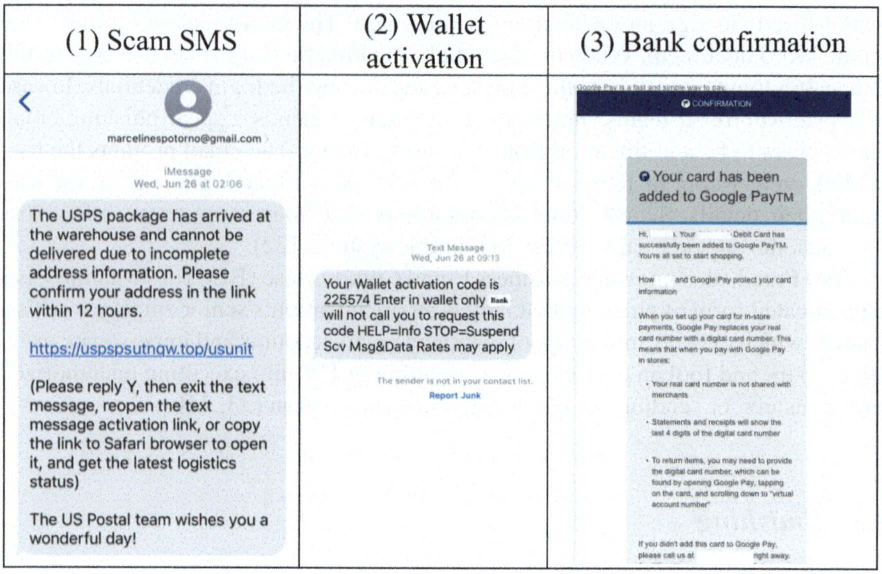

Fig. 2 Stages of real Smishing. *Source* Authors

money due to possible purchases through malicious Wallet to call the bank timely.

Lesson learnt. The most common mistake of the victim was to go to the scam link and put real debit card data and address. To avoid this, consumers must not access the link to any inaccurate and unknown website and, of course, do not fill out any personal data.

Smishing has emerged as a significant threat in the cybersecurity landscape due to the widespread adoption of text messaging. In addition to SoEng tactics, smishing attacks can also leverage malicious software or fraudulent websites to further their objectives [26].

In case of a malware, the smishing URL link deceive the user to download malware (malicious software) that installs itself on the smart device (phone, tablet, etc.).

The link in the smishing message directs the user to a false site (malicious website) that requests consumer to type sensitive personal data [27, 28].

3.4 Vishing

Vishing ("voice phishing") is a cyberattack initiated by phone calls [29, 30]. During a vishing attack, a cybercriminal attempts to obtain personal data and financial details, such as bank account numbers and passwords. Vishing scam make use of verbal tricks to cheat people into doing things they believe are in their best interests. Vishing thrives when cybercriminals have a bit of information about the consumer's interests. Vishing is the next step in the phishing attack chain. After attackers have successfully harvested sensitive information or credentials through phishing emails or text messages, they may use this data to initiate a vishing call. Posing as a representative of a trusted organization, such as a bank or government agency, they try to convince the victim to take an urgent action, such as providing additional information, making a payment, or authorizing a fraudulent transaction. The sense of urgency and perceived legitimacy of the call can cause victims to fall prey to vishing attacks [31].

Both smishing and vishing are serious threats in today's digital age. It's important to stay alerted, never share sensitive information unless recipient's identity is known, and keep the devices and accounts secure with the latest security updates and strong, unique passwords [32].

3.5 Page Hijacking

Page hijacking, also known as is a form of search engine index spamming and involves redirecting consumers to malicious websites or exploit kits through the compromise of legitimate web pages, often using cross site scripting [33].

hijacking, which encompasses a variety of tactics to trick users into visiting malicious or fake websites. At its core, page hijacking aims to exploit weaknesses in web pages or search engines to deceive web crawlers and genuine users, thereby redirecting them to compromised sites or serving up malware. This malicious technique encompasses the hijacking of entire web pages, as well as the insertion of malicious code into legitimate web pages through inline frames or other means. The impact of page hijacking can be severe, particularly in corporate settings, where attackers may target specific websites to launch watering hole attacks on corporate victims. Hackers might insert kits as MPack into fake website to bate legitimate consumers visiting the server. The insertion of malicious inline frames, or iframes, enables attackers to load exploit kits, which can deliver malware or execute other harmful actions on the victim's device. This method is often deployed in conjunction with watering hole attacks, which target specific websites or web pages that corporate users frequent. By hijacking a page or leveraging glitches in search engines, attackers can redirect users to malicious pages that appear to be authentic, bypassing traditional security measures. [34]. It works through the use of a script that temporarily redirect site visitors. A "temporary page cloaking," is often used in conjunction with page hijacking to boost the ranking of a malicious site. By employing methods such as *robots.txt* files or meta tags, a hijacker can mark the original page as temporary or non-indexable, reducing its ranking in search engine results pages (SERPs). This allows the attacker's duplicate page, which is designed to deceive users into divulging sensitive data or downloading malware, to rank higher and appear more prominently in search results, increasing its potential for success [35].

The main goal in page hijacking is to make the second page more prominent than the first one. This can be used for various purposes such as generating fraudulent ad revenue, installing spyware to steal personal data, or installing ransomware to lock a individual's browser until the hacker is paid to unlock it [36].

3.6 Calendar Phishing

Involves sending fake calendar invitations with phishing hyperlinks. These invitations usually imitate public event requests and can be simply added automatically to calendars. It is a type of scam where attackers abuse calendar services to trick victims into giving away their personal sensitive data. This can include passwords, card details, or account IDs. The scam is highly effective as it comes from trusted, legitimate services, and can even trick users who might not fall for other forms of attacks [37].

3.7 Quishing

QR codes are used maliciously in phishing attacks. The term "quishing" encompasses misleading users into thinking a QR code is harmless while the true intent is malicious, aiming to access sensitive personal data. Quishing, represents a growing concern in the world of cybersecurity, as hackers increasingly leverage the widespread use of QR codes to launch attacks on unsuspecting victims. By disguising malicious links as harmless QR codes, attackers can easily bypass traditional anti-phishing measures designed to detect and block malicious URLs. This makes Quishing attacks particularly dangerous, as users may accidentally scan a malicious QR code, leading to device compromise or data theft [38].

3.8 Juice Jacking

Juice jacking is a concerning security risk associated with charging devices via USB ports, particularly in public places such as airports, malls, or cafes. The concept revolves around the idea of compromised charging stations or cables that can be used to either install malware on the connected device or secretly copy sensitive data. The term "juice jacking" was first coined in 2011. The concept was first showcased when researchers created a compromised charging kiosk to bring awareness to the problem [39]. This happened during an event at Defcon, where the Wall of Sheep, an informational juice jacking kiosk, was set up. The kiosk would change its screen from "Free charging station" to a warning message that the user "should not trust public charging stations with their devices" when a USB device was plugged in.

The P2P-ADB, was a groundbreaking attack framework introduced by Kyle Osborn in 2012. Utilizing USB On-The-Go technology, the vulnerability of smartphones to attacks was demonstrated using devices connected via USB. P2P-ADB showcased the potential for attackers to gain unauthorized access to a victim's device, unlock locked phones, and even steal sensitive data such as Google Account credentials, bypassing traditional security measures. This groundbreaking research served as a wake-up call to the cybersecurity community, highlighting the need for increased protection against USB-based attacks targeting mobile devices. [40]. In 2013, researcher graduated and students from Georgia Tech released a proof-of-concept malicious tool "Mactans" that utilized the USB charging port on Apple mobile devices. They constructed a malicious wall charger that could infect an iPhone with the current (at that time) version of iOS with malicious software while it was being charged. As of April 2023, there have been no substantiated reports of juice jacking attacks occurring in the wild, despite the potential for this type of cyberattack to wreak havoc on unsuspecting individuals. Despite the lack of reported incidents, it is crucial to exercise caution when using public charging stations, as the threat of juice jacking remains a possibility. Researchers have demonstrated the feasibility of this attack, highlighting the importance of implementing safeguards such as using charging-only cables or portable power banks to prevent the potential exploitation of mobile devices through USB charging ports. [41].

3.9 USB Attacks

Although the USB security test has been designed especially for employees, it can provide valuable insights into users' behavior, highlighting areas where additional training may be necessary. USB drives can pose a significant security risk, as they can be used to introduce malware or exfiltrate sensitive personal data. To assess user behavior when faced with unknown USB drives, a USB Security Test, consists in placing specially crafted "beaconized" files on a USB drive and leaving it in a high-traffic area.

The USB-based attacks, have caused destruction on numerous high-profile targets worldwide. In recent years, USB drives have proven to be a potent vector for infiltrating networks and compromising sensitive data. The incidents, such as the Stuxnet attack on Iran's nuclear program, the Ukraine power grid attack, and the Sony Pictures Entertainment breach, demonstrate the devastating consequences of targeted USB-based attacks. In these cases, attackers exploited human nature, planting malware on USB drives or plugging in infected devices directly into the target systems to steal data or cause widespread disruption [42, 43].

3.10 *Malware Infection*

Malware, brief for malicious software, includes worms, viruses, spyware, Trojans, and ransomware. These programs can infect devices and compromise data security, leading to data loss, identity theft, or financial fraud.

3.11 *Social Engineering Attacks*

SoEng attacks [12] performing actions that compromise security or manipulate consumers into revealing confidential data. These attacks often exploit human weakness and are based on psychology to cheat victims.

3.12 *Identity Theft*

Identity theft happens when cybercriminals snip personal data, such as social security numbers, personal ID data, bank account details, or driver's license numbers, to impersonate consumers for fraudulent purpose, including accessing financial accounts or applying for loans or credit cards.

3.13 *Weak Passwords and Credential Theft*

Weak passwords are like inviting cybercriminals to a buffet of personal information and access to sensitive accounts. By choosing easily guessable passwords, such as "123,456", password", "admin" or common words with simple number substitutions, users make it effortless for hackers to bypass security measures and gain unauthorized access to online accounts, potentially resulting in devastating consequences such as identity theft, financial losses, and data breaches. The importance of using strong, unique passwords for each online account cannot be overstated. Moreover, utilizing MFA can provide an extra layer of protection against password-related cyberattacks. Credential theft, such as through data breaches or phishing attacks, can also lead to compromised accounts and identity theft.

3.14 Unsecured Wi-Fi Networks

Individuals are often connecting their devices to unsecured Wi-Fi networks, especially in public places like coffee shops or airports. Therefore, they expose themselves to various security risks, including eavesdropping, man-in-the-middle (MITM) attacks, and Wi-Fi spoofing.

3.15 Fake Software Updates and Downloads

Cybercriminals may distribute fake software updates or malicious applications disguised as legitimate software. Unsuspecting users who download and install these programs may unknowingly infect their devices with malware.

3.16 Online Scams and Fraud

Various online scams and fraudulent schemes aim to deceive consumers into sending money or personal data under false facades. Examples can be tech support scams, romance scams, and lottery scams.

3.17 Data Breaches

Data breaches can be associated to a digital burglar breaking into a vault and stealing its most valuable contents. These security incidents happen when cybercriminals use a variety of tactics, such as phishing attacks, malware infections, or exploiting security vulnerabilities, to gain unauthorized access to confidential information stored by companies or other organizations. Sensitive data, such as personal information, intellectual property, financial records, and trade secrets, can then be compromised, exposing the organization to potential reputational damage, financial losses, and regulatory penalties. Consumers may become victims of identity theft or financial fraud if their personal data is compromised in a data breach.

3.18 IoT Vulnerabilities

IoT devices, such as smart home gadgets and wearable technology, may have security vulnerabilities that cybercriminals can exploit to gain unauthorized access to networks or compromise user privacy.

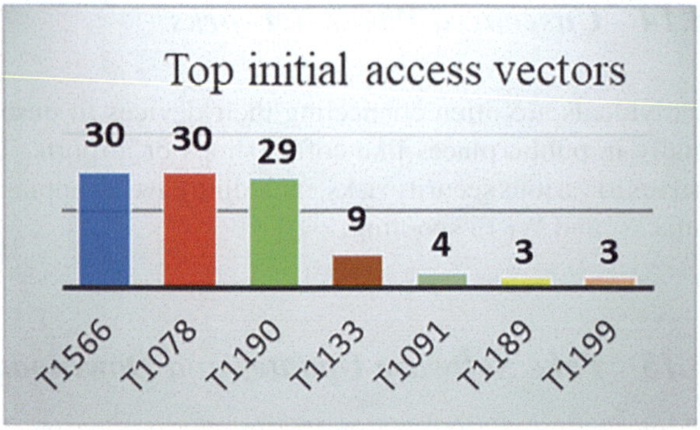

Fig. 3 Top initial access vectors during 2023 (adapted from [44])

Findings from IBM's X-Force Threat Intelligence Index 2024 paint a disturbing picture of the evolving cyberthreat landscape, highlighting a worrying trend of cybercriminals seeking to exploit vulnerabilities in authentication systems, steal and monetize sensitive data, and target emerging AI technologies. The staggering 71% increase in attacks leveraging stolen or compromised credentials underscores the importance of multifactor authentication and other security measures to mitigate credential-based attacks. Similarly, the 32% increase in data theft and leakage incidents is a stark reminder of the ongoing threat posed by data-driven cybercrime. Similarly, out of the five geographic regions, Europe experienced the highest percentage of incidents (32%). The most observed malware action (Fig. 3) was reported as 44% incidents [44].

Phishing—T1566—Cybercriminals send (to targeted victims) emails with malicious addons or hyperlinks, aiming to implement malicious code on victims' systems. Social media platforms have become a fertile ground for phishing attacks, as they provide an ideal ground for SoEng, enabling criminals to masquerade as trusted entities, build rapport with their victims, and earn their trust. This may be achieved by hijacking social media accounts, altering message headers or metadata to appear as legitimate communications, or using convincing language and messaging to persuade users to click on malicious links or divulge sensitive data. Another way to accomplish this is by counterfeiting or spoofing the ID of the sender which can be used to deceive both the user as well as the automated security apps [45].

Valid accounts—T1078—Compromised credentials pose a significant threat to an organization's security posture, as they can be leveraged by attackers to avoid access controls and gain unauthorized entry into various resources, systems, and services. The abuse of compromised credentials is a prevalent and significant cybersecurity threat faced by consumers and organizations alike [46].

Exploit public facing application—T1190—Cybercriminals usually exploit a weakness (a temporary malfunction or error, a software bug, or a wrong configuration) in an Internet-facing host or system to access a network. Exploited applications can be websites or web servers, including databases (like SQL—Structured Query Language), standard services (for example SSH—Secure Shell Protocol), network device administration and management protocols (as SNMP—Simple Network

Management Protocol), and any other system with open sockets accessible by Internet connection [47].

External remote services—T1133—Cybercriminals gain control over an external-facing remote service in order to gain access and/or remain into a network. Users are able to connect to internal organization network resources from an outside location throughout remote services (such as VPNs—Virtual Private Networks or the Citrix platform). Remote service gateways manage connections and credential authentication for these services. Windows Remote Management and VNC—Virtual Network Computing can also be services used externally.

Credential phishing and pharming, a combination of phishing and farming, involves techniques such as keylogging, SoEng, or malware to harvest login credentials and breach enterprise networks. Alternatively, once inside the network, attackers may use valid accounts to access remote services, enabling them to maintain persistent and redundant access to sensitive data and systems. This underscores the importance of robust security measures, such as multi-factor authentication, to protect against credential-based attacks and limit unauthorized access to enterprise resources [48].

Replication through removable media—T1091—Cybercriminals can access IT networks, by copying malware into a removable media by using Autorun features when the media is introduced into a system and runs. Mobile devices (even USB charging cables) can be used to contaminate PCs with malware when connected via USB. If Autorun features are enabled, malware on the mobile device can infect the PC or laptop [49].

Drive-by compromise—T1189—an insidious technique known as "drive-by download," is a type of malware attack that leverages malicious code embedded in websites to infect unsuspecting visitors. By disguising the malicious code as benign content, such as an image, audio file, or JavaScript, attackers can take advantage of vulnerabilities in web browsers and browsers plugins to silently install malware onto the victim's system, often without any user interaction required. In addition to web browser exploitation, compromised websites may also be used to acquire sensitive data, such as application access tokens, which can provide attackers with unauthorized access to corporate resources and sensitive information.

There are multiple ways of delivering exploit code to a browser (i.e., Drive-by Target), as follows:

A valid website is compromised where cybercriminals have injected malicious code such as cross-site scripting, JavaScript, and/or iFrames. Script files served to a legitimate website from a publicly writeable cloud storage container are altered by a cybercriminal. Malicious ads are paid for and served through legitimate ad providers (i.e., Malvertising). The "client-side exploitation," or an insidious technique employed by cybercriminals to gain unauthorized access to internal systems by exploiting vulnerabilities in web application interfaces and user-generated content. Unlike traditional exploit-based attacks that target servers and public-facing applications, client-side exploitation targets web browsers and client software to deliver malicious payloads directly to unsuspecting users, thereby bypassing external security measures. This technique is often used to steal sensitive information, such as

OAuth tokens, which can grant attackers access to protected applications and data within an organization's internal network.

Trusted relationship—T1199—In a trusted relationship attack, bad actors take advantage of the relationship between a target and a trusted third party (like an IT contractor, infrastructure service contractor, and managed service/security provider) that might have access to an otherwise unreachable target network [50].

Research has shown that a combination of security awareness training and simulated tests can reduce the risk of becoming victim to cyber-attacks. By empowering consumers with the knowledge and tools to identify and report potential threats, a more resilient and secure digital environment can be created for consumers.

4 Generative AI: A New Frontier in Cybersecurity Attacks and Defense

The advent of AI technology, particularly Large Language Models (LLMs) like ChatGPT or Copilot has generated significant interest and excitement within the all sectors. These tools offer promising opportunities to enhance operational efficiency, improve learning experiences, and support institutional growth.

In 2023, AI chatbots based on large language models, such as OpenAI's ChatGPT, revolutionized the tech landscape. ChatGPT became ubiquitous by making accessible worldwide the power of generative AI. This encouraged other businesses (that had been working on comparable AI in labs for years) to introduce their own public LLM services, hence thousands of tools based on these LLMs became popular and available online to millions of users as fast as a storm light [51]. There have been an emerging arms race between AI cybersecurity defenders and cybercriminals seeking to exploit AI tools for malicious purposes. As mainstream LLMs like ChatGPT became publicly available, cybercriminals wasted no time in harnessing these powerful tools to craft more sophisticated and convincing phishing emails. However, the robust guardrails embedded within these mainstream LLMs proved to be a formidable barrier for cybercriminals, prompting them to seek out alternative AI tools specifically designed for malicious activities. This highlights the ongoing battle between cybersecurity practitioners and cybercriminals, both vying to gain an upper hand in this increasingly complex and dynamic digital landscape [51].

Cybercriminals are skilled to power AI as an attack vector, and as an attack surface as well. We consider here SoEng attacks, deep fakes, document-scraping malware, ransomware attacks, automated phishing attacks, evasion of image recognition and voice biometrics, data pollution, impersonation attacks, and fake customer support chatbots [52].

4.1 Deep Fake

Deep fake is at this time the best-known use of AI as an attack vector, being used for disinformation campaigns and extortion. The concept of deepfakes, first popularized in 2017, involves the use of AI and machine learning techniques to generate highly realistic manipulated images and videos that can be nearly impossible to distinguish from authentic ones. The potential applications of deepfakes are vast and troubling, with significant implications for cybersecurity and the spread of disinformation. Cybercriminals have already started to leverage deepfake technology to impersonate individuals and organizations for phishing and scamming purposes, and there is increasing concern that deepfakes could be used for more sophisticated forms of disinformation and misinformation campaigns, particularly in the political sphere. They are often realistic enough to deceive consumers and automated detection systems. This can lead to a variety of malicious activities, such as disinformation campaigns, identity theft, and fraud [53].

The term "deepfake" is created by mixing "deep learning" and "fake." In an era where digital content is easily manipulated, we can no longer blindly trust the authenticity of images and videos. The age-old adage "seeing is believing" has lost its relevance in the context of AI-powered synthetic media, where truth and falsehood can often blur beyond recognition. As deepfakes become more sophisticated, discerning fact from fiction will become increasingly challenging, making it more important than ever to critically evaluate information from multiple sources and apply skepticism in the digital world. Deepfakes have raised significant concerns due to their potential misuse [54].

These tools exploit AI algorithms to study and spot deepfakes with remarkable precision.

StyleGAN is a successful deepfake software that generates realistic-looking images uses generative autoencoders instead of GANs. It learns high-level attributes (e.g., pose and identity) without supervision and allows scale-specific control of image synthesis. StyleGAN's generator architecture enables intuitive modifications to styles, resulting in impressive visual output [55]. Generative adversarial networks (GANs) consist of two neural networks: a generator that produces fake content and a discriminator that evaluates its authenticity. The generator and discriminator learn from each other iteratively, resulting in increasingly realistic output [56].

Deepfakes have been used in various contexts, including entertainment, for misinformation, in pornography or political manipulation. Movie studios use AI to create realistic visual effects, such as de-aging actors or resurrecting deceased ones, malicious actors create fake videos or audio clips to deceive viewers, celebrities have unintentionally become the stars of deepfake pornographic content and they can be used to create videos of politicians saying things they never actually said [57].

Detecting deepfakes is challenging because they are becoming increasingly realistic. Legislation and platform policies aim to curb the spread of deepfakes, but enforcement remains a challenge and researchers, responsible institutions as well as organizations are developing tools to identify manipulated content, but the arms race continues [58].

There are tools and methods being developed to detect deep fakes, such as analyzing inconsistencies in lighting or facial expressions, or looking for signs of editing in the pixels of an image or video [59].

Sentinel is an AI-based shield platform used by democratic governments, defense agencies, and businesses that allows users to upload digital media (such as images or videos) through their website or API. The system automatically analyzes the media for AI-forgery, determining whether it's a deepfake or not. It provides the imagining of the manipulation, helping consumers to understand where and how the media has been altered [60].

FakeCatcher, is real-time deepfake detector from Intel. It has been conceived in collaboration with the State University of New York at Binghamton and can identify fake videos with an impressive 96% accuracy rate. It looks for authentic hints in authentic videos, including delicate "blood flow" in the pixels. When the hearts pump blood, veins change color, and these signals are detected from various parts of the face. Deep learning algorithms translate these signals into spatiotemporal maps, instantly determining whether a video is real or fake [61].

WeVerify is a project focused on evolving intelligent human-in-the-loop content proof and misinformation investigation methods. It analyzes and contextualizes social media and web content to depict fake content within the larger online ecosystem [62].

Microsoft has also developed a deepfake detection tool called the Video Authenticator that aims to identify manipulated videos and provide a level of confidence in their authenticity [63].

The field of deepfake detection is constantly evolving, and researchers continue to explore new techniques to combat this threat.

4.2 *SoEng and AI*

AI has significantly impacted SoEng, both in terms of attack techniques and defense mechanisms. As AI technology advances, the methods employed by criminals such as SoEng attacks, evasion of image recognition and voice biometrics, intelligent targeting for ransomware attacks, are more and more sophisticated and difficult to be recognized. AI can be used to support convincing SoEng attacks at scale. The growing sophistication of deepfakes has also given rise to the threat of AI-powered evasion techniques. Cybercriminals can use deepfakes to circumvent security measures like image recognition and voice biometrics, which have traditionally been relied upon to authenticate users and prevent fraud. By generating synthetic media that mimics a person's face or voice, attackers can bypass facial recognition or voice biometric systems to gain unauthorized access to systems and accounts. Advanced research conducted by Darktrace revealed a 135% rise in SoEng attacks using generative AI [64].

AI is changing the landscape of SoEng as for example, the deepfake detection, and AI-generated voice calls. Phishing attacks, traditionally relying on human creativity and research, are now supercharged by AI. Generative AI tools can craft personalized, compelling messages that mimic legitimate communications from trusted entities [65]. Attackers use AI-generated voice calls that are alarmingly convincing, enhancing the credibility of SoEng. These voice synthesis tools gather audio data from targets, often from public sources like interviews, customer service calls, or video clips posted on social media sites [66].

As part of the ethical structure, the AI for positive SoEng technology can be used positively to counter disinformation campaigns and raise awareness about phishing scams. It can also determine when an image or video is counterfeit, helping identify deepfakes. Personalized educational and training materials, incorporating AI-generated content, can make them more engaging and effective [67]. Incorporating AI-generated content in phishing simulations familiarizes consumers with AI-generated communication styles [68].

4.3 *Document-Scraping Malware*

AI can make attacks more efficient by using *document-scraping malware*. Document scraping is the process of parsing and extracting data from documents being focused on extracting data from structured documents, such as *.pdf, *.docx, *.xlsx, and other file formats (Fig. 4). Document-scraping malware, on the other hand, also known as memory-scraping malware or RAM-scraping malware, is a type of malicious software that targets and extracts sensitive information from the RAM memory of the infected devices. This malware is particularly dangerous because it can collect valuable data, such as credit card numbers, personal IDs and identification numbers (PINs), and other sensitive personal data, which are then used for fraudulent activities or sold on the dark web [69].

Document-scraping malware often targets Point-of-Sale (POS) systems, ATMs, and other devices that process payment card data. It works by scanning the memory of the infected device for specific data patterns related to payment card information. Once the malware identifies the desired information, it extracts and exfiltrates it to a remote server controlled by the attacker. One of the reasons document-scraping malware is challenging to detect is that it does not interfere with the normal operation

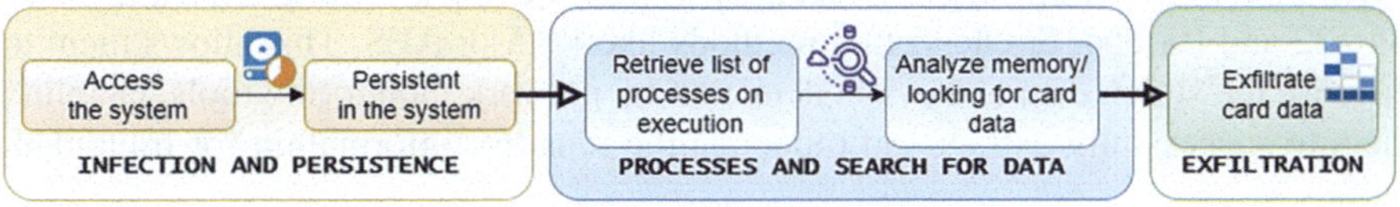

Fig. 4 Flowchart sample of a POS RAM scrapping malware [70]

of the infected device. This means that the device continues to function as usual, and the malware can go unnoticed for extended periods [70].

Malware often employs various techniques to access the system and achieve persistence on a compromised machine. Persistence means that the malware can maintain its presence on the infected system even after a reboot or an attempted removal.

This ensures that the malware can continue to execute its malicious activities over an extended period.

Malware's persistence can be achieved by modifying the system registry, creating scheduled tasks on the infected system, placing executable files or shortcuts in the system's startup folders, creating new services or modify existing services to load the malicious code during system startup, using fileless techniques to persist on a system without relying on traditional executable files, infecting the Master Boot Record (MBR) or Volume Boot Record (VBR), allowing the malicious code to be loaded before the operating system or replacing or modifying legitimate Dynamic-Link Library (DLL) files with malicious versions.

Upon installation, the malware initiates a comprehensive analysis of all executing processes within the system. These processes are then meticulously filtered to examine their allocated memory. The selection of processes to be explored can follow (i) a non-selective condition, which involves the indiscriminate exploration of all processes, or a (ii) selective one, that focuses solely on specific processes. The selective process list can be generated using either an inclusion (whitelisting) or exclusion (blacklisting) approach. The presence of whitelist-selective POS malware in a system may indicate a prior knowledge of the targeted system, suggesting a heightened probability of an Advanced Persistent Threat (APT) attack in progress.

Various exfiltration techniques have been identified in the behavior of POS malware. Some perform host-based exfiltration rather than establishing a connection to a Command & Control (C&C) server. These malware variants deposit scraped data into specific files within the compromised system's own disk, shared network folders, or designated USB devices. The latter scenario suggests that the attacker may have some form of physical access to the compromised system. Other POS malware strains complete the data extraction process by connecting through the Internet using different methods such as HTTP(S) GET/POST requests, FTP requests, and DNS requests, among others. [71]. Exfiltration of stolen data is a crucial component of POS RAM scraping attacks. POS RAM scraping malware typically collects sensitive information from the memory of compromised POS systems, including payment card data, names, and addresses. Cybercriminals often encode or cipher the exfiltrated data to avoid detection by security systems, using encoding algorithms like Base32 and Base64, or encryption methods like RSA or AES. This allows them to disguise the stolen data and evade detection by network monitoring tools, enabling them to successfully extract and monetize the sensitive information for fraudulent purposes.

4.4 Evasion of Image Recognition and Voice Biometrics

Refers to techniques where attackers manipulate data or use AI to avoid detection by these systems [72].

Image recognition systems use deep learning models to identify objects, faces, or patterns in images, therefore attackers can evade these systems by adversarial attacks, wearing disguises, obfuscating features, data poisoning, and intelligent targeting. AI crafts vague modifications to input images that cause misclassification. These perturbations are often imperceptible to humans but can fool the model. The wearing disguises technique employs accessories, masks, or makeup to alter facial features and avoid recognition. AI can manipulate images to hide key features (e.g., blurring or adding noise) to confuse the system or inject malicious samples into training data to bias the model's decision boundaries. Intelligent targeting allows AI to identify blind spots in detection rules and exploit them [73].

Voice biometrics systems authenticate users based on their unique voice patterns. Attackers can evade voice biometrics by voice cloning voice morphing, background noise, and speaker adaptation. Using AI to clone someone's voice is realized within minutes allowing an attacker to impersonate the victim.

They alter voice characteristics (pitch, tone, etc.) to sound different during authentication, introduce noise during voice recording that confuses the system, and train AI models to mimic a specific user's voice for unauthorized access [74].

4.5 Ransomware Attacks

Ransomware attacks are a type of malicious cyberattack where files are encrypted or computers (servers) are locked, rendering them inaccessible to the rightful users.

Ransomware has evolved into a sophisticated and devastating form of cybercrime, with significant ramifications for both organizations and individuals. Financial losses, data breaches, reputational damage, and even business disruptions can occur as a result of successful ransomware attacks [75].

The attackers then claim for a ransom payment, typically in cryptocurrency, in exchange for decrypting the files or unlocking the system. Ransomware typically enters a system through phishing emails, malicious attachments, or compromised websites. Once a user interacts with the malicious content, the ransomware is installed on their device. After gaining access to the system, the ransomware encrypts files on the infected device or network, effectively blocking access to them. Once the files are encrypted, the attackers display a ransom note on the victim's screen, informing them of the situation and demanding payment in exchange for a decryption key. The ransom note serves as the "proof of life" for the cybercriminals, ensuring that the victims are aware of the attack and the ransom demand. The note typically includes instructions on how to make the payment, often via cryptocurrency (such as Bitcoin or Monero) to avoid detection, and may set a deadline for payment, threatening to delete the encrypted files if the deadline is not met [76]. This adds a sense of urgency and pressure to the victim's already stressful situation, increasing the likelihood that they may comply with the attacker's demands.

In case that the victim chooses the payment, they may receive a decryption key from the attackers. However, there is no guarantee that paying will result in the files being decrypted, as some attackers may fail to provide a working decryption key, or the decryption process may not fully restore the files [77].

AI can support ransomware attacks through intelligent targeting and evasion. Some notable ransomware attacks that occurred in 2023 were the LockBit attack on the UK's Royal mail in January 2023, ESXiArgs attacks on VMware ESXi servers in February 2023, and Clop Group Exploits Zero-Day in GoAnywhere MFT in March 2023 [78].

The Royal Mail ransomware attack in January 2023 was a significant incident that underscores the devastating impact of ransomware on critical infrastructure. The LockBit ransomware gang, known for their ruthless approach, successfully infiltrated Royal Mail's systems, disrupting international mail delivery and crippling vital services, such as parcel tracking and online payment. As a result, millions of letters and parcels were left in limbo, causing significant disruptions to the postal service. The refusal of Royal Mail to pay the ransom, despite the threat of data leaks, illustrates the difficult choices organizations face when dealing with ransomware attacks [79].

In February 2023, an automated ESXiArgs ransomware attack exploited a vulnerability (CVE-2021–21,974) in VMware ESXi servers. Despite a patch being available, over 3000 servers were encrypted. The attackers demanded around 2 BTC (approximately $45,000 at the time). They even created new Bitcoin wallets for each victim, making recovery more challenging [80].

The Clop group widely exploited a zero-day vulnerability in Fortra's GoAnywhere MFT (managed file transfer) tool. Over 100 organizations, including Procter & Gamble, Toronto, and Community Health Systems, suffered attacks on vulnerable GoAnywhere MFT servers [81].

4.6 Data Pollution

Data pollution, also known as adversarial data injection, is a technique where an attacker intentionally manipulates the training data used by machine learning models to introduce vulnerabilities or biases. In the context of cybersecurity, data pollution can be particularly concerning because it can lead to false negatives (missed detections) or false positives (false alarms) in security systems.

Adversarial attacks involve crafting input data (e.g., images, text, network traffic) to deceive machine learning models. For example, an attacker can generate adversarial samples by adding imperceptible perturbations to legitimate data.

These perturbations are designed to mislead the model into making incorrect predictions. In cybersecurity, this can lead to evasion of intrusion detection systems (IDS), malware classifiers, and other security tools.

AI can be used for data pollution by identifying *blind spots* in detection rules. Data pollution is a relatively new technique in the world of cybercrime, and AI can be a powerful weapon in the hands of threat actors looking to exploit this technique. By analyzing and identifying blind spots in data detection rules, such as those used by security systems and data analytics tools, attackers can craft specifically designed "noise" that can evade detection and cause erroneous conclusions, effectively polluting the data. This could result in the bypassing of security systems or the compromise of sensitive data, leading to significant damage for the affected organizations [52]. AI algorithms based on historical data can discover patterns that were previously overlooked by rule-based systems. These blind spots may arise due to incomplete feature sets, biased training data, or limitations in the detection logic.

Data pollution also refers to data poisoning that involves injecting malicious or misleading samples into the training dataset. An attacker can manipulate training data to bias the model's decision boundaries or cause it to misclassify certain inputs. For example, in network intrusion detection, an attacker might inject benign network traffic samples that resemble malicious patterns, leading the model to misclassify future traffic.

4.7 Automated Phishing Attacks

Automated phishing attacks with AI support represent a significant evolution in cybercrime tactics. Traditional phishing attacks involve sending deceitful emails or messages to trick individuals into disclosing sensitive personal/organizational information, such as login credentials or financial details. However, with the integration of AI technologies, these attacks have become more sophisticated, scalable, and difficult to detect [82].

AI algorithms can analyze vast amounts of data to identify potential targets for phishing attacks. These algorithms may consider factors such as online activity, social media profiles, and previous breaches to tailor the phishing message to specific consumers or organizations, then, AI-powered tools can generate highly convincing phishing emails or messages by mimicking the writing style, tone, and formatting of legitimate communications. This makes it harder for recipients to distinguish between genuine and fraudulent messages [83].

AI can customize phishing messages with personal information obtained from publicly available sources or previous data breaches. By including details such as the recipient's name, job title, or recent activities, attackers increase the likelihood of successful deception. AI algorithms can optimize the timing and delivery method of phishing messages to maximize their impact. For example, messages may be sent during periods of high activity or urgency to increase the likelihood of immediate response. AI-powered phishing tools can adapt and learn from interactions with recipients to improve their effectiveness over time. By analyzing feedback, open rates, and response rates, attackers can refine their tactics and evade detection by security measures [84].

With AI support, cybercriminals can automate various aspects of the phishing campaign, including target selection, message generation, and distribution.

This enables them to launch large-scale attacks against multiple targets simultaneously with minimal manual intervention [85].

4.8 Impersonation Attacks

Impersonation attacks, involve attackers assuming the identity of a trusted individual or entity to manipulate victims into exposing sensitive data, granting unauthorized access, or performing actions that the attacker can take advantage of. These attacks exploit human trust and credulity, often using psychological tactics to deceive and mislead victims.

AI-driven impersonation is a rapidly evolving threat with alarming consequences. As mentioned above, the ability to mimic a victim's voice, also known as "voice deepfakes," can be used in a variety of malicious activities, including cybercrime and SoEng attacks. In the case of a kidnapping, the cybercriminals could potentially use AI to mimic the victim's voice in phone calls to the victim's family, causing immense distress and potentially misleading law enforcement efforts. Such uses of AI highlight the urgent need for effective regulations and safeguards to prevent the abuse of this powerful technology [86].

4.9 Fake Customer Support Chatbots

Fake customer support chatbots are a growing concern in the realm of cybersecurity. Hackers have been known to create fake chatbots to trick users into handing over their personal and payment. The victim receives an email, often appearing to be from a reputable company like DHL, stating that they have a pending parcel and further instructions are needed. If the victim accesses the hyperlink in the email, is redirected to a false customer support platform that appears to be running a chatbot. This chatbot is not a "real" chatbot, but rather an app with limited options and predefined responses. The victim is guided through a series of steps, including passing a fake captcha page, and is eventually asked to provide sensitive data such as their login credentials (email and password) and credit card data (card number, cardholder name, expiration date, CVV code). Once the victim enters the card data, the payment gateway checks the legitimacy of the card. The consumer is then forwarded to a one-time password (OTP) page, where is asked to enter a code received via SMS [87].

Therefore, by posing as a customer support representative, the attackers trick victims into providing sensitive data, including login credentials and credit card information, by utilizing a fake chatbot. The captcha page and payment gateway add a layer of legitimacy to the fraudulent process, further convincing victims of the scam's authenticity. By extracting OTPs via SMS, the attackers can bypass two-factor authentication, which makes this technique a particularly dangerous threat to individuals and organizations alike.

Although the AI-related threats are still in their early stages, the volume and sophistication of AI-driven attacks are increasing every day.

5 Cyber Security Strategy Based on Consumers' Behavior in AI Era

Cybersecurity is not just about the technology and systems, but also about understanding human behavior and psychology. By recognizing psychological traits and individual differences, cybersecurity professionals can develop more effective strategies for protecting computer systems against attacks. For instance, some people are naturally more trusting or risk-taking, which makes them more susceptible to phishing emails or social engineering attacks. Understanding these traits can help organizations tailor security awareness training and implement safeguards that cater to the unique vulnerabilities of their workforce [88]. Attackers often employ SoEng tactics (such as spreading false information) to manipulate consumers. Behavioral training equips consumers with techniques to recognize and resist such tactics, reducing the impact of cognitive hacking methods [89]. Behavioral training helps consumers comply with security policies. By addressing cognitive capabilities and situational awareness, consumers are empowered to actively participate in safeguarding network and information security. This involves pushing users toward security-conscious actions, making risk mitigation an instinctive habit [88]. Thus, behavioral cybersecurity training bridges the gap between technical solutions and human behavior, ultimately strengthening defense against cyber threats [88]. A cybersecurity strategy (Fig. 5) should develop extra steps, certain procedures for specific actions, but all businesses have to be today cyber-responsible and workers must understand that these procedures help to protect the business' assets.

Rather than slowing down or restricting progress, cybersecurity measures are designed to empower organizations to navigate the digital landscape with confidence, knowing that their assets, data, and systems are protected against cyber threats. As the business landscape continues to evolve at breakneck speed, cybersecurity must evolve alongside it, constantly adapting to changing threats and adopting cutting-edge technologies to ensure that innovation and security go hand in hand [90].

As AI continues to evolve, it's crucial for consumers to adopt a consumer-focused cybersecurity strategy (Fig. 5) for the following reasons:

- Personalized protection; consumers face unique cybersecurity risks based on their individual online behavior, including the websites they visit, the information they

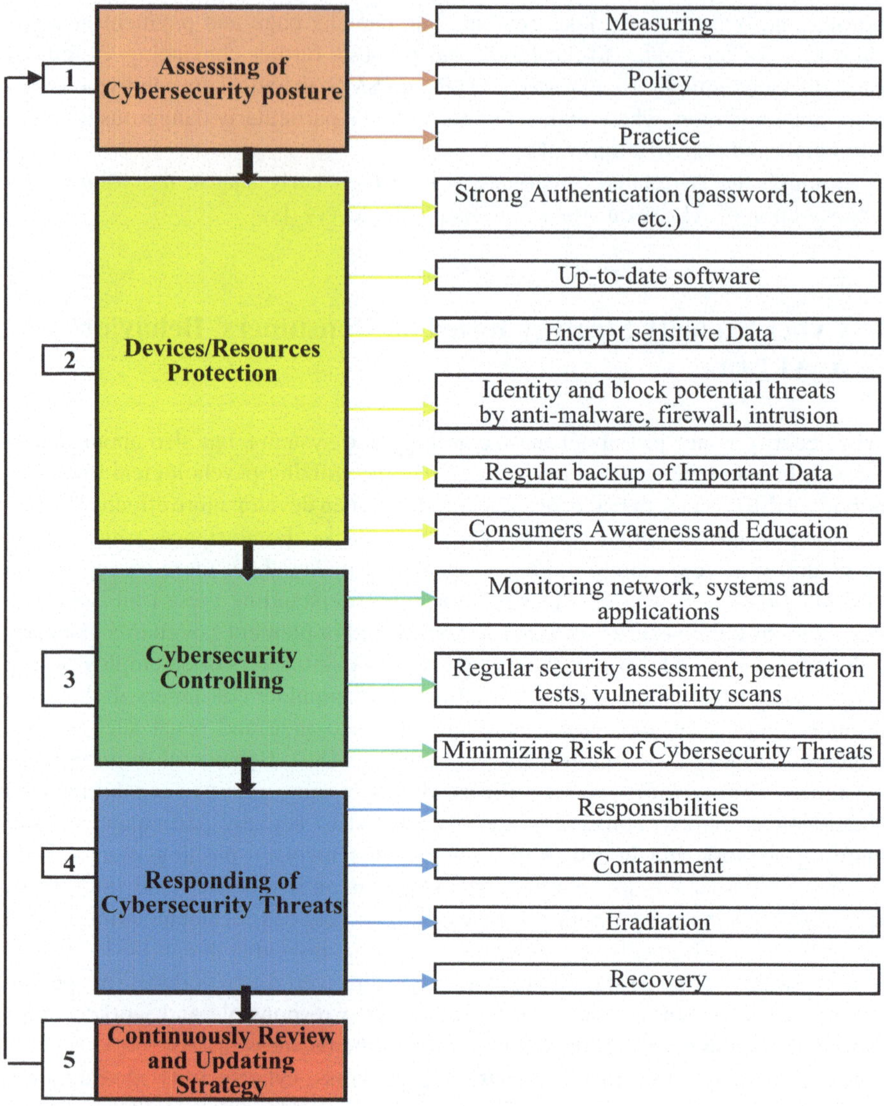

Fig. 5 Consumer-focused cybersecurity strategy. *Source* Authors' design

share, and the devices they use. A consumer-focused strategy can help identify and address these individual risks, providing personalized protection.

- Increased awareness; by understanding the various cybersecurity threats and best practices, consumers can become more aware of the potential risks associated with their online activities, allowing them to make more informed decisions when engaging with digital technologies.

- Ethical considerations; consumers play a vital role in shaping the development of AI and the future of cybersecurity. By advocating for transparency, fairness, and ethical considerations in AI, consumers can help promote a more secure and trustworthy digital environment.
- Building resilience; cyberattacks are a reality of our digital world, but consumers who are equipped with the necessary knowledge and tools can build resilience and mitigate the impact of these attacks, ensuring that their personal information and financial assets remain secure.

6 Conclusion

Cybersecurity threats advance continue and become more sophisticated, especially with the emergence of generative AI. It is therefore essential for consumers to stay vigilant and adopt a multi-layered approach to cybersecurity. The chapter identified 18 cybersecurity threats for consumers and defined a new frontier in cybersecurity attacks and defense by generative AI.

Based on analysis of cybersecurity threats and new frontier in cybersecurity attacks and defense authors designed a clear consumer-focused strategy. When consumers adopt a consumer-focused cybersecurity strategy, they can enjoy the following benefits:

Protection against identity theft can be achieved by implementing secure passwords, using a VPN, and regularly monitoring their online accounts, consumers can reduce the risk of identity theft and fraud.

Prevention of phishing attacks refers to the consumers who understand the tactics of phishing attacks, such as suspicious emails and pop-up ads, can avoid becoming victims of these scams, which can lead to financial losses or stolen personal information.

Online shopping experience can be improved by consumers who prioritize cybersecurity when shop online more safely, without fear of fraudulent activity or data breaches. This can lead to greater confidence and peace of mind when making online purchases.

Using secure Wi-Fi networks, keeping device software up-to-date, and avoiding public charging stations, consumers can protect their devices from malware and other security threats, ensuring that their personal data remains secure.

References

1. https://www.digitalocean.com/resources/chapter/ai-and-privacy
2. https://hai.stanford.edu/news/privacy-ai-era-how-do-we-protect-our-personal-information
3. https://aisecuritycentral.com/what-is-ai-privacy/
4. https://soofastaei-publications.com/battle-bytes-warfare-in-the-age-of-artificial-intelligence/
5. https://www.brookings.edu/chapters/protecting-privacy-in-an-ai-driven-world/

6. https://hgs.cx/blog/customer-focused-cybersecurity-remains-a-top-focus/
7. https://www.csoonline.com/chapter/644796/best-practices-for-an-effective-cybersecurity-strategy.html
8. Antipova, T., Riurean, S.: Managing cyber resilience literacy for consumer. Int. J. Inform. Commun. Technol. (IJ-ICT) **14**(1), 122–131 (2025). https://doi.org/10.11591/ijict.v14i1.pp122-131
9. Antipova, T.: Cyberattacks on business website: case study. In: Antipova, T. (eds.) Digital Science. DSIC 2021. Lecture Notes in Networks and Systems, vol. 381, pp. 505–512. Springer, Cham (2022). https://doi.org/10.1007/978-3-030-93677-8_44.
10. Moldovan, D., Riurean, S.: Cyber-security attacks, prevention and malware detection application. J. Digit. Sci. **4**(2), 3–23 (2022). https://doi.org/10.33847/2686-8296.4.2_1
11. Riurean, S.: Concepte şi tehnologii noi de comunicaţii în arhitecturi de reţele/ Novel Communications Concepts and Technologies in Network Architectures, 528p. Universitas Petroşani (2023). ISBN:978-973-741-948-4
12. Manukyan, L., Gevorgyan, M.: SoEng attacks: how to prevent. JDS **6**(1), 28–35 (2024). https://doi.org/10.33847/2686-8296.6.1_3
13. https://www.stationx.net/phishing-statistics/
14. https://www.verizon.com/business/resources/chapters/s/the-history-of-phishing/
15. https://www.techtarget.com/searchsecurity/definition/ILOVEYOU-virus
16. https://www.graphus.ai/blog/history-of-phishing/
17. https://www.fortinet.com/resources/cyberglossary/types-of-phishing-attacks
18. https://www.dnsc.ro/citeste/alerta-tentative-de-frauda-promovate-prin-anunturi-sponsorizate-social-media
19. https://securitytoday.com/chapters/2023/04/04/generative-ai-changes-everything-you-know-about-email-cyber-attacks.aspx
20. https://docs.apwg.org/reports/apwg_trends_report_q4_2023.pdf?
21. https://apwg.org
22. https://www.tessian.com/blog/5-real-world-examples-of-phishing-attacks/
23. https://www.knowbe4.com/ceo-fraud
24. https://www.cisco.com/site/us/en/learn/topics/security/what-is-a-whaling-attack.html
25. Shweta, K.: Main, what is Smishing? Definition, examples & protection. https://www.forbes.com/advisor/business/what-is-smishing/
26. https://www.kaspersky.com/resource-center/threats/what-is-smishing-and-how-to-defend-against-it
27. https://www.malwarebytes.com/blog/news/2021/04/what-is-smishing-the-101-guide
28. https://www.trendmicro.com/en_us/what-is/phishing/smishing.html
29. https://www.proofpoint.com/us/threat-reference/vishing
30. https://us.norton.com/blog/online-scams/vishing
31. https://www.kaspersky.com/resource-center/definitions/vishing
32. https://www.fortinet.com/resources/cyberglossary/vishing-attack
33. https://www.techopedia.com/definition/27967/page-hijacking
34. https://www.microsoft.com/en-us/microsoft-365-life-hacks/privacy-and-safety/what-is-browser-hijacker-how-remove
35. https://www.searchenginegenie.com/101-chapters/Page-Hijacking.html
36. https://webmasters.stackexchange.com/questions/122945/what-is-page-hijacking
37. https://blog.knowbe4.com/alert-your-users-about-calendar-scams-and-what-to-do-about-them
38. https://heimdalsecurity.com/blog/quishing/
39. https://us.norton.com/blog/mobile/what-is-juice-jacking
40. https://checkpeople.com/blog/what-is-juice-jacking-dangers-of-using-public-usb-ports/
41. https://www.comparitech.com/blog/information-security/juice-jacking/
42. https://www.avast.com/c-stuxnet
43. https://www.cfr.org/cyber-operations/compromise-power-grid-eastern-ukraine
44. https://www.ibm.com/reports/threat-intelligence?adoper=234412_1_PB1

45. https://attack.mitre.org/techniques/T1566/
46. https://attack.mitre.org/techniques/T1078/
47. https://attack.mitre.org/techniques/T1190/
48. https://attack.mitre.org/techniques/T1133/
49. https://attack.mitre.org/techniques/T1091/
50. https://www.senseon.io/resource/mitre-attck-initial-access-techniques-how-attackers-gain-access-to-corporate-networks/#:~:text=In%20a%20trusted%20relationship%20%28T1199%29%20attack%2C%20bad%20actors,have%20access%20to%20an%20otherwise%20unreachable%20target%20network
51. Elgan, M.: What to know about new generative AI tools for criminals. https://securityintelligence.com/chapters/what-to-know-about-new-generative-ai-tools-for-criminals/. Accessed 4 Oct 2023
52. https://www.europol.europa.eu/media-press/newsroom/news/new-report-finds-criminals-leverage-ai-for-malicious-use-%e2%80%93-and-it%e2%80%99s-not-just-deep-fakes
53. https://spectrum.ieee.org/what-is-deepfake
54. https://www.wipo.int/wipo_magazine/en/2022/02/chapter_0003.html
55. https://deepmake.com/
56. https://theaisummer.com/deepfakes/
57. https://metaroids.com/learn/ai-generated-media-a-guide-to-understanding-deepfakes/
58. https://www.scientificamerican.com/chapter/detecting-deepfakes/
59. https://www.analyticsinsight.net/artificial-intelligence/top-10-tools-to-detect-ai-deep-fakes
60. https://www.unite.ai/best-deepfake-detector-tools-and-techniques/
61. https://www.fastcompany.com/90957928/intel-fakecatcher-deepfake-detector
62. https://weverify.eu/tools/deepfake-detector/
63. Suratkar, S., Sharma, P.: A simple and effective way to detect DeepFakes: using 2D and 3D CNN. In: Rao, U.P., Patel, S.J., Raj, P., Visconti, A. (eds.) Security, Privacy and Data Analytics. Lecture Notes in Electrical Engineering, vol. 848. Springer, Singapore (2022). https://doi.org/10.1007/978-981-16-9089-1_19
64. Leibler, Y.: The rising threat of generative AI in SoEng cyber attacks — what you need to know. https://www.entrepreneur.com/science-technology/how-cyber-criminals-are-weaponizing-generative-ai/455896. Accessed 27 July 2023
65. https://www.compassitc.com/blog/beyond-phishing-understanding-ai-powered-social-engineering-attacks
66. https://blog.knowbe4.com/generative-ai-automation-social-engineering
67. https://www.infosecinstitute.com/resources/machine-learning-and-ai/how-artificial-intelligence-is-changing-social-engineering/
68. Rosca, S., Riurean, S., Leba, M., Ionica, A.: An educational model of graduation project for students at automation and computer engineering. J. Digit. Sci. **1**(1), 34–42 (2019). https://doi.org/10.33847/2686-8296.1.1_4
69. Bauer, J., Gruhn, M., Freiling, F.C.: Lest we forget: cold-boot attacks on scrambled DDR3 memory. Digit. Investig.Investig. **16**, S65–S74 (2016). https://doi.org/10.1016/j.diin.2016.01.009
70. Rodríguez, R.J.: Evolution and characterization of point-of-sale RAM scraping malware. J. Comput. Virol. Hack. Tech. **13**(3), 179–192 (2016). https://doi.org/10.1007/s11416-016-0280-4
71. Siddiqui, A.S., Lee, C.-C., Saqib, F.: Hardware based protection against malwares by PUF based access control mechanism. In: 2017 IEEE 60th International Midwest Symposium on Circuits and Systems (MWSCAS), Boston, MA, USA, pp. 1312–1315 (2017). https://doi.org/10.1109/MWSCAS.2017.8053172
72. Arora, S., Mittal, R., Kukreja, H., et al.: An evaluation of denoising techniques and classification of biometric images based on deep learning. Multimed. Tools Appl. **82**, 8287–8302 (2023). https://doi.org/10.1007/s11042-021-11573-w
73. https://www.idrnd.ai/faq-nist-presentation-attack-detection-pad-evaluation-nist-ir-8491/

74. https://www.theguardian.com/technology/2023/mar/16/voice-system-used-to-verify-identity-by-centrelink-can-be-fooled-by-ai
75. Alvee, S.R.B., Ahn, B., Kim, T., Su, Y., Youn, Y., Ryu, M.: Ransomware attack modeling and artificial intelligence-based ransomware detection for digital substations. In: 2021 6th IEEE Workshop on the Electronic Grid (eGRID), New Orleans, LA, USA, pp. 01–05 (2021). https://doi.org/10.1109/eGRID52793.2021.9662158
76. Kareem, K.M.: The intelligence technology and big eye secrets: navigating the complex world of cybersecurity and espionage, 10 May 2024 (2024). https://doi.org/10.31234/osf.io/v36ys
77. Lee, S.-J., Shim, H.-Y., Lee, Y.-R., Park, T.-R., Park, S.-H., Lee, I.-G.: Study on systematic ransomware detection techniques. In: 2021 23rd International Conference on Advanced Communication Technology (ICACT), PyeongChang, Korea (South), pp. 297–301 (2021). https://doi.org/10.23919/ICACT51234.2021.9370472
78. https://www.kaspersky.com/blog/ransowmare-attacks-in-2023/50634/
79. https://www.blackfog.com/2023-ransomware-attack-report/
80. https://www.cnet.com/tech/services-and-software/ransomware-cost-us-schools-3-56-billion-in-2021-study-says
81. https://www.fortinet.com/content/dam/fortinet/assets/reports/report-2023-ransomware-global-research.pdf
82. https://www.forbes.com/sites/forbestechcouncil/2023/08/04/defending-against-ai-based-phishing-attacks/
83. https://www.f5.com/company/blog/generative-ai-automation-of-spear-phishing
84. https://www.automation.com/en-us/chapters/may-2023/how-defend-ai-powered-phishing-attacks-data
85. De La Torre, R.: How AI is shaping the future of cybercrime. https://www.darkreading.com/vulnerabilities-threats/how-ai-shaping-future-cybercrime. Accessed 21 Dec 2023
86. https://www.techradar.com/news/hackers-are-spinning-up-fake-chatbots-to-trick-you-into-handing-over-your-data
87. Moustafa, A.A., Bello, A., Maurushat, A.: The Role of user behaviour in improving cyber security management. Front. Psychol. **12** (2021). https://www.frontiersin.org/journals/psychology/chapters/10.3389/fpsyg.2021.561011/full
88. https://www.frontiersin.org/journals/psychology/chapters/10.3389/fpsyg.2021.561011/full
89. https://www.weforum.org/agenda/2022/11/how-user-experience-and-behavioural-science-can-guide-smart-cybersecurity/
90. https://www.pwc.com.au/digitalpulse/cyber-behaviours-security-strategy.html

Security, Safety and Sustainability in Digital Era

Enhancing Data Security and Sustainability Using a Modified AES for an Energy-Efficient Cryptography

Bogdan Fustei and Monica Leba

Abstract This chapter presents a detailed description of the steps required for encrypting and decrypting information using a modified Advanced Encryption Standard (AES) algorithm, suitable for transmission over wireless communication channels. As a cryptographic algorithm, that can be implemented with various optimizations that help conserve energy and reduce the carbon footprint of IT systems, ultimately contributing to environmental sustainability. This includes minimizing power consumption, reducing waste heat, and optimizing the use of renewable energy sources. As a symmetric encryption algorithm, is widely used for securing data, being notable for its speed, security, and efficiency, making it a popular choice for both software and hardware encryption. It is a basis of modern encryption technology, providing a robust and efficient method for securing digital information. The modifications made to the AES algorithm seek to improve the data transfer rate by reducing the complexity of the required hardware resources and increasing the transfer speed without sacrificing the security of the transmitted data. This makes the modified algorithm an ideal choice for implementation in innovative wireless communication technologies, such as visible light communication (VLC), that prioritize high transfer rates but often have limited hardware resources.

Keywords Symmetric encryption algorithm · AES · Complex calculations · High speed

1 Introduction

The AES cryptographic algorithm, is commonly used for securing data across numerous platforms, including servers, IoT devices, and communication networks. The relationship between AES and sustainability primarily revolves around the algorithm's energy efficiency and its role in supporting secure IT practices. Optimized

---

B. Fustei (✉) · M. Leba
University of Petrosani, Petrosani, HD, Romania
e-mail: mihai_bogdan666@yahoo.com

T. Antipova (ed.), *Digital Technology Platforms and Deployment*, Information Systems Engineering and Management 36, https://doi.org/10.1007/978-3-031-86547-3_17

implementations of AES can lead to reduced energy consumption, lower carbon emissions, and contribute to the overall goal of environmental sustainability [1].

The efficiency of AES in terms of computational resources is crucial for minimizing energy consumption, especially when implemented in hardware or software that processes large volumes of data. Optimizing AES implementations can lead to significant reductions in energy usage, contributing to lower carbon footprints in data centers and other IT infrastructures [2].

Table 1 provides a clear and concise comparison of the key aspects of AES and other encryption algorithms, highlighting their respective strengths, weaknesses, and typical use cases. These encryption algorithms are Data Encryption Standard (DES), Triple DES (3DES), Rivest-Shamir-Adleman (RSA), and Elliptic Curve Cryptography (ECC).

Research has shown that certain hardware implementations of AES, particularly on Field-Programmable Gate Arrays (FPGAs) and Application-Specific Integrated Circuits (ASICs), can achieve high energy efficiency [3]. By designing low-power AES circuits, manufacturers can produce devices that consume less power,

Table 1 Comparison of the key aspects of AES and other encryption algorithms

Feature	AES	DES	3DES	RSA	ECC
Type	Symmetric Key Encryption	Symmetric Key Encryption	Symmetric Key Encryption	Asymmetric Key Encryption	Asymmetric Key Encryption
Key length [bits]	128, 192, 256	56	Effectively 112 (168 bits)	2,048, 4,096	256 (equivalent to 3,072—bit RSA)
Security	High secure, resistant to known attacks	Insecure by today's standards	More secure than DES, less secure than AES	Secure for key exchange and digital signatures	Strong security with shorter keys
Performance	Fast, efficient for bulk data encription	Slower, less efficient	Slower due to truple encription	Slower, complex mathematical operations	More efficient than RSA, especially on limited devices
Use Cases	Encrypting data-at-rest and data-in-transit	Largely absolute replaced by AES and 3DES	Legacy systems, being phased out	Secure key exchange digital signatures	Secure communications, mobie, IoT
Strenghts	Speed, strong encryption, modern design	Simplicity, historical significance	Compatibility with DES, increased security	Robust security for authentication and encription	Efficient security with smaller keys
Weknesses	Requires secure key management	Vulnerable to brute-force attacks	Slower, higher computational cost	Slower for large data encryption	Complexity in implementation

thus extending battery life in portable electronics and reducing overall energy demand. This reduction in energy consumption directly correlates with a decrease in greenhouse gas emissions associated with power generation [4, 5].

Secure communication, enabled by AES, reduces the likelihood of data breaches and cyber-attacks, which can have indirect environmental benefits. For instance, secure data management reduces the need for repeated data transfers or redundant data storage, both of which consume additional energy. By maintaining the integrity and confidentiality of data, AES helps in the efficient use of digital resources, thereby supporting environmental sustainability [6].

As part of Green IT initiatives, organizations are increasingly focusing on energy-efficient computing. The choice of encryption algorithms like AES, which can be optimized for low energy consumption, aligns with these initiatives. AES's efficiency ensures that secure data encryption does not significantly contribute to energy waste, thereby supporting broader environmental protection goals [7].

By reducing the complexity of hardware requirements and increasing data transfer speeds without compromising security, the modified AES algorithm offers a powerful and practical solution for the unique needs of modern VLC systems [8].

AES is widely adopted for various applications. It offers a good balance of security and performance compared to older standards like DES and 3DES. While RSA and ECC are asymmetric algorithms used for different purposes, such as key exchange and digital signatures, AES remains the preferred choice for encrypting data at rest and in transit due to its speed and security. The choice between these algorithms depends on the specific security needs and constraints of the application [9].

2 Overview of the AES with Key Features, Security Implications, and Applications in Modern Cryptography

The AES is widely used for securing data. It has been developed by Vincent Rijmen and Joan Daemen, and, in 2001 was selected as the encryption standard by the U.S. National Institute of Standards and Technology (NIST), replacing the older Data Encryption Standard (DES) [10].

The key structures of AES are symmetric key encryption, block cipher, key lengths, rounds, and efficiency.

Both encryption and decryption with AES have the same key, meaning that the sender and receiver must share the key securely. It encrypts data in fixed-size blocks of 128 bits. If the data to be encrypted isn't a multiple of 128 bits, padding is added. AES supports three key lengths: 128, 192, and 256 bits. The different key lengths provide different levels of security; therefore AES-128 is considered secure for most purposes, AES-192 offers a higher level of security, and AES-256 provides the highest security level and is often used for military-grade encryption.

AES encryption involves multiple rounds of processing, where each round includes several operations like substitution, permutation, and mixing. The number of rounds depends on the key length; hence AES-128 has 10 rounds, AES-192 has 12 rounds, and AES-256 has 14 rounds [11, 12].

It is designed to be efficient in both hardware and software implementations. It can be executed quickly on a variety of processors, making it suitable for a wide range of applications, from securing personal data on mobile devices to encrypting data in cloud storage.

Due to its large key size, AES is resistant to brute force attacks. Even with the fastest computers available today, cracking AES encryption by brute force would take an impractical amount of time.

As of now, AES has proven to be secure against known practical attacks, such as differential and linear cryptanalysis. However, like any cryptographic algorithm, it is subject to scrutiny, and continuous research is conducted to ensure its security [13].

Its widespread adoption and continued relevance underscore its effectiveness in protecting sensitive data across various platforms and applications. AES is used in a wide range of applications, including secure communications, data encryption, and wireless security, being used in protocols like HTTPS, VPNs, and other secure communication channels. It is also commonly used to encrypt sensitive data stored on devices, in databases, and in cloud storage. Wireless communications, such as Wi-Fi networks are secured by AES (WPA2).

In processing the plaintext, each round is divided into 4 stages:

(a) Byte Substitution (S-box) in the input block, where each element undergoes a nonlinear transformation using a lookup table with special mathematical properties.
(b) ShiftRows (row permutation) operates at the byte level.
(c) MixColumns is a matrix operation that combines 4 bytes each.
(d) Key Addition is the XOR logical operation between the current block and the key.

As with the DES algorithm, each round of AES key generation uses a subkey. The operations in the AES processing stage use calculations in Galois fields. The finite field contains 256 elements, using the notation GF (28) [14]. This field was chosen because each element can be represented by a byte. For the byte and column substitution steps, AES treats each byte in the data block as an element in the GF (28) field and performs arithmetic operations in this field. In the 10th round of AES-128, the mix columns transformation is not performed. Figure 1 shows the logical schema of the classic AES algorithm [15].

3 The Proposed Modified Algorithm

The AES algorithm has been extensively studied by specialists over time, and although it is not the only existing algorithm, it is considered one of the most efficient, if not the most efficient, algorithm for symmetric encryption [14]. Despite its excellence, due to the large volume of calculations, AES affects storage capacity

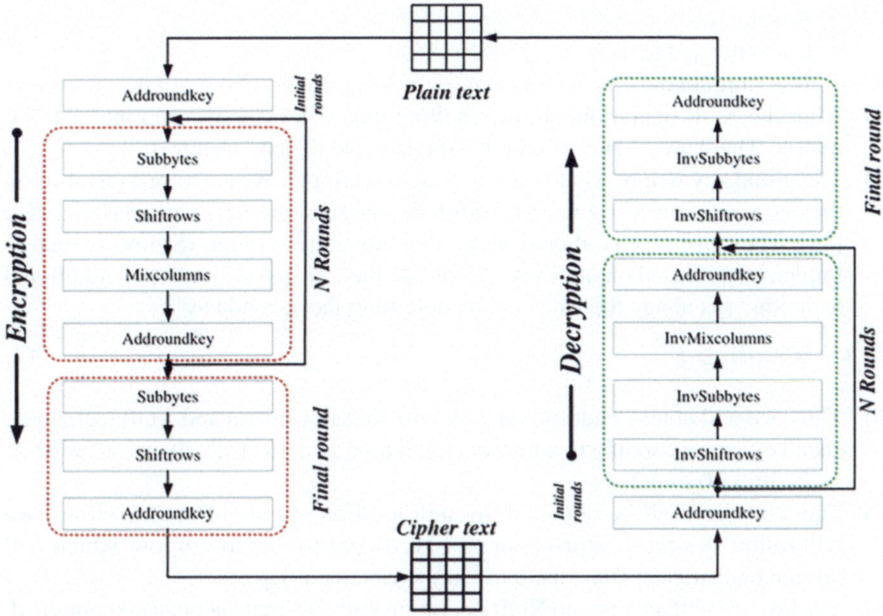

Fig. 1 Logic scheme of the classical AES algorithm. Adapted from [16]

and consumes a significant portion of processing hardware resources. The proposed method will bring modifications to the encryption/decryption process.

3.1 The Encryption Process

We consider the function as f_{ct} being the function of the encoded text, with the following formula:

$$f_{ct} = P \oplus K \quad (1)$$

where:
P—Plain text;
K—key.

(1) We choose the plaintext for the encryption process and a 128-bit key and divide the text and the key into a matrix of 4 × 4.
(2) The plaintext will undergo an XOR operation with the initial key.

$$C_0 = P \oplus K_0 \quad (2)$$

where:

P—plaintext;
C_0—encoded text.
K_0—initial key.

(3) After the XOR operation, all the resulting data will be rearranged into a 4 × 4 matrix. There are 16 bits of data for the key and the encoded text.
(4) The initial key will be modified for the second round. We perform a left shift on the first row of the 4 × 4 matrix. Similarly, the second row is shifted right twice [15]. The third row is altered using the substitution block (S-box), which is implemented based on our lookup table. In the fourth row, we use an additional technique to change the values, thus obtaining the second key K_1.
(5) $$C_1 = C_(0) \oplus K_1 \quad (3)$$
(6) Thus, we will obtain 9 additional keys with this adjustment and XOR technique, each key corresponding to a new encoded text. K = (0, 1, …, 9) and C = (0, 1, …,8) after 10 rounds.
(7) The tenth key will be obtained through a different type of modifications: we will follow the entire adjustment process except for the fourth row, which will remain unchanged. Thus, the tenth key is obtained: K_{10}.
(8) The last key will undergo an XOR operation with the final encoded text obtained, C_9, and we obtain the main encoded text: $C_{10} = C_9 \oplus K_{10}$.

In Table 2 the source (S-box) for byte substitution in xy (hexadecimal format) for the encoding phase is presented.

3.2 The Decryption Process

For the decryption process, we consider the function f_{pt}

$$f_{pt} = C \oplus K \quad (4)$$

where:
f_{pt} is plaintext function
C—the ciphertext; K—key
The steps for the decryption process are as follows:

(1) We have the main ciphertext C_{10} and the received key, which we will consider as the 11th key K_{10}. These will be converted into 4 × 4 matrices for use in the decryption process.
(2) We apply the XOR operation to the two matrices obtained in the previous step. This will yield the ciphertext C_9 (the tenth ciphertext). Therefore, we can write that:

$$C9 = C10 \oplus K10. \quad (5)$$

Table 2 S-box for byte substitution

	0	1	2	3	4	5	6	7	8	9	A	B	C	D	E	F
0	63	7C	77	7B	F2	6B	6F	C5	30	01	67	2B	FE	D7	AB	76
1	CA	82	C9	7D	FA	59	47	F0	AD	D4	A2	AF	9C	A4	72	C0
2	B7	FD	93	26	36	3F	F7	CC	34	A5	E5	F1	71	D8	31	15
3	04	C7	23	C3	18	96	05	9A	07	12	80	E2	EB	27	B2	75
4	09	83	2C	1A	1B	6E	5A	A0	52	3B	D6	B3	29	E3	2F	84
5	53	D1	00	ED	20	FC	B1	5B	6A	CB	BE	39	4A	4C	58	CF
6	D0	EF	AA	FB	43	4D	33	85	45	F9	02	7F	50	3C	9F	A8
7	51	A3	40	8F	92	9D	38	F5	BC	B6	DA	21	10	FF	F3	D2
8	CD	0C	13	EC	5F	97	44	17	C4	A7	7E	3D	64	5D	19	73
9	60	81	4F	DC	22	2A	90	88	46	EE	B8	14	DE	5E	0B	DB
A	E0	32	3a	0A	49	06	24	5C	C2	D3	AC	62	91	95	E4	79
B	E7	C8	37	6D	8D	D5	4E	A9	6C	56	F4	EA	65	7A	AE	08
C	BA	78	25	2E	1C	A6	B4	C6	E8	DD	74	1F	4B	BD	8B	8A
D	70	3E	B5	66	48	03	F6	0E	61	35	57	B9	86	C1	1D	9E
E	E1	F8	98	11	69	D9	8E	94	9B	1E	87	E9	CE	55	28	DF
F	8C	A1	89	0D	BF	E6	42	68	41	99	2D	0F	B0	54	BB	16

(3) The 11th key undergoes a reversible process. We will shift the first row to the right. Similarly, the second row will be shifted to the left twice, reversing the encryption process. The third row will be substituted with a value chosen from the inverted S-box, which has been implemented as a database in the program. This will yield the 10th key, K_9.

(4) The ninth ciphertext C_8 is obtained as follows:

$$C8 = C9 \oplus K9. \tag{6}$$

(5) We will further modify the tenth key by applying the steps from point 3 to the first three rows. For the fourth row, we will add the value of decrements. This will yield the ninth key. K_8.

(6) We apply XOR between C_8 (ciphertext 9) and K_8 (ninth key) to obtain the coded text C_7 (eighth ciphertext). Similarly, the subsequent keys will follow the previously described steps (recurrence relation) until we obtain the ciphertext and key numbered 0. This will yield the following:

$$\text{K} = (10, 9 \ldots 0) \text{ and } \text{C} = (10, 9 \ldots 0).$$

(7) We apply XOR to the ciphertext and the remaining key, thus obtaining the desired plaintext. Therefore, we can write the following: $P = C_0 \oplus K_0$, where P is plain text.

Table 3 S-box for byte substitution

	0	1	2	3	4	5	6	7	8	9	a	b	c	d	e	f
0	52	09	6a	d5	30	36	a5	38	bf	40	a3	9e	81	f3	d7	fb
1	7c	e3	39	82	9b	2f	ff	87	34	8e	43	44	c4	de	e9	cb
2	54	7b	94	32	a6	c2	23	3d	ee	4c	95	0b	42	fa	c3	4e
3	08	2e	a1	66	28	d9	24	b2	76	5b	a2	49	6d	8b	d1	25
4	72	f8	f6	64	86	68	98	16	d4	a4	5c	cc	5d	65	b6	92
5	6c	70	48	50	fd	ed	b9	da	5e	15	46	57	a7	8d	9d	84
6	90	d8	ab	00	8c	bc	d3	0a	f7	e4	58	05	b8	b3	45	06
7	d0	2c	1e	8f	ca	3f	0f	02	c1	af	bd	03	01	13	8a	6b
8	3a	91	11	41	4f	67	dc	ea	97	f2	cf	ce	f0	b4	e6	73
9	96	ac	74	22	e7	ad	35	85	e2	f9	37	e8	1c	75	df	6e
a	47	f1	1a	71	1d	29	c5	89	6f	b7	62	0e	aa	18	be	1b
b	fc	56	3e	4b	c6	d2	79	20	9a	db	c0	fe	78	cd	5a	f4
c	1f	dd	a8	33	88	07	c7	31	b1	12	10	59	27	80	ec	5f
d	60	51	7f	a9	19	b5	4a	0d	2d	e5	7a	9f	93	c9	9c	ef
e	a0	e0	3b	4d	ae	2a	f5	b0	c8	eb	bb	3c	83	53	99	61
f	17	2b	04	7e	ba	77	d6	26	e1	69	14	63	55	21	0c	7d

In Table 3 the values of the Inverse S-Box substitution for bytes are presented in xy coordinates.

3.3 *Key Expansion*

Key expansion refers to the process in which all keys are transformed and combined into a single key.

The key will undergo three different stages of transformation, which will be implemented sequentially. These steps are as follows:

(1) The row permutation operation (Shift Rows) is performed. The rows are shifted from one block to another. When a key is represented by a 4×4 matrix, the first 2 rows of this matrix go through this stage.
(2) The byte substitution process in the input block (S-BOX implementation) is performed. All calculations will be carried out in row number 3 of the matrix. We will use the S-BOX and Inverse S-BOX tables. The S-BOX will be used in the encryption process, while the Inverse S-BOX will be used for decryption.
(3) In the final stage, the Increment and Decrement processes occur, with calculations performed in row number 4 of the matrix corresponding to the keys. Bytes will be added/removed in every fourth row of the matrix. The Increment will be used for encryption, while the Decrement will be used for decryption. The last

key in the encryption process and the first key in the decryption process will NOT be subjected to these calculations.

In Figs. 2 and 3 are presented logical diagrams for encrypting and decrypting data according to the algorithm with the proposed modifications.

Fig. 2 Diagram for data encryption

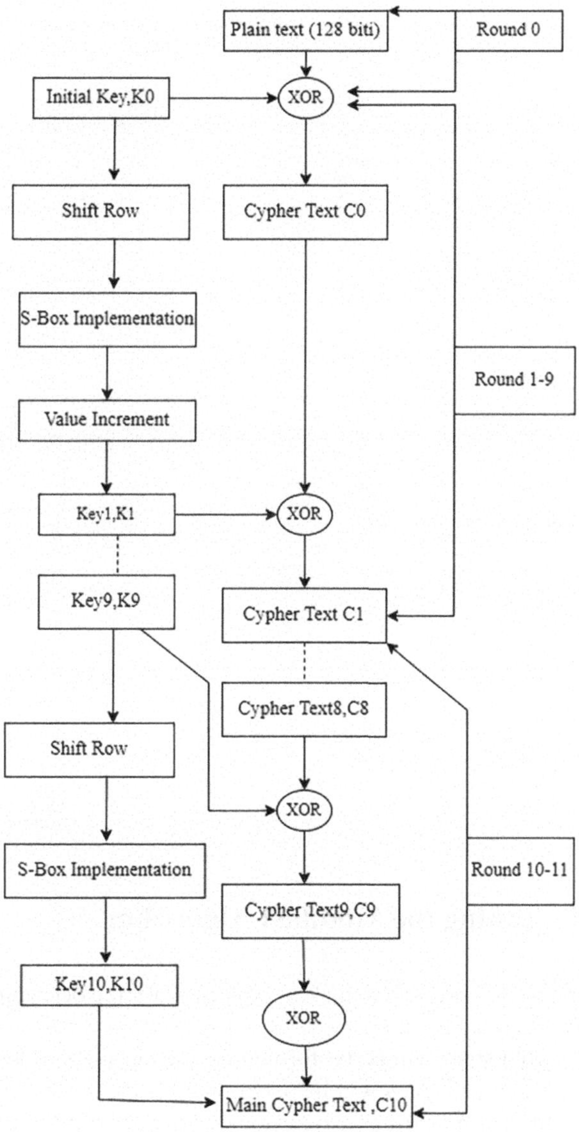

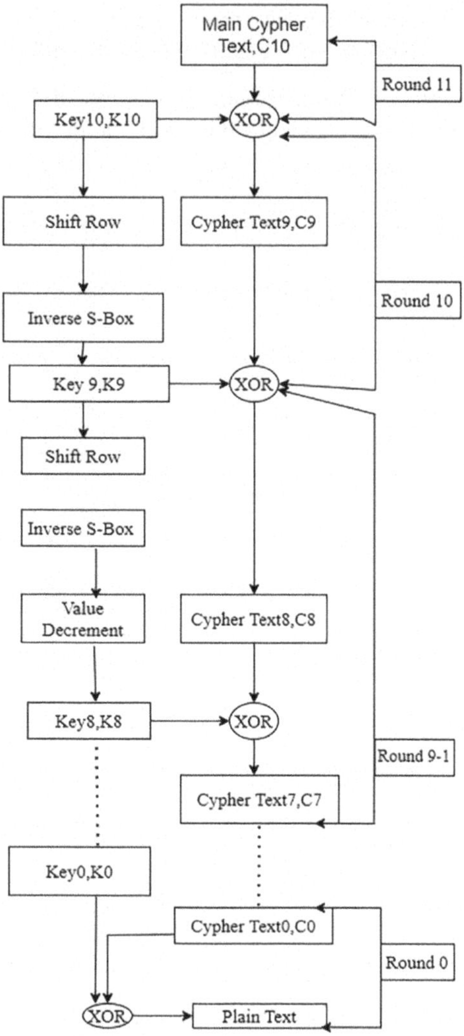

Fig. 3 Diagram for data decryption

4 Testing the Modified Algorithm

In Fig. 4 is presented the key expansion for the decryption process (characterized by the decrement value).

All 4 steps will be performed for the keys: K9 to K0.

The first 3 steps will be performed for the keys: from: de la K10 to K9.

In Fig. 5, the key expansion for the encryption process (characterized by the increment value) is presented.

All 4 steps are performed for the keys: K0 to 9.

The first 3 steps are performed for the keys: K9 to K10.

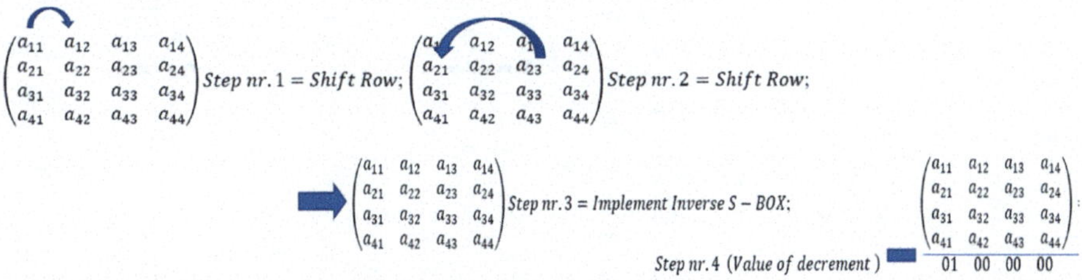

Fig. 4 Key expansion for decryption process

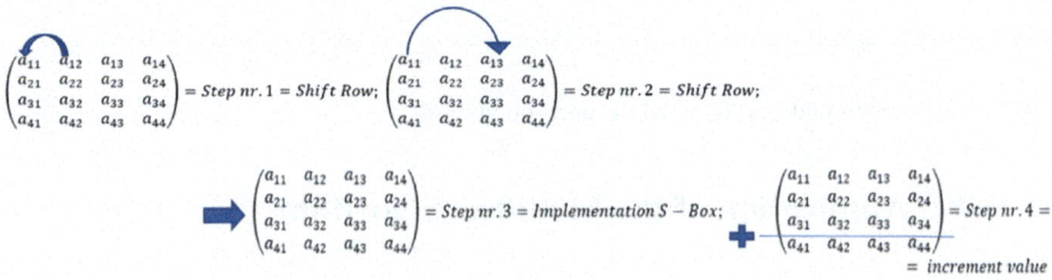

Fig. 5 Key expansion for the encryption process

In Figs. 6 and 7 the methods for implementing the AES algorithm with the previously presented modifications are shown.

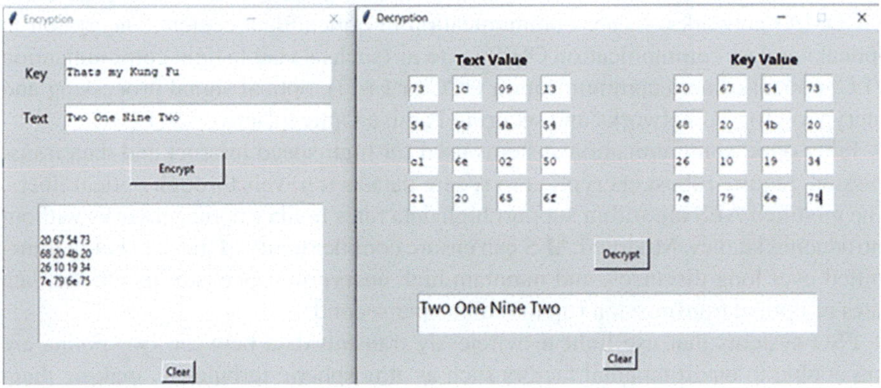

Fig. 6 Encryption and decryption of the text

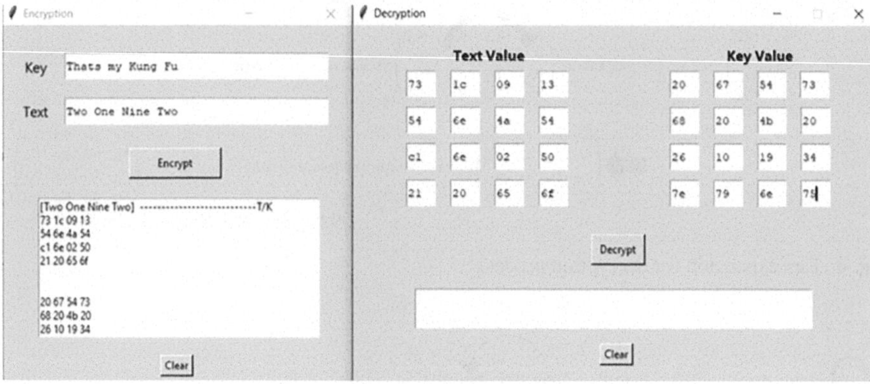

Fig. 7 Text encryption/decryption and the decryption result

5 OWC Application of the Modified Algorithm

When applied to optical communication systems, modifications to the AES algorithm are optimized to achieve the unique characteristics of optical systems, such as high bandwidth, low latency, high data rates and high security.

There are several key applications of the modified AES algorithm in optical communication systems such as secure data transmission in fiber optic networks, free-space optical communication (FSO) systems, Quantum Key Distribution (QKD) and classical encryption in hybrid systems, Optical Code Division Multiple Access (OCDMA) networks, secure communication in optical data centers, encryption in optical wireless communication OWC systems (such as visible light communication VLC, optical camera communication OCC or Li-Fi), optical signal processing and encryption for 5G networks and securing Passive Optical Networks (PON).

Fiber optic communication systems used for high-speed internet and data transmission, require robust encryption to secure data as it travels through optical fibers. The modified AES algorithm support high data rates in fiber optic networks without introducing latency. Modified AES can ensure confidentiality of the data being transmitted over long distances, and maintain high encryption speeds to match the data rates of optical transmission (up to terabits per second).

FSO systems that use light to wirelessly transmit data between two points are susceptible to environmental factors such as atmospheric turbulence, making them vulnerable to eavesdropping. AES modifications secure high-speed, high-bandwidth communication channels in FSO systems, handling rapid data rates without introducing significant delays or computational overhead, and enhancing encryption to adapt to varying signal strengths and potential interruptions in free-space environments (such as weather or physical obstructions) [17].

In quantum communication systems, QKD provides unconditionally secure encryption keys. However, once keys are established, classical encryption algorithms, such as AES, are often used to secure the actual data transmission [18].

In hybrid systems that combine QKD and classical optical communication, a modified AES algorithm can work in tandem with quantum key protocols to ensure the confidentiality of data, and utilize QKD-derived encryption keys while optimizing AES performance to match the high throughput of optical systems.

OCDMA as a method for transmitting multiple data signals simultaneously over a single optical channel by assigning each signal a unique code can integrate modified AES to encrypt the data for each unique code [19].

Data centers that use optical interconnects for high-speed communication between servers and storage systems need to ensure the confidentiality and integrity of the data. AES is a common choice for encryption, but its modified version is essential for supporting high-speed optical links within data centers, where data throughput is extremely high, and minimizing encryption-related delays.

AES modifications applied in VLC, OCC and Li-Fi systems enable real-time encryption and decryption of optical signals, resilience to various forms of interference, such as signal reflections and obstructions, as well as optimized performance for high-speed light-based data transmission in environments such as smart homes, offices, and industrial systems [20–22].

The integration of 5G networks with optical communication systems (for backhaul and fronthaul connections) requires secure encryption mechanisms due to the sensitivity and high data rates involved. Modified AES can secure the high-speed optical data links used in 5G infrastructure, ensure low-latency encryption/decryption, a critical requirement for 5G applications such as autonomous vehicles, IoT devices, and real-time video streaming and provide robust encryption against potential cyberattacks targeting the optical communication backbone of 5G networks.

A PON is a telecommunications technology used to deliver high-speed broadband internet, TV, and voice services to end-users through optical fiber. PON are widely used in telecom infrastructure to deliver broadband services. Since these networks distribute the same signal to multiple end-users, it is critical to encrypt the transmitted data to prevent unauthorized access [23].

In the medical field, especially with the rise of eHealth services, telemedicine, and healthcare networks, data security and privacy are critical due to the sensitive nature of patient data. The use of AES encryption in the medical field ensures patient data privacy, secure transmission of medical imaging, telemedicine and remote monitoring and compliance with regulations [24].

AES encryption in underground environments ensures secure communication for critical infrastructure being vital for managing and monitoring utilities like gas pipelines, electrical grids, and transport systems. AES ensures that any data transmitted between underground sensors, monitoring systems, and control centers is encrypted, protecting it from tampering or unauthorized access [25].

6 Conclusion

This chapter details the elimination of the Mix Columns step in the modified AES-128 algorithm, which decrease complex calculations without compromising its security. The resulting algorithm is faster and less demanding on hardware resources than its classical counterpart, making it suitable for implementation in a wide range of applications. Despite some degree of vulnerability to brute-force attacks, the modified.

AES-128 provides the same level of security as the original AES-128, while offering improved performance. The modifications made to the AES algorithm strike a critical balance between enhancing the data transfer rate and maintaining the security of transmitted data, making it an optimal choice for implementation in novel optical wireless communication technologies. These modifications reduce the complexity of hardware resources required, thereby increasing the data transfer speed without compromising security. This makes the modified AES algorithm an ideal candidate for use in cutting-edge optical systems that prioritize high transfer rates and efficient resource utilization.

Modifying the AES algorithm for optical communication systems allows it to handle the unique demands of high-speed, high-throughput, and low-latency environments that define modern optical networks. Applications range from securing fiber optic and FSO networks to safeguarding data in Li-Fi, quantum hybrid systems, 5G networks, and beyond. The flexibility of a modified AES algorithm in such environments ensures robust encryption without compromising the efficiency of optical communication. Modifying the AES algorithm for optical communication systems allows it to handle the unique demands of high-speed, high-throughput, and low-latency environments that define modern optical networks. Applications range from securing fiber optic and FSO networks to safeguarding data in VLC, OCC and Li-Fi, quantum hybrid systems, 5G networks, and beyond. AES in optical communication plays a pivotal role in securing sensitive data, whether it's patient health information in medical settings or mission-critical data in underground infrastructure. Given the high-speed and large-capacity benefits of optical networks, combining them with robust encryption methods like AES ensures that the data transmitted remains both private and secure, which is especially important in environments where the stakes are high, such as healthcare and underground communication systems [26].

References

1. Z'aba, M.R., Maarof, M.A.: A survey on the cryptanalysis of the advanced encryption standard (2006)
2. Moldovan, D., Riurean, S.: Cyber-security attacks, prevention and malware detection application. J. Digit. Sci. **4**(2), 3–23 (2022). https://doi.org/10.33847/2686-8296.4.2_1
3. Deshpande, H.S., Karande, K.J., Mulani, A.O.: Efficient implementation of AES algorithm on FPGA. In: 2014 International Conference on Communication and Signal Processing, Melmaruvathur, India, pp. 1895–1899 (2014). https://doi.org/10.1109/ICCSP.2014.6950174

4. Joan, D., Vincent, R.: The Design of Rijndael. The Advanced Encryption Standard (AES). Editura Spinger (2020)
5. Yingzong, L., et al.: An adaptive CO2 Brayton-Rankine power cycle for efficient utilization of low environment temperature: a thermodynamic analysis and optimization study. J. Clean. Prod. 140547. https://doi.org/10.1016/j.jclepro.2023.140547, ISSN 0959-6526
6. Antipova, T.: Cyberattacks on business website: case study. In: Antipova, T. (eds.) Digital Science. DSIC 2021. Lecture Notes in Networks and Systems, vol. 381. Springer, Cham. https://doi.org/10.1007/978-3-030-93677-8_44
7. Forhad, M.S.A., Riaz, S., Hossain, M.S., Das, M.: An improvement of advanced encryption standard. Int. J. Comput. Sci. Netw. Secur. **18**(11), 159–166 (2018)
8. Simona, R., Dobre, A., Marcu, A.-E.: Security and propagation issues and challenges in VLC and OCC systems. In: Proceedings, Advanced Topics in Optoelectronics, Microelectronics and Nanotechnologies , vol. X, p. 117182B (2020). https://doi.org/10.1117/12.2572029
9. Riurean, S., Leba, M., Crivoi, L.: Enhanced security level for sensitive medical data transmitted through visible light. In: 2021 International Symposium on Networks, Computers and Communications (ISNCC), pp. 1–6 (2021). https://doi.org/10.1109/ISNCC52172.2021.9615732
10. William, S.: Cryptography and Network Security, 4th edn. Prentice Hall (2005). ISBN 0-13-187319-3
11. NordLayer: AES encryption: how does it safeguard your data? (2023). https://nordlayer.com/blog/aes-encryption/
12. Tehranipoor, S., Karimian, N., Edmonds, J.: Breaking AES-128: machine learning-based SCA under different scenarios and devices. In: 2023 IEEE International Conference on Cyber Security and Resilience (CSR), Venice, Italy, pp. 564–571 (2023). https://doi.org/10.1109/CSR57506.2023.10225009
13. Zobaed, S., Salehi, M.A., Zomaya, A., Sakr, S.: Big data in the cloud. In: Sakr, S., Zomaya, A.Y. (eds.) Encyclopedia of Big Data Technologies (2019)
14. Navneet, J.R., Patil, R., Sawant, O., Madasamy, S. and Sakthivel, R.: AES algorithm with dynamic shift rows and bit permuted mix column. In: 2023 International Conference on Next Generation Electronics (NEleX), Vellore, India, pp. 1–6 (2023). https://doi.org/10.1109/NEleX59773.2023.10421625
15. Kumar, K.R.R., Kumar, K.J.J., Partibane, B.: Reverse cellular automata-based AES S-BOX. In: 2017 International Conference on Innovations in Information, Embedded and Communication Systems (ICIIECS), Coimbatore, India, pp. 1–4 (2017). https://doi.org/10.1109/ICIIECS.2017.8276006
16. Mulani, A., Mane, P.B.: High-speed area-efficient implementation of AES algorithm on reconfigurable platform (2019). https://doi.org/10.5772/intechopen.82434
17. Erdogan, E., Altunbas, I., Kurt, G.K., Yanikomeroglu, H.: The secrecy comparison of RF and FSO eavesdropping attacks in mixed RF-FSO relay networks. IEEE Photonics J. **14**(1), 1–8, Art no. 7901508 (2022). https://doi.org/10.1109/JPHOT.2021.3127397
18. Noirie, L.: From existing quantum key distribution systems towards future quantum networks. In: 2024 13th International Conference on Communications, Circuits and Systems (ICCCAS), Xiamen, China, pp. 339–344 (2024). https://doi.org/10.1109/ICCCAS62034.2024.10652815
19. Kaur, S., Singh, S.: Highly secured all optical DIM codes using AND gate. In: 2020 Indo – Taiwan 2nd International Conference on Computing, Analytics and Networks (Indo-Taiwan ICAN), Rajpura, India, pp. 64–68 (2020). https://doi.org/10.1109/Indo-TaiwanICAN48429.2020.9181369
20. Arman, S., Rehnuma, T., Rahman, M.: Design and implementation of a modified AES cryptography with fast key generation technique. In: 2020 IEEE International Women in Engineering (WIE) Conference on Electrical and Computer Engineering (WIECON-ECE), Bhubaneswar, India, pp. 191–195 (2020). https://doi.org/10.1109/WIECON-ECE52138.2020.9397992
21. Riurean, S., Leba, M., Ionica, A., Nassar, Y.: Technical solution for burnout, the modern age health issue. In: 2020 IEEE 20th Mediterranean Electrotechnical Conference (MELECON), Palermo, Italy, pp. 350–353 (2020). https://doi.org/10.1109/MELECON48756.2020.9140516

22. Riurean, S.: A study on the VLC security at the physical layer for two indoor scenarios. In: MATEC Web of Conferences, vol. 342. EDP Sciences (2021)
23. Begam, A., Sekh, M.A.: Design of secured hybrid passive optical networks for FSO system. In: 2024 Fourth International Conference on Advances in Electrical, Computing, Communication and Sustainable Technologies (ICAECT), Bhilai, India, pp. 1–4 (2024). https://doi.org/10.1109/ICAECT60202.2024.10469086
24. Rîurean, S.M., Leba, M., Ionică, A.: VLC embedded medical system architecture based on medical devices quality requirements. Qual.-Access Success **20** (2019)
25. Riurean, S., et al.: Li-Fi embedded wireless integrated medical assistance system. In: World Conference on Information Systems and Technologies. Springer International Publishing, Cham (2019)
26. Stoicuta, O., Riurean, S., Burian, S., Leba, M., Ionica, A.: Application of optical communication for an enhanced health and safety system in underground mine. Sensors **23**, 692 (2023). https://doi.org/10.3390/s23020692

Enhancing Cybersecurity and Environmental Sustainability Through Active Directory Integration

Sorin Andrei Sav and Simona Riurean

Abstract This chapter explores the integration of environmental sustainability and cybersecurity within IT infrastructures using Active Directory (AD). It highlights the benefits of enhancing security through systematic policy enforcement and monitoring. It emphasizes the critical role of cybersecurity in today's digital landscape, focusing specifically on AD networks. The chapter also aims to assist security responsible in being updated with the current cybersecurity threats and defense techniques, while also presenting real-world scenarios for AD and infrastructure administration to assess potential risks from internal adversaries. As AD expands to manage many objects and resources, it becomes increasingly essential to have clear and regularly maintained policies in place. Without these safeguards, forgotten or misconfigured objects may slip through the cracks, turning into security vulnerabilities. A lack of attention to detail and ongoing maintenance in AD management can expose an organization to serious risks, highlighting the criticality of thorough and consistent maintenance, inventorying, and monitoring of AD domains. Technical terminology and details of cybersecurity within networks and AD, addressing the gap in public knowledge regarding modern at-tack techniques and the methods employed by bad actors, are also addressed here.

Keywords Green IT · Cybersecurity · Active directory · SIEM

S. A. Sav (✉)
Cyber Threat Defense, Cluj, CJ, Romania
e-mail: ssorinandreig@gmail.com

S. Riurean
University of Petroşani, Petroşani, HD, Romania

T. Antipova (ed.), *Digital Technology Platforms and Deployment*, Information Systems Engineering and Management 36, https://doi.org/10.1007/978-3-031-86547-3_18

1 Introduction

Integrating AD into IT infrastructures can lead to more efficient and manageable operations, addressing environmental sustainability and cybersecurity challenges. By consolidating resources and centralizing management, a hardened AD deployment reduces the number of potential entry points for cyberattacks.

This consolidation enhances security and contributes to more efficient resource utilization. Implementing security measures such as multi-factor authentication (MFA) and regular audits can prevent costly security breaches that might otherwise require extensive resources to address. This proactive approach aligns with sustainable IT practices by minimizing the need for reactive measures [1].

The substantial role of cybersecurity in today's world, with a particular focus on AD networks is discussed in this chapter. Integrating AD into IT infrastructures can effectively address both environmental and security challenges in several ways that bring important (i) environmental and (ii) security benefits [2].

The environmental benefits are resource optimization, efficient operations, and virtualization support. AD enables centralized management of IT resources, reducing the need for redundant hardware and software. This consolidation leads to lower energy consumption and reduced electronic waste.

By streamlining user and resource management, AD minimizes the operational overhead associated with managing multiple systems. This efficiency translates to less energy usage and a smaller carbon footprint. AD integration supports virtualization technologies, which allow multiple virtual machines to run on a single physical server. This reduces the number of physical servers required, leading to significant energy savings [3].

The security benefits are centralized authentication and authorization, group policies, and monitoring and auditing.

AD provides a centralized platform for managing user identities and access permissions. This centralization enhances security by ensuring consistent application of security policies across the organization. AD can integrate with identity management solutions such as Keycloak or Kerberos, enforce policies such as MFA, and grant or deny access to resources with various levels of authorization based on identity to provide an additional layer of security. Additionally, policies such as MFA, or strong password policies enforced throughout the domain can significantly reduce the risk of unauthorized access by requiring multiple or more complex forms of verification. Further, AD can be complemented by Endpoint Detection and Response (EDR) and Device/Endpoint Management (DM/EM) solutions which only allow devices with certain configurations to join the domain; thereby, only if a workstation matches the security policies adhered to by the domain, such as having software like Sysmon and Defender installed and active, and strong password policies it can then join the domain.

Administrators can also enforce security policies across all devices and users in the network. These policies can include password complexity requirements, account lockout policies, and software restrictions, thereby enhancing overall security [4].

AD provides robust tools for monitoring and auditing user activities. This capability helps in detecting and responding to security incidents promptly, ensuring that any breaches are quickly contained [5].

There is a certain synergy between environmental and security benefits by reduced attack surface and sustainable security practices. By consolidating resources and centralizing management, AD reduces the number of potential entry points for cyberattacks. This consolidation not only enhances security but also contributes to more efficient resource utilization. Implementing security measures such as MFA and regular audits can prevent costly security breaches that might otherwise require extensive resources to address. This proactive approach aligns with sustainable IT practices by minimizing the need for reactive measures [6].

By seamlessly integrating AD into IT infrastructures, organizations can streamline operations, reducing costs, energy consumption, and risks associated with cybersecurity and environmental sustainability. AD provides a central management system for resources, permissions, and identities, ensuring compliance with environmental policies and security standards. This not only supports sustainable business practices but also enhances the overall security posture of the organization by mitigating risks related to misconfigurations and human error [7].

This study aims to enhance responsible current knowledges on threats and defense methods, while also presenting scenarios for AD and security administrators to help in assessing the risks posed by adversaries present within internal networks. It also covers technical terminologies and details of cybersecurity within networks and AD, dealing with what is a large knowledge gap in public education on how modern attacks are performed and what a bad actor does. Increasing public knowledge in these areas can help safeguard most victims from successful attacks and identify such incidents at a much earlier stage.

Another motivation of the research and the practical solution presented is based on the fact that security personnel commonly neglect internal infrastructure security, thus underestimating how an incident may affect them. Security incidents may result in undesired outcomes since AD technicalities may be used against an organization to gain access to resources.

Considering the fact that AD is resident in approximately 90% of organizational networks and infrastructures worldwide, the need for developing a routine for fast inquiry into the security state of an organization would guarantee the security maturity and data safety of an organization.

2 Navigating Online Safety and the Evolution of Cyber Threats

The landscape of online safety and cybersecurity has evolved significantly over the years, driven by technological advancements and the increasing sophistication of cyber threats [8]. This section presents general cybersecurity concepts, focusing on social engineering attacks, the particulars of malicious artificial intelligence use, and protective measures that can be established to ensure a robust organizational perimeter.

2.1 Social Engineering and AI

In today's digital landscape, cybersecurity is no longer an afterthought or niche concern. Rather, it has become a central concern for businesses, individuals, and governments alike. With increasing reliance on technology for daily activities, the risks posed by cybersecurity threats have grown exponentially, making it essential for everyone to understand and prioritize security measures. As such, cybersecurity is rapidly evolving from a niche discipline into a critical component of the digital world, affecting virtually all aspects of modern life.

In the past decades, the use of devices, and networks has grown exponentially in many sectors such as healthcare, finance, the government, education, and the private sector. However, its counterpart, cybersecurity, has only recently become a priority due to the increasing threat and potential damage that malicious actors could pose. As a result, attackers are more likely to target these old and vulnerable systems, which leads to a rise in cybersecurity incidents [9].

Of the many different approaches used by threat actors, it would probably be accurate to say that the most common and successful vector is social engineering. These attacks can take many forms, the most common of which is phishing and all its variations, such as smishing, vishing, and spear phishing. Smishing is a form of a social engineering attack that involves sending fake mobile text messages with malicious intent [10].

Phishing is the most common and known form of social engineering attack. Phishing generally involves multiple measures through which the victim is coerced or manipulated to perform various actions, as desired by the attacker, usually resulting in access to confidential data, or monetary loss. These attacks are carried out by pretexting a plausible scenario and implying a vector to manipulate the feelings of the victim to divulge confidential information, or to get the victim to perform an act that hinders the security of critical assets. Vishing or voice phishing, is a form of social engineering attack that involves the usage of fraudulent phone calls aiming to manipulate the victim [11, 12].

One of the most recent additions to an attacker's arsenal is artificial intelligence (AI). The analysis ability and generational aspect of a GPT algorithm have exacerbated the threat landscape due to the fast creation of deepfake videos presenting seemingly real videos with modulated voice or video content, making it increasingly difficult for an unseasoned individual to distinguish between real and fraudulent content. The latest forthcoming attacks have been discovered to impersonate reputable individuals, enticing victims to engage in malicious behavior by instilling a sense of authority, fear, trust, or security, or fear of missing out on the victim (Fig. 1) [13].

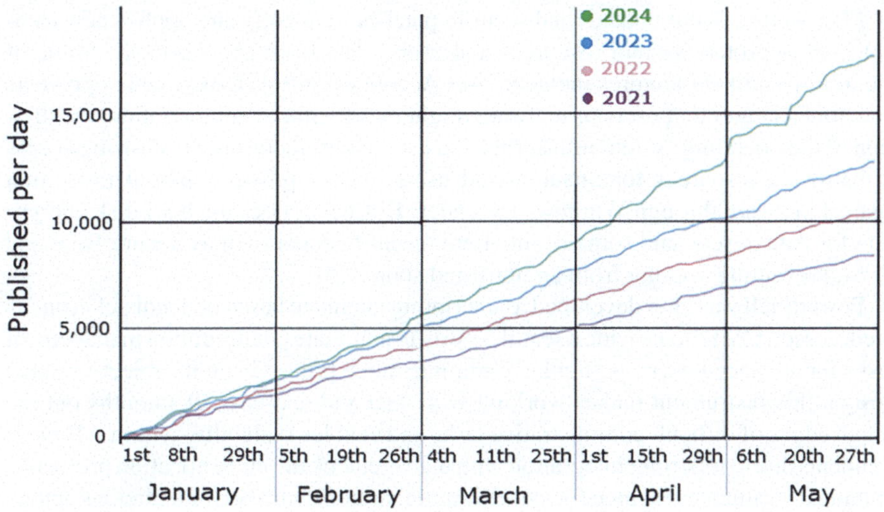

Fig. 1 Growth of CVE reported numbers per day in 2024 (Green), in contrast to past years [13]

2.2 Evolution of the Common Vulnerabilities and Exposures (CVE) Landscape

CVE is a standardized identifier for publicly known cyber security vulnerabilities affecting software or hardware. Due to the usage of AI, CVE numbers have been seen to grow exponentially over the past two years, as such inadvertently affecting an organization's ability to handle, share, and address these issues effectively. This is due to AI's ability to analyze vast amounts of code and identify potential vulnerabilities more quickly than its human counterpart.

As such, this behavior showcases the need for a novel method of addressing the CVE matter, a few takes in this direction have been performed by NIST and CISA with their specific projects and additions, both of which have yet to see room for improvement [14].

3 Technical Particulars of the Windows Ecosystem. An Attacker's Perspective and the Importance of Security Controls

Windows appears to be the dominant system of choice for both personal and enterprise environments. The architecture of Windows is designed to provide the user with a consistent experience, intuitive UI, and tight integration with various tools and services.

Microsoft rolls out updates and security patches frequently and applies new technologies to protect system resources and users' data from cyber-attacks. Some of the most prominent features include: User Account Control (UAC), which prevents potential changes that users or malware might make without authorization, the division of the operating system into kernel and user modes therefore disallowing access to hardware resources for unauthorized users; it also provides inbuilt protection against malware through Windows Defender; BitLocker does the hard disk encryption for data at rest; and some of the more recent features such as Secure Boot and TPM, preventing booting from an untrusted source [5].

PowerShell was later developed as a scripting language based on feedback from its predecessor, CMD, to provide a seamless way to automate configuration management tasks for advanced users, particularly administrative tasks. Given its object-oriented design, this instrument makes working with data and text easy. It smooths out the rough edges of difficult administrative tasks and enables scalability within AD environments due to its strong integration with management and authentication protocols, making it a valuable resource for system administrators, but also for malicious actors.

Building on these foundations, AD was created to serve as a central management and control center for operational resources and identities. AD promotes business growth by providing a central control center for managing small to large networks of Windows computers from a single "vantage" point. It enables con-figuration distribution across a variety of machines by defined rules, thereby improving efficiency and control over organizational networks [15].

In the interconnected digital world, the increased integration and connectivity between nodes have led to a new challenge: the greater the level of integration, the greater the potential for cyber-attacks to spread across the network. Unlike in the past, where a compromise of one computer typically limited the damage to a small number of systems, today's interconnected networks create vast attack surfaces, enabling a single breach to rapidly spread to other nodes, causing widespread damage. This presents a critical trade-off that cybersecurity professionals must carefully balance—the advantages of increased connectivity versus the elevated security risks that come with it.

This was because only a few protocols usually linked a user's access to another computer, such as credentials for file-sharing services, such as FTP, or session-establishing protocols such as SSH, or services provided over HTTP, or other LAN resources. In such cases, inside access for the attacker is limited. In most scenarios, it rarely gave access to multiple or all resources, due to network configurations such as segregation, and deployment of defensive controls such as an antivirus solution on end-points, and perhaps in some cases an Intrusion Prevention/Detection System (IPS/IDS) [16].

However, this simple approach is not the case with AD anymore. The complexity and linkage between the objects of AD deployments can often enable an attacker to compromise the entire domain using Microsoft signed mechanisms and tools (as such making non-behavioral based defense solutions such as string/rule-based antivirus useless) when only a single domain machine has been compromised.

This is due to trust-based relationships between domain-joined objects. Mechanisms such as session management and environment information, including user sessions and credentials for various services stored on each machine, contribute to this risk. Additionally, users with domain-joined accounts can query the Domain Controller (DC) or global catalog to find network information.

When analyzed, this information can reveal vectors for privilege escalation, such as Kerberoast-able domain administrator accounts, or delegations to sensitive services such as easily accessible file-sharing service provider users or database users. Kerberoasting is a type of cyber-attack that targets a misconfiguration of the Kerberos authentication protocol used in AD. It can be an effective method for extracting service account credentials from AD as a regular user without sending any packets to the target system. This attack is effective since people tend to create poor passwords. It enables the extraction of a hashed string, which can then be subjected to offline brute-force attacks, potentially exposing weak passwords in plain text if successfully brute-forced, allowing for the impersonation of the given user [17].

Once a malicious actor acquires access to a domain administrator account, it can access all resources within the AD domain. Moreover, it can establish a stronger foothold on the domain through various methods, such as dumping the Local Security Authority Subsystem Service (LSASS.exe), which contains credentials for all users of the machine or the entire domain in the case of a domain controller. It can also create privileged backups or Kerberos tickets using the *krbtgt* domain user's hash. The krbtgt user is a built-in AD account that is used by the Kerberos authentication protocol [18]. The extraction of this user's NTLM hash could allow the forgery of tickets which can grant access to any resource within the domain. More silent methods include dumping the Directory Services Restore Mode (DSRM) password, creating a skeleton key, or tampering with the AdminSDHolder object to change the privileges of the domain admin objects. The skeleton key attack is a persistence attack that changes the master password on one or multiple domain controllers. The Administrative Security Descriptor object controls what objects can modify the permissions of the Domain Administrators Group. [19].

Given the numerous options for attackers, simple protection and detection methods are no longer sufficient. A more thorough and security-mature approach is instead required. This approach should include a per-category defense mechanism with multiple controls for each category, abiding by the principle of least privilege, and relevant per-industry standards, such as Center for Internet Security (CIS), National Institute of Standards and Technology (NIST), Payment Card Industry Data Security Standard (PCI DSS), etc. CIS is a non-profit organization that aims to safeguard public and private organizations against cyber threats through the provision of hardening recommendations for various systems and software. NIST is a US-based government agency that provides multiple security recommendations and standards across various sectors. PCI DSS represents a set of security standards designed to ensure that all companies handle and possess payment card data securely. Additionally, implementing thorough network segregation, inventorying AD objects and resources, and constantly cleaning up and rechecking user access and privileges is advised.

Moreover, deploying a Security Information and Event Management (SIEM) solution to aggregate all data and logs from the network and machines, in conjunction with various Endpoint Detection and Response (EDR) and endpoint logging systems such as Sentinel or Microsoft Sysmon, can provide more detailed information on actions performed on each system, providing of great help during a digital forensics investigation. Performing frequent security audits, creating policies for event and incident management, and establishing an internal digital forensics and incident response team are also advised [20].

All of these measures are intended to provide a robust approach, with multiple security controls if one fails. This makes it more difficult for a malicious actor to gain and maintain access to a domain, while also allowing for quick and documented incident response procedures, cutting off malicious access quickly and assisting in investigative and recovery steps.

4 Tools for Rapid Assessment of AD Network Security

There are several tools and templates for analyzing and managing security baselines in Windows environments, including AD. It helps ensure that security settings comply with Microsoft's recommended security configurations, providing guidance on best practices for securing AD environments.

4.1 Available Specialized Tools

For quickly assessing the security AD networks, several specialized tools can provide comprehensive security audits, vulnerability assessments, and configuration analysis such as BloodHound for mapping attack paths, PingCastle for comprehensive AD security reports, ADAudit Plus for real-time monitoring, Purple Knight for identifying weaknesses, and Microsoft Security Compliance Toolkit for ensuring adherence to best practice.

BloodHound is an open-source tool that uses graph theory to reveal the hidden and often unintended relationships within an Active Directory environment. It maps out the attack paths an adversary could take and helps security professionals identify weaknesses and misconfigurations in AD permissions and group memberships. This tool is especially useful for visualizing potential privilege escalation paths [21].

PingCastle is a lightweight, easy-to-use tool specifically designed for AD security assessment. It provides a thorough analysis of AD health and security posture, including detailed reports on vulnerabilities, best practice compliance, and potential attack paths. It also includes a scoring system to help prioritize remediation efforts [22].

ManageEngine ADAudit Plus is a comprehensive AD auditing solution that tracks all changes in real time. It provides detailed insights into user activity, security group modifications, and more. This tool is useful for monitoring and reporting, helping organizations ensure compliance and detect potential security incidents [23].

Purple Knight is a free AD security assessment tool developed by Semperis. It checks for vulnerabilities and misconfigurations that could be exploited by attackers. The tool provides a score and detailed findings, making it easier for organizations to understand their security posture and prioritize remediation actions [24].

Microsoft Security Compliance Toolkit includes a collection of tools and templates for analyzing and managing security baselines in Windows environments, including AD. It helps ensure that security settings comply with Microsoft's recommended security configurations, providing guidance on best practices for securing AD environments [25].

Each of these tools offers unique features and advantages, making them suitable for different aspects of AD security assessment. Depending on the specific needs and scope of the security assessment, organizations may choose one or a combination of these tools to ensure comprehensive coverage.

4.2 Personalized Tool

This section presents a personalized tool for quickly assessing the security of AD networks. It is intended to provide a detailed analysis, inventorying network objects and identifying the vulnerabilities present using various security methodologies. The ultimate goal is to increase the resilience of AD networks and protect critical data of organizations.

An AD environment was created to generate a lab necessary to develop and test the capabilities of the script that is our personalized tool for quickly assessing the security of AD networks. For this purpose, several solutions such as using Cloud environments (Microsoft Azure, Digital Ocean, AWS), or using a Raspberry Pi microcontroller with an SSD to have a portable laboratory, are available. The most convenient solution is to use a virtualization solution on a personal system equipped with a 1 TB SSD and 32 GB of RAM. VMware is used to create and configure the virtual machines. Having ISOs from the official Microsoft page, two virtual machines with Windows 10 and one with Windows Server 2019 are created, each with separate virtualization instances. When running concurrently, these virtual machines take up about 200 GB of SSD space and 6 GB of RAM.

Regarding the implementation and writing of the script, the Microsoft Active Directory PowerShell module is used, to reduce the complexity of the queries. Then, tests are made according to a personalized methodology, based on examples such as the cheat sheet created by Cas Van Cooten [26], resources like HackTricks [27], and ired.team [28], but also Nikhil Mittal's Certified Red Team Professional course from Altered Security [29].

The tool is aimed to provide a quick overview of the operational AD environment while keeping a verbose output as showcased in Fig. 2.

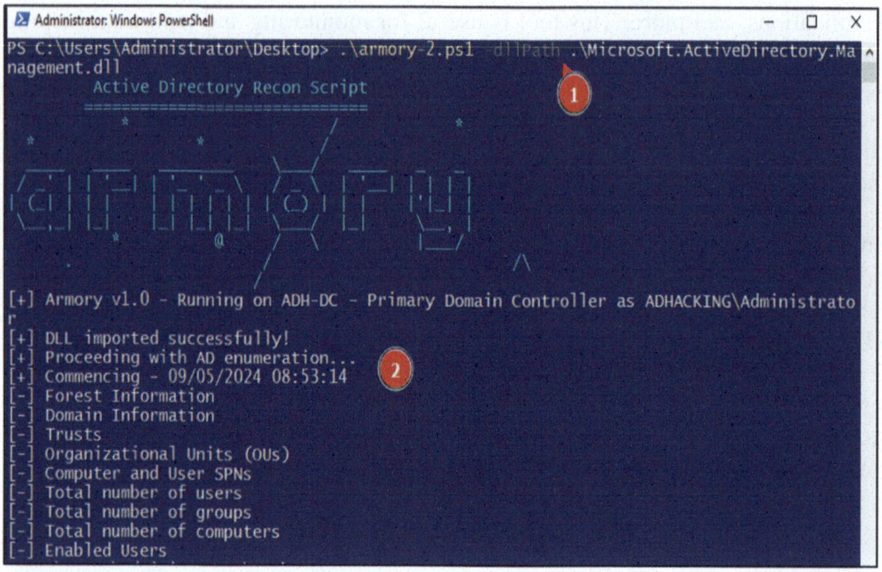

Fig. 2 Execution of the tool, showcasing the tests performed and details about the environment. *Source* Authors' elaboration

The result of this analysis is later saved within multiple individual files for each of the findings, to offer granularity for the auditor, as shown in Fig. 3.

It is important to highlight that this approach has room for improvement. Enhancements could include the addition of customization options to provide multiple output metrics or integration with existing tools such as PingCastle or Bloodhound. Moreover, the script could potentially be rewritten in a language that can be compiled to offer a more portable solution and improve execution speed.

As such the benefits of automating various administrative tasks are evident, particularly in terms of reducing time and resource consumption, leading to faster and more efficient outcomes. While this solution currently focuses on analysis, it could be further developed to comply with organizational policies or specific benchmarks, enabling automatic modifications to the environment [30].

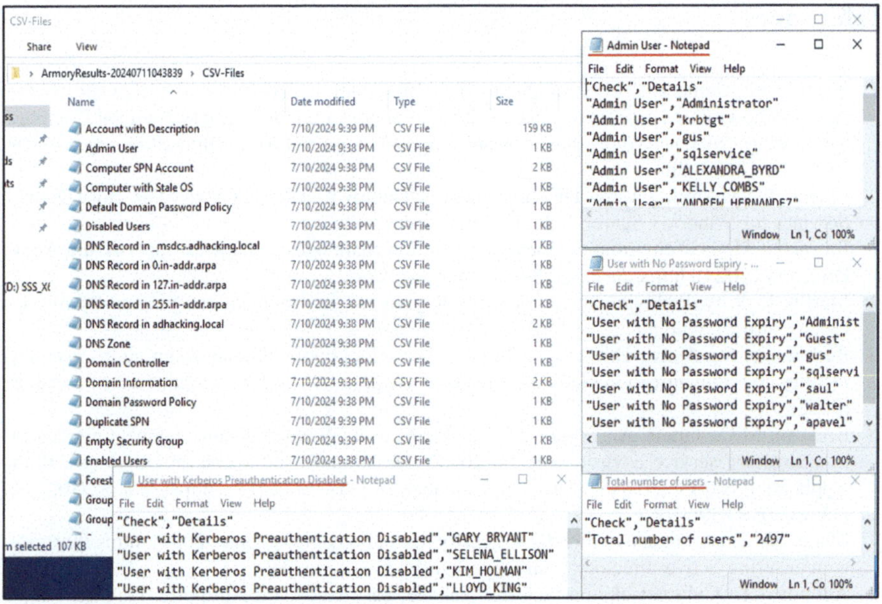

Fig. 3 Audit results are granularly saved within multiple CSV files. *Source* Authors' elaboration

5 Conclusion

The implementation of AD provides multiple benefits for organizations, including enhanced security, efficiency, and scalability. A centralized management of user identities and permissions helps improve overall security, reducing the risk of unauthorized access and malicious activities. Consolidating user and resource management activities in one location saves time and effort, allowing administrators to focus on other important tasks. AD can easily adapt to changing organizational requirements, from small businesses to large enterprises, ensuring smooth operations as the business grows. By implementing systematic policies and ongoing monitoring, organizations can leverage AD to achieve twofold benefits: reduced carbon footprints and enhanced security measures. General users must be equipped with an understanding of existing cybersecurity threats and defense techniques, while scenarios that assess potential risks posed by internal adversaries are presented.

Ensuring AD networks remain secure and resilient requires regular assessment and monitoring. A variety of specialized tools that help administrators to quickly evaluate their security posture are also presented. These tools provide comprehensive security audits, vulnerability assessments, and configuration analysis, allowing administrators to identify and address potential security risks promptly. Such proactive measures help ensure that AD networks remain secure and resilient against potential threats, mitigating the risk of costly and damaging cyber-attacks. A personalized, handy, and easy-to-use tool for quickly assessing the security of AD networks is also presented.

References

1. Binduf, A., Alamoudi, H.O., Balahmar, H., Alshamrani, S., Al-Omar, H., Nagy, N.: Active directory and related aspects of security. In: 21st Saudi Computer Society National Computer Conference (NCC), Riyadh, Saudi Arabia, pp. 4474–4479 (2018). https://doi.org/10.1109/NCG.2018.8593188
2. Francis, D.: Mastering Active Directory: Design, Deploy, and Protect Active Directory Domain Services for Windows Server 2022. Packt Publishing (2021)
3. Rasin, R.: 3 hybrid environment identity protection challenges and how to apply Azure Active Directory (2023). https://techcommunity.microsoft.com/
4. Martin, J.: Why deploying active directory in your OT environment can be a costly mistake (2023). https://cyolo.io/
5. Berkouwer, S.: Active Directory Administration Cookbook: Proven solutions to everyday identity and authentication challenges for both on-premises and the cloud. Packt Publishing (2022)
6. Breaban, M.C., Graur, A., Potorac, A.D., Balan, D.G.: Bandwidth management application in a directory service environment. In: 2018 International Conference on Development and Application Systems (DAS), Suceava, Romania, pp. 88–92 (2018). https://doi.org/10.1109/DAAS.2018.8396077
7. Marat, M.: Ensuring active directory security in hybrid environments (2024). https://infrasos.com/ensuring-active-directory-security-in-hybrid-environments/
8. Antipova, T.: Cyberattacks on business website: case study. In: Antipova, T. (eds.) Digital Science. DSIC 2021. Lecture Notes in Networks and Systems, vol. 381. Springer, Cham (2022). https://doi.org/10.1007/978-3-030-93677-8_44
9. Moldovan, D., Riurean, S.: Cyber-security attacks, prevention and malware detection application. J. Digit. Sci. **4**(2), 3–23 (2022). https://doi.org/10.33847/2686-8296.4.2_1
10. Bishnoi, A., Bishnoi, S., Gupta, N.: Comprehensive assessment of reverse social engineering to understand social engineering attacks. In: 2023 5th International Conference on Smart Systems and Inventive Technology (ICSSIT), Tirunelveli, India, pp. 681–685 (2023). https://doi.org/10.1109/ICSSIT55814.2023.10061054
11. Sharma, A.K., Rakesh, N., Verma, P.K.: An evaluation and comparison for phishing attack detection using machine learning approaches. In: 2024 3rd International Conference on Power Electronics and IoT Applications in Renewable Energy and its Control (PARC), Mathura, India, 2024, pp. 464–468. https://doi.org/10.1109/PARC59193.2024.10486547
12. Hegde, A.M., Kumar, S.B., Bhuvantej, R., Vyshak, R., Sarasvathi, V.: Spear phishing using machine learning. In: Singh, M., Tyagi, V., Gupta, P., Flusser, J., Ören, T. (eds.) Advances in Computing and Data Sciences. ICACDS 2023. Communications in Computer and Information Science, vol. 1848. Springer, Cham (2023). https://doi.org/10.1007/978-3-031-37940-6_43
13. Ansari, M.F., Panigrahi, A., Jakka, G., Pati, A., Bhattacharya, K.: Prevention of phishing attacks using AI algorithm. In: 2022 2nd Odisha International Conference on Electrical Power Engineering, Communication and Computing Technology (ODICON), Bhubaneswar, India, pp. 1–5 (2022). https://doi.org/10.1109/ODICON54453.2022.10010185
14. Bressers, J.: Open-source security, why are vulnerabilities out of control in 2024? (2024). https://opensourcesecurity.io/2024/06/03/why-are-vulnerabilities-out-of-control-in-2024/
15. Warner, T.L.: Managing identity and access with azure active directory. In: Microsoft Azure for Dummies, pp. 229–249. Wiley (2020)
16. Akacki, D.: Detecting Kerberoasting Activity (2024) https://adsecurity.org/?tag=kerberoast
17. Cordis, G.A., Costea, F.M., Pecherle, G., Győrödi, R., Győrödi, C.: Considerations in mitigating Kerberos vulnerabilities for active directory. In: 2023 17th International Conference on Engineering of Modern Electric Systems (EMES), Oradea, Romania, pp. 1–6 (2023). https://doi.org/10.1109/EMES58375.2023.10171623
18. Alexeevskaya, Y.A., Molodtsova, Y.V., Alexeevsky, R.A.: Forensic search for traces of unauthorized access using the Kerberos authentication protocol. In: 2023 International Russian Smart

Industry Conference (SmartIndustryCon), Sochi, Russian Federation, pp. 404–409 (2023). https://doi.org/10.1109/SmartIndustryCon57312.2023.10110781
19. Mittal, N.: Certified Red Team Professional (CRTP) Course Material (2024)
20. Bakshi, A., Sawant, T., Thakare, P., Dandawala, A., Hanawal, M.K., Kabra, A.: Improving threat detection capabilities in windows endpoints with Osquery. In: 2023 15th International Conference on COMmunication Systems & Networks (COMSNETS), Bangalore, India, pp. 432–435 (2023). https://doi.org/10.1109/COMSNETS56262.2023.10041379
21. https://bloodhoundenterprise.io/
22. https://www.pingcastle.com/
23. https://www.manageengine.com/products/active-directory-audit/active-directory-auditing.html
24. https://www.semperis.com/purple-knight/
25. https://www.microsoft.com/en-us/download/details.aspx?id=55319
26. https://casvancooten.com/posts/2020/11/windows-active-directory-exploitation-cheat-sheet-and-command-reference
27. https://book.hacktricks.xyz/windows-hardening/active-directory-methodology
28. https://www.ired.team/
29. https://www.alteredsecurity.com/post/certified-red-team-professional-crtp/
30. Sav, S., Riurean, S.: Rolul Securității Cibernetice în Rețele de tip Active Directory. Unpublished Bachelor's thesis, University of Petroșani (2024)

Mobile Application Based on Ransomware Technology to Improve the Security of Smartphones

Lidice Haz, Jaime Moises Minchala Marquino, Jenny Garzón Balcazar, and José Mero Morán

Abstract In the digital age in which we live, computer security is one of the most important issues. The use of Internet and the growing number of devices connected to the Internet, as well as trends towards teleworking or business policies such as BYOD, have increased the possibilities of being victims of technological threats and vulnerabilities. Which mainly puts at risk unauthorized access to personal data. Therefore, it is important to implement computer security mechanisms that minimize IT risks, and ensure the confidentiality, integrity and availability of data, whether for organizations or end users. This work presents the development of a mobile application based on ransomware technology. The software was developed applying the cascade development methodology using tools such as Java Code Conventions, and Android Studio IDE. The functionality and performance tests of encryption and data backup algorithm were carried out on mid- and high-range mobile devices with Android platforms from version 4.0. In general, the results showed high effectiveness in data encryption, being compatible with all file types. Users perceived the mobile application with a high level of ease of use and handling.

Keywords Mobile application · Ransomware · Encryption · Data backup · Computer security

1 Introduction

Nowadays, the use of mobile devices has changed the way people communicate and interact. In past decades it was unthinkable to be able to carry out multiple activities on a small screen in the palm of your hand. This scenario changes with the so-called smartphones. Organizations, and employees in general, take advantage of the

L. Haz (✉) · J. M. M. Marquino · J. M. Morán
Peninsula of Santa Elena University State, Santa Elena, Ecuador
e-mail: lhaz@upse.edu.ec

J. G. Balcazar
University Ecotec, Guayaquil, Ecuador

T. Antipova (ed.), *Digital Technology Platforms and Deployment*, Information Systems Engineering and Management 36, https://doi.org/10.1007/978-3-031-86547-3_19

capabilities and tools that can be exploited on these devices. In this same sense, the concept of BYOD (Bring Your Own Device) arises, which refers to the IT policy that determines that workers can use their own mobile devices to carry out company activities [1]. During 2020, this mechanism and teleworking were exploited because the Covid19 pandemic modified the way of working and doing business, moving us to a much more digital context [2].

In this context, the advantages of these forms of work can also be misused. History has shown that there is a relationship between crime and technology. Thus, computer security predictions suggest exposure to various cyber threats caused by the materialization of technological risks [3].

Worldwide statistics record a constant increase in the number of threats to mobile devices. In 2023, the number of attacks on these devices reached almost 33.8 million attacks, which represents an increase of more than 50% compared to figures from the previous year. In the period of 2023, the threat with the highest percentage of detection on mobile devices was adware with 40.8% [4].

In this scenario, the most latent risk is the theft of personal data and loss of confidential information. This is because users store a large amount of personal, work and academic information on their mobile devices. Which is exposed to being misused by cybercriminals who generally use malicious code and social engineering techniques to obtain unauthorized access to devices and generate illicit profits [5].

Therefore, it is necessary to implement cybersecurity mechanisms that guarantee the integrity, availability and confidentiality of data.

In this work, we try to answer the question: How effective is it to use ransomware as a security mechanism embedded in a mobile application? To answer this question, we investigated the behavior and operation of ransomware, and developed a computer application based on this method to encrypt sensitive data. In addition, we evaluate the level of knowledge of users who telework, regarding the security mechanisms they implement on their mobile devices, and we carry out experimental tests of the use and functionality of the application.

The chapter is organized as follows: Sect. 2 describes recent works and approaches related to the operation of ransomware and information encryption mechanisms. Section 3 describes the project development methodology. Section 4 presents the details of the results obtained in the experimentation, and the surveys carried out with the users who participated in the study. Finally, Sect. 5 summarizes the research conclusions, and presents suggestions for future work.

2 Literature Review

Technological advances and the internet promote the use of mobile devices. This reality implies the exposure of risks and threats that may affect the confidentiality of the information [6]. Information is the most valuable asset for organizations and people, and consequently, it must be protected to guarantee the continuity of operations for companies, and for users to avoid being victims of the commission of computer crimes, such as theft of data, identity theft, extortion, fraud, among others [7].

Cybercriminals have adopted various techniques to profit from information; one of the most current and dangerous is malware infection [8]. Among the different types of malware, there is ransomware. This is a type of malware that is used to digitally extort victims in exchange for payment of a specific fee. Typically, malicious code makes specific files or an entire system unavailable to the victim through encryption or change-of-use rights [9].

In this context, various studies promote the development of computer security mechanisms and tools to secure information. The work proposed by [10] implements an image encryption application in the Android system. Another study develops a native Android mobile application for the encryption of information files on mobile devices by applying symmetric cryptographic algorithms. Researchers use the development methodology based on prototyping to build a mobile application under the object-oriented programming paradigm [11]. In another investigation, they develop an anti-espionage application by encrypting calls on mobile devices with Android system [12]. The works described above use cryptographic techniques whose objective is to transform a message so that it is unintelligible except for those who have the key to decipher it. Thus, [13] focus their research on the use of the RSA algorithm between mobile devices. The objective is to encrypt the data to send it through communication channels called threads that, through formulas and processes executed on the server, will help execute the encryption and decryption of data.

A similar investigation, presented by [14] which includes a brief introduction to cryptography, without delving into the mathematics that supports cryptographic algorithms. The work describes the different techniques, symmetrical and asymmetrical. The researchers apply both techniques through the use of the algorithms: RSA and 3DES, implemented in Visual Basic. NET. Similarly [15], propose a symmetric encryption algorithm that takes as input the original information of length L and upon encoding it generates the ciphertext of greater length LM. The algorithm is implemented in three orbits. The first is used to apply a diffusion technique in order to mix the original information. The second orbit combines the mixed information and the length from L to LM is increased and with the third the confusion technique is implemented. The encryption algorithm was applied to encode an image.

In this context, during the research, a mobile application prototype based on ransomware technology was designed to improve the security of smartphones with the aim of encrypting data.

3 Data and Methodology

This work is developed under the modalities of descriptive, bibliographic and experimental research. The data collection technique applied to the participants was the survey with a 10-question multiple choice questionnaire. The participants were 20 users who belong to a public institution in the province of Santa Elena, Ecuador.

The same ones who use their mobile devices for the daily development of their work activities, that is, they apply BYOD [1]. The objective of this survey was to evaluate the level of knowledge regarding the security mechanisms they apply on their mobile devices, and the technological risks to which they are exposed. The questionnaire was enabled online for a period of 10 days through the Google platform using the Google Form service. The information was processed on this same platform, evidencing 100% participation of those surveyed. The results are presented using statistical pie charts. These data were used as input to delimit some functional and non-functional requirements of the mobile application.

To develop the project, we use the cascade methodology defined as a sequence of five phases [16]. Below, we describe each of the phases:

- **Phase 1: Approach**. We applied a survey to know the level of use of the computing services available in the cloud. The information collected allows you to define the functional and non-functional requirements of the application.

Functional requirements:

1. Key registration to protect stored data.
2. Enter the personal account associated with the Google Drive cloud storage service.
3. Make backup copies by sending them to the account associated with the service.
4. Encrypt the data using a key preconfigured by the user.

Non-functional requirements:

1. Simple and user-friendly graphical interface.
2. Code development using the Android Studio IDE using the Java programming language.
3. Encrypt mobile device information using Cipher.

- **Phase 2:** Analysis. We carried out a bibliographical investigation on the characteristics and operation of Slocker and WannaLocker ransomware. Also, we study and select the APIs that allow you to use cloud storage services. We evaluate the compatibility and efficiency of the Java and JavaScript programming languages, the operation of source code editors, the requirements for using the Google Assistant API, and finally the architecture of the Android platform.

In this phase, we structure the operating architecture of the mobile application described in Fig. 1. The first step of the process is when the user logs into the rebel application. The second step is to create the encryption key. The third step is to encrypt the data using the ransomware algorithm. The fourth step is to upload and store the encrypted data in the cloud repository. The fifth and final step is to decrypt the data using the encryption key from the rebel app.

- **Phase 3:** Design. We developed design of graphical interface of the mobile application as shown in Fig. 2. For this, we used a wireframe that allows us to simulate the functionalities that the application would have.

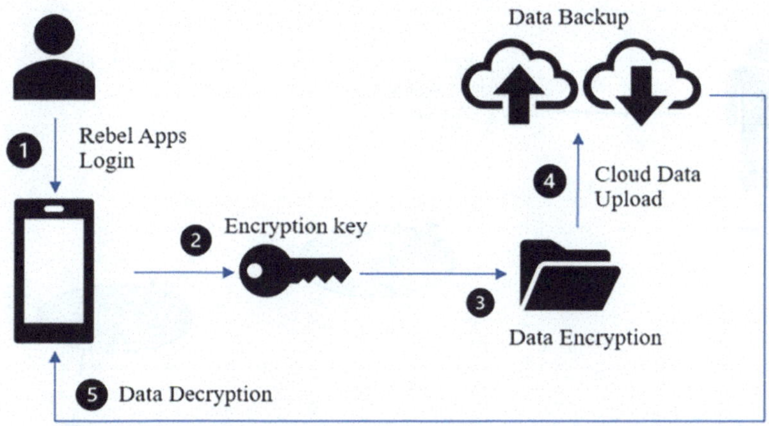

Fig. 1 Mobile application architecture

Fig. 2 Interface design of Rebel 1.1 app

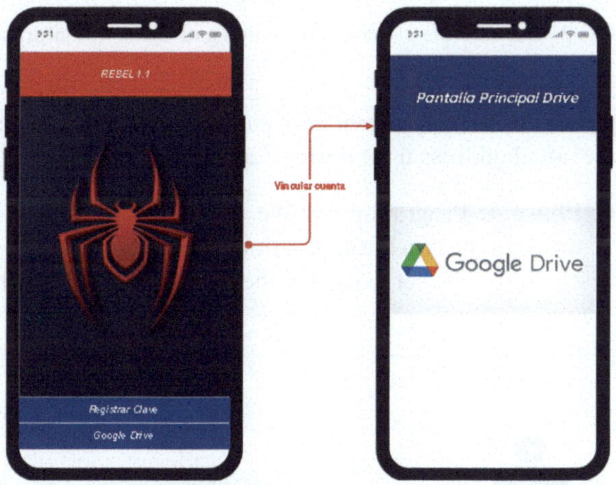

Additionally, we use case diagrams to describe the behavior of software. A use case contains a textual description of all the ways that the intended actors could work with the software [17].

Figure 3 describes the use case diagram for registration and backup of information from the rebel app. The user must previously install the rebel and Google Drive applications on their mobile device. Subsequently, the user manually selects the files they want to back up and then generates the backup from the rebel app and stores it in the Google Drive repository that is associated with their account.

Figure 4 shows the use case diagram for data encryption from the rebel app. The user activates the data encryption process by assigning an encryption password. Later, this key will be used for the decryption process. In case you lose or forget this

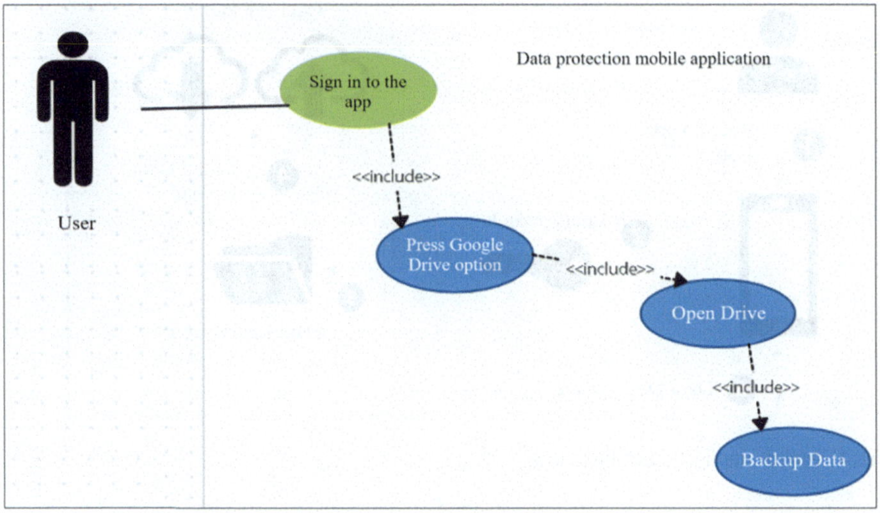

Fig. 3 Use case diagram for registration and backup of information from the rebel app

password, the app provides a password recovery procedure. Which will be sent to the email address used during registration.

- **Phase 4:** Programming. We coded the prototype of mobile application implementing the two basic functionalities. The first is to generate a backup copy using cloud storage services, and the second is to encrypt the data using the ransomware

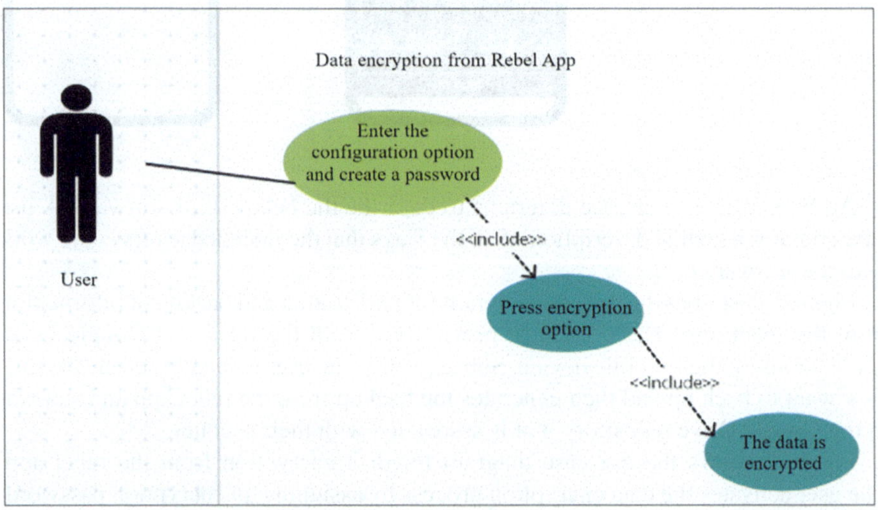

Fig. 4 Use case diagram for data encryption from rebel app

algorithm. The development tools used were the Java programming language, the Gradle framework, and the Android Studio integrated development environment (IDE). Data encryption is performed by four tools, namely AES Encryption, Initialization Vector IV, Cipher Block Chaining (CBC), and PKCS5Padding.

- **Phase 5:** Testing. We developed three test scenarios to evaluate the functionality of application.

Table 1 shows the description of test scenario 1 to validate the key registration process. In this test two possible cases are evaluated. The first is the creation of encryption key, and the second is the replacement of the previously assigned key.

Table 2 shows the description of test scenario 2 to validate the data encryption activation and execution process. This test evaluates the correct and incorrect entry of encryption key. If it is successful, the encryption proceeds, otherwise an error message is displayed, and the encryption operation is aborted.

Table 1 Test scenario 1 for key record validation

Test 1: key registration	
Aim	Key creation and registration to control the data encryption and decryption process
Complexity level	Low
Case 1: check the key registration procedure	
Input data	Encryption key
Condition	Minimum character length and use of special characters
Process	Encryption key creation and registration
Output data	Authorization to encrypt information
Case 2: verify existing key	
Input data	Encryption key
Condition	Shows a notification asking if you want to replace the old key
Process	Encryption key update and registration
Output data	Authorization to encrypt information

Table 2 Test scenario 2 for encryption execution validation

Test 2: encryption execution	
Aim:	Turn data encryption on and off
Complexity level	Low
Case 1: data encryption using key	
Input data	Encryption key
Condition	Key validation
Output data 1	Correct key. Successful authorization to perform data encryption
Output data 2	Wrong key. Error message due to incorrect password entry. Aborts the encryption process

Table 3 Test scenario 3 for running and uploading backup data to Google Drive

Test 3: access to Google Drive	
Aim	Access Google's cloud storage service
Complexity level	Low
Case 1: uploading data to Google Drive	
Input data	Google Drive access key
Condition	Internet connection and access
Output data 1	Successful authorization to execute and load backup data
Output data 2	Warning message. Mobile data disabled. Impossible to load data from the backup

Table 3 shows the description of test scenario 3 for execution and uploading backup data to Google Drive. In this test, the device has an internet connection or mobile data is enabled. If successful, the encrypted data is backed up and subsequently uploaded and stored in the Google Drive account associated with the Smartphone.

4 Results

To evaluate the data encryption method using the ransomware algorithm, we use the metrics described in Table 4 that allow us to evaluate the performance of mobile application as a security mechanism.

Next, we present the results of survey applied to users of public institution who participated in the research, and who carry out teleworking. This survey made it possible to evaluate the level of knowledge regarding the security mechanisms implemented on their mobile devices.

Table 4 Performance metrics of data encryption and backup process using mobile application

Metrics	Average time
Data encryption	5 s. per file (up to 8 s depending on file type, size and mobile device range)
Data decryption	3 s. (up to 5 s depending on file size, and range of mobile device)
Deleting the original files	6 s
Encryption speed under load	10 s
Encryption initialization time	3 s
Data backup speed	1 s. per file (up to 5 s depending on file type and internet connection)
Backup download speed	1 s. per file (up to 8 s depending on file type and internet connection)

Fig. 5 Methods used to recover information stored on the mobile device

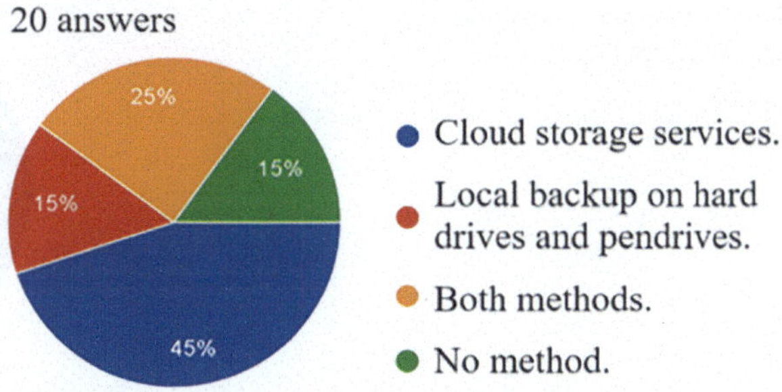

Figure 5 shows the methods used to recover the information stored on the mobile device. 45% of respondents mentioned that they use storage services available in the cloud. This means that users know and apply iCloud SaaS services. Which is important to know, since the mobile application uses this medium as a backup and data storage method.

The next question, regarding the use of iCloud services for data storage. Figure 6 shows that 90% of users mainly use the services offered by Google.

Another question related to the constant connection for Internet access. Figure 7 shows that 65% of respondents have a permanent internet connection via mobile data or Wi-Fi, while 35% do not. This means that there is a limitation to permanent use of app as a security mechanism. Due to intermittent internet connection.

Finally, Fig. 8 describes the percentage of use of data encryption tools. 95% indicated not using this mechanism as part of the security their data. Which favors the proposal of this project.

In response to our research question, regarding the effectiveness of using ransomware as a security mechanism embedded in the mobile application, and the evaluation of users regarding the level of knowledge and use of security mechanisms that they implement on their mobile devices. The results obtained show, first

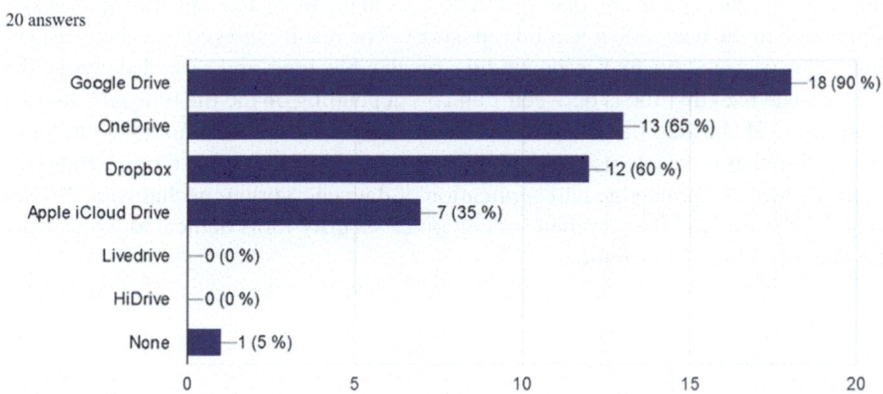

Fig. 6 Using iCloud services for data storage

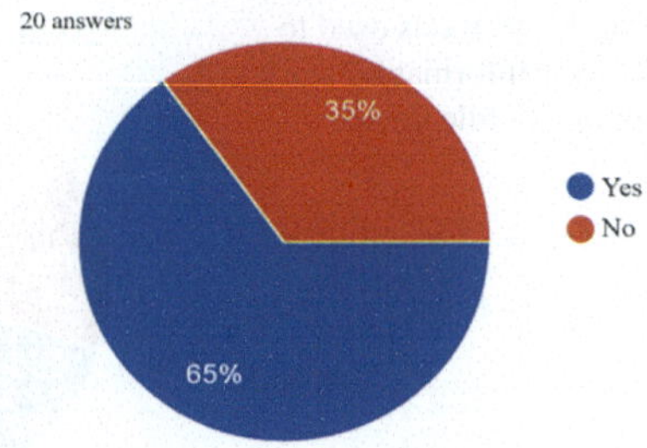

Fig. 7 Internet connection via mobile data or Wi-Fi

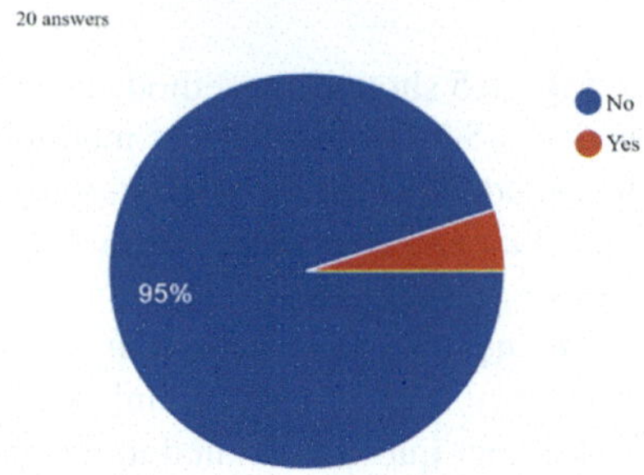

Fig. 8 Use of data encryption tools

of all, Sara and WannaLocker Ransomware are data encryption mechanisms with a high level of effectiveness. WannaLocker implements a function inherited from Utils.java to perform the encryption process. The WannaLocker algorithm does not affect images, that is, it only allows you to encrypt documents and even.rar files. On the other hand, Sara Ransomware encrypts all types of files, and includes an additional function that consists of sending the data to a repository in the cloud, specifically to the remote server https://api.anonfile.com/upload (currently offline). Sara uses libraries and her own code developed in Python.

In this way, the advantages of both Ransomware are taken advantage of to implement the data encryption and decryption process in the rebel app, and the data backup is included in the user's own iCloud repository. The results showed that data encryption time averages up to 8 s depending on the file type and size. Likewise, the average data backup time is between 1 and 5 s depending on the file type and internet connection. Regarding the usability of the application, it has a minimalist and intuitive design that is easy to use. The findings obtained in the surveys demonstrate that there is a lack of knowledge and application of data encryption mechanisms. Which suggests the use and development of computer security tools dedicated to ensuring the integrity of the information.

5 Conclusion

In this work, we develop a mobile application to encrypt data based on Ransomware, and back up the data using an iCloud service. The encryption procedure is based on the algorithms of Sara and WannaLocker Ransomware. Encryption is performed on all file types and sizes with 100% effectiveness. The mobile application is developed using the Java programming language, which is cross-platform, and is compatible with the Cipher library. Which facilitates the encryption process due to its AES encryption scheme, and its File library that facilitates the management of the mobile directories. The data backup process is done through an iCloud service specifically Google Drive. The backup is performed semi-automatically, which means that the user must select the most important files that they want to back up to the iCloud service drive. The experimental tests were carried out in real environments, that is, on mobile devices of the study participants. The results demonstrated high performance and ease of use of the application. However, the limitations are the use of a stable internet connection through mobile data or Wi-Fi for the data backup process, and the use of mid- and high-range smartphones, since these characteristics affect encryption and decryption times of data. Likewise, the mobile application is compatible with Android platforms from version 4.0, requiring root mode to be enabled. Finally, encourage users to use computer security tools on their mobile devices.

For future work, the data backup process should be performed automatically for all files stored on the mobile device that have been encrypted. Also, it is necessary to automate root mode to use the rebel application, that is, this process is validated during the installation of the application. In addition, API permissions must be acquired to improve and implement more functionalities in the rebel app, such as password management, software installation permissions.

References

1. Ratchford, M., El-Gayar, O., Noteboom, C., Wang, Y.: BYOD security issues: a systematic literature review. Inform. Secur. J.: Glob. Perspect. **31**(3), 253–273 (2022)
2. Peiró, J., Soler, A.: The push for teleworking during COVID-19 and the challenges it poses (2020). https://www.ivie.es/wp-content/uploads/2020/05/11.Covid19IvieExpress.El-impulso-al-teletrabajo-durante-el-COVID-19-y-los-retos-que-plantea.pdf. Accessed 8 Sept 2024
3. Fitzpatrick, D., Griffin, D.: Cyber-extortion losses skyrocket, says FBI (2016)
4. Kivva, A.: IT threat evolution in Q3 2023. Mobile statistics (2023). https://securelist.com/it-threat-evolution-q3-2023-mobile-statistics/111224/. Accessed 10 Sept 2024
5. Qayyum, R., Ejaz, H.: Data security in mobile cloud computing: a state of the art review. Int. J. Mod. Educ. Comput. Sci. **12**(2), 30–35 (2020)
6. Kumar, M., Meena, J., Singh, R., Vardhan, M.: Data outsourcing: a threat to confidentiality, integrity, and availability. In: 2015 International Conference on Green Computing and Internet of Things (ICGCIoT), pp. 1496–1501. IEEE (2015)
7. Martinez-Garcia, H.A., Medina, M.M.: Encrypted data recovery by versioning in cloud, an alternative against ransomware: Case Study (2016)
8. Aslan, Ö.A., Samet, R.: A comprehensive review on malware detection approaches. IEEE Access **8**, 6249–6271 (2020)

9. Liska, A., Gallo, T.: Ransomware: defending against digital extortion (2017)
10. López, R.: Development of an image encryption application on the android system (2014)
11. Laguado, M.: Android mobile application to protect information through symmetric cryptography (2020)
12. Hernández, R., Morales, P.: Anti-espionage application by encrypting calls on mobile devices with Android system (2014)
13. Solís, F., Pinto, D., Solís, S.: Information security in data exchange between mobile devices with Android system using the RSA encryption method (2017)
14. Mendoza, J.: Demonstration of symmetric and asymmetric encryption. Magazine of Science and Technology, Ingenius (2008)
15. Maricela, J.R., Octavio, F.S., Guadalupe, G.N.M.: Sistema para codificar información implementando varias órbitas caóticas. Ingeniería, Investigación y Tecnología [Internet]. **16**(3), 335–343 (2015). http://www.revistaingenieria.unam.mx/numeros/v16n3-02.php
16. Pico, M.A.P., Torres, E.F.L.: Implementation of agile methodologies in the development of mobile applications for inexperienced programmers. Univ. Soc. **15**(S2), 208–218 (2023)
17. López, R., Ramiro, L.: Software engineering. Center for Financial Studies (2011)

Robots' Team Controlling in Emergencies

Gennady Ross and Valery Konyavsky

Abstract The chapter proposes an approach to solving the problem of controlling robot teams in emergency situations, which involves the distribution of territorial areas between robots for the search of different objects. It takes into account the nature of the tasks performed by the robots, their software and hardware capabilities, and the specifics of the environment under study. It is shown that an increase in efficiency in solving real-world problems can be achieved by taking into account important logical conditions that are often encountered in practice. A method for describing possible logical relationships between robots and territorial areas is presented. A modified formulation of the classical allocation task and the methods for its execution are proposed, within the framework of which the analyst can choose in a dialogue any option for the distribution of robots by territorial areas, relying only on his own intuition, experience, etc. The examples of solving some problems are presented, taking into account the situations often encountered in practice.

Keywords Robot distribution model · Intelligent robot · Robot control · Multi-agent system · Team of robots

1 Introduction

Technological progress in the field of emergency rescue operations is connected with the use of robot complexes - Mobile robots for emergency use (topuch.ru) - which are capable of autonomously carrying out search and rescue operations in the case of both man-made and natural accidents (remote survey of the area, examination of a suspicious object in the case of a terrorist attack, carrying out repair works

G. Ross (✉)
Plekhanov Russian University of Economics, Moscow, Russia
e-mail: ross-49@mail.ru

V. Konyavsky
Moscow Institute of Physics and Technology (National Research University), Moscow, Russia

T. Antipova (ed.), *Digital Technology Platforms and Deployment*, Information Systems Engineering and Management 36, https://doi.org/10.1007/978-3-031-86547-3_20

in conditions dangerous for people, etc.). In addition, a team of robots (TR) must effectively search for various objects.

The object is considered to be the occurrence of a series of events, such as a fire, a traffic accident, a breach of public order, the discovery of a suspicious object, an emergency. This object performs the following functions: surveying and patrolling various areas, environmental monitoring, providing radio retransmission, radio direction finding, surveying areas of major accidents and disasters, etc.

In order to organise the work of a team of robots based on multi-agent principles, it is necessary to solve a number of problems, one of which is to draw up a plan for the distribution of territorial areas between the robots, taking into account the nature of the tasks they perform, the functionality of each robot and the environment in which they operate. The robot is characterised by the speed of movement, the level of battery power, the availability of special equipment (photo, video, air, water, soil analysis tools) with varying degrees of accuracy. The robot's software and hardware are related to its functional orientation in the studied territorial area.

In the context of a Multi-Agent System (MAS), which takes over the functions of goal setting, the TR control system should have the property of self-organisation, i.e., after receiving a task from the "centre", it should ensure the planning of the behaviour of a team of robots, coordinating or coordinating the actions of each of them for some upcoming period of time [1]. A robot with more powerful software and hardware acts as the "scheduler", carrying out the distribution of the team of robots by territorial areas. The "scheduler" robot has a powerful computing complex, a database, a knowledge base designed to schedule the activities of a group of other robots. On this basis, each robot independently plans its activities in a given territorial area [2]. The robot's schedule can be characterised by a certain set of quantitatively measurable characteristics, for example, the robot's movement within the territorial area is characterised by location coordinates, energy consumption, the weight of the load being moved, the speed of movement, and so on. In the practical implementation, it is necessary to take into account the specifics of the logic of the functioning of the whole TR and the logic of the relationships between the robots [3].

In general, not all robots of the team can be used to perform a specific target task of the "centre", but a certain set of robots is formed to achieve a specific goal. In case of failure of part of the set of robots, it is necessary to continue to perform the tasks of the "centre" by redistributing the efficient robots to the unexplored areas. It is possible to reduce some indices of the quality of the functioning of the TR within acceptable limits.

Within this framework, the problems of controlling a team of robots focused on surveying a given area in emergency situations are considered. The "scheduler" robot of the TR [4] forms a complex task for robots to survey a certain area, which is divided into separate territorial areas, which are within the "functional orientation sphere" (self-organisation) of robots of one or another type.

Various options for setting the task of surveying given territorial areas by special robots and methods of controlling them are studied, taking into account various conditions of their functioning. It is shown that the application of the classical formulation of the task setting and the algorithms for its performance in real control systems is not efficient. However, the modification of the classical formulation of the task setting, taking into account the logic of the functioning of the TR and the logic of the relationships between the robots, allows to formulate more efficient solutions for the control of robots.

The structure of the chapter includes a research of the classical formulation of the assignment task setting and the proposed modification of it, taking into account the specific features of the logic of the functioning of the TR, the examples of various options for modelling the functioning of robots within the framework of the TR using "Combinatorics" module of Decision Tool System are given. An example of the solution of the problem of time optimisation of all the areas to be surveyed by a team of robots is given.

2 The Classical Formulation of Assignment Task Setting and Its Modification

In the case of discrete search tasks, where a given area is to be explored, the task of allocation is implemented by defining the set of areas, and the process of searching for the corresponding object is performed by the robots in the areas allocated to them. It is assumed that the whole search area is divided into several areas. Then, the assignment of certain search areas should be related to the topographical conditions and terrain elements, such as the structure of the terrain and local peculiarities, i.e. natural objects (rivers, lakes, green areas, wetlands, etc.) and artificial objects (settlements, roads, engineering structures, etc.). To search for an object in territorial areas, special survey robots are used. A survey robot is a robot designed to observe and detect an event (or group of events). The search itself can be performed by a single robot or by a group of robots.

The assignment task [5, 6] is a special type of distribution task in planning theory, the model of which can be reduced to the task of distributing robots over territorial areas. Any search area can be assigned to any robot. The resource expenditure matrix (indices such as total energy consumption, time spent exploring territorial areas, etc. can be used as resources) is presented as the initial data for solving the search tasks. The task is to distribute the robots over the areas in such a way that the search work is carried out with minimum effort.

Let us consider the mathematical formulation of the task: I will be the total number of robots, J—the total number of areas to be searched; αi, j—stands for a Boolean variable that is equal to 1 if the ith robot is assigned to search the jth area, and equal to 0 otherwise. The assignment task then has the following form:

$$\sum_{\mathrm{i}=1}^{\mathrm{I}} \alpha_{\mathrm{ij}} = 1, j = \overline{1,\mathrm{J}}; \tag{1}$$

$$\sum_{j=1}^{J} \alpha_{ij} = 1, i = \overline{1, I}; \quad (2)$$

$$\sum_{i=1}^{I} \sum_{j=1}^{J} r_{ij} \alpha_{ij} \rightarrow \min \quad (3)$$

Equality (1) expresses the requirement that all robots be distributed, and equality (2) expresses the requirement that each area be surveyed. To solve problem (1)–(3), assuming that the number of robots is equal to the number of areas, a simple Hungarian method (Flood's algorithm) is proposed in the literature to find the optimal solution [5].

Example 1. Assume that 5 areas are to be surveyed by 5 robots with minimised expenditures. The initial data for solving the problem is as follows: the list of areas to be surveyed; the list of possible robots; expenditures matrix $r_{i,j}$ (Table 1) required for the *j*th area to be surveyed by the *i*th robot $\left(j = \overline{1, J}; i = \overline{1, I}\right)$. It is assumed that each area is to be surveyed by one robot, i.e. J = I.

The allocation in bold with an asterisk in Table 1 corresponds to the optimal allocation and has the following expenditure: $1 + 5 + 9 + 7 + 9 = 31$.

In order to find the maximum of the objective function of this task using Flood's algorithm, it is necessary to transform the original matrix rij into a matrix r0ij in the following way.

r0ij = (max rij) – rij, now the problem is solved similarly to the minimisation i, j problem.

Problems (1)–(3) and the algorithm for solving them contain some rather strict and not very realistic requirements, and at the same time they do not take into account such an important aspect as the logic of TR operation [7], which has to be considered in practice. In particular, it is not always necessary to survey all areas or to engage all robots. Some areas can be postponed to other dates, and robots can be selected from the list of extras. At the same time it has to be taken into account that the survey of some areas requires a certain order of their survey, or they are alternative. This does not fit into the formulation (1)–(3). The consideration of these requirements can be formalised as follows:

Table 1 Expenditures matrix

j / i	1	2	3	4	5
1	**1***	4	6	9	5
2	2	**5***	7	6	9
3	3	10	13	**7***	8
4	4	11	**9***	11	12
5	7	8	10	8	**9***

$$\sum_{i \in \Omega s} \alpha_{ij} \geq 1, s = \overline{1,\ S}; \tag{4}$$

$$\sum_{j \in \Omega t} \alpha_{ij} = 1, t = \overline{1,\ T}; \tag{5}$$

$$\sum_{j \in Vz} \alpha_{ij} \geq \sum_{j \in Wz} \alpha_{ij}, z = \overline{1, Z} \tag{6}$$

$$\sum_{j \in Uk} \alpha_{ij} \leq 1, k = \overline{1, K}; \tag{7}$$

$$\sum_{j \in He} \alpha_{ij} \leq 1, e = \overline{1, E}; \tag{8}$$

$$\sum_{i} \sum_{j} r_{ijp} \alpha_{ij} \leq M_p, p = \overline{1, P} \tag{9}$$

$$\sum_{i=1}^{I} \sum_{j=1}^{J} r_{ij} \alpha_{ij} \to \min \tag{10}$$

Formulations (4)–(9) are a generalisation of formulations (1)–(3). In particular, condition (4), in contrast to condition (1), expresses the requirement that at least one robot from a group of robots Ωs is involved. There can be any number of such groups $s = (1, S)$. Condition (1) on the involvement of all robots is a special case of this condition. Contrary to condition (2), condition (5) expresses the requirement that at least one region from the set of regions Gt is to be explored. The number of sets of areas $t = (1, T)$ is not limited. The requirement (2) that all areas must be surveyed is a special case of this requirement.

In addition, model (4)–(9) allows (but does not require) the consideration of conditions such as the logical interdependence of areas. In particular, condition (6) expresses the requirement that the areas of set Vz are surveyed before or together with the areas of set Wz, and that the number of surveyed areas of set Vz should not be less than the number of surveyed areas of set Wz.

Conditions (7) and (8) allow (but do not require) taking into account that some groups of areas or some groups of robots may be incompatible for some reason.

Condition (9) allows (but does not require) taking into account the constraint on the amount of expenditure for performing the survey of all areas. In addition, this condition allows (but does not require) taking into account the presence of several limited resources Mr, i.e. funds, energy, etc.

Condition (10) is similar to condition (3).

Let us formulate the requirements for the description of the logic of the operation of the TR, which determines the complex of logical connections between robots and zones.

3 Logical Conditions and Technology for Their Implementation in "Combinatorics" Module

3.1 *The Representation of Logical Interrelations Between Robots and Areas to Be Surveyed*

In terms of meaning, logical conditions can be very diverse. However, all logical conditions, including very complex ones, can be represented using only 2 types of logical links, for example: negation and conjunction (i.e. "AND"), or negation and disjunction (i.e. "OR").

Let P be a propositional variable, i.e. a variable that can be replaced by any proposition, e.g.: "Sampling the soil by a robot 1", "Robot 7 is equipped with high-resolution photo and video equipment", etc. Let us consider the most convenient set of 3 types of logical connections: or (exclusive disjunction); if…then… (implication); at least one of … must be present [8].

The following symbols are used to represent logical connections: $\cup$, $\rightarrow$,{}. Notation $P_1 \cup P_2$ means that P_1 and P_2 sentences contradict each other (mutually exclude each other). Notation $P_3 \rightarrow P_4$ means that sentence P_4 is a consequence of sentence P_3. Notation $\{P_5, P_6\}$ means that any of sentences P_5 or P_6 has a certain property, for example, it is true. Logical connections represented by symbols $\cup$, $\rightarrow$,{} are called *logical formulas*. The independent propositional variable P_7 is a logical formula as well.

Each sentence P_i can be true (T) or false (F). Complex sentences obtained by means of logical connections can also be true or false. This is determined using the so-called truth tables, which are combined in the form of Table 2.

Table 2 shows that complex sentence $P_1 \cup P_2$ is false only if both sentences P_1 and P_2 are true. For example, if P_1 = "Send robot 2 to area 1", P_2 = "Send robot 2 to area 2". These sentences are mutually exclusive as it is impossible to send robot 2 to area 1 and area 2 simultaneously. Sentence $P_3 \rightarrow P_4$ is false if P_4 is true and P_3 is false. The implication assumes the truth of sentence P_4 only if sentence P_3 is true. For example, P_3 = "Equip robot 2 with photo and video tools", P_4 = "Conduct the survey of area 3 using photo and video tools". It is clear that area 3 can be surveyed only if robot 2 is equipped with photo and video tools. Finally, $\{P_5, P_6\}$ connection is false if both sentences in it (P_5 and P_6) are false. For example, P_5 = "Set 50%

Table 2 Truth table

$P_1 \cup P_2$	P_1	P_2	$P_3 \rightarrow P_4$	P_3	P_4	$\{P_5, P_6\}$	P_5	P_6
F	T	T	T	T	T	T	T	T
T	T	F	T	T	F	T	T	F
T	F	T	F	F	T	T	F	T
T	F	F	T	F	F	F	F	F

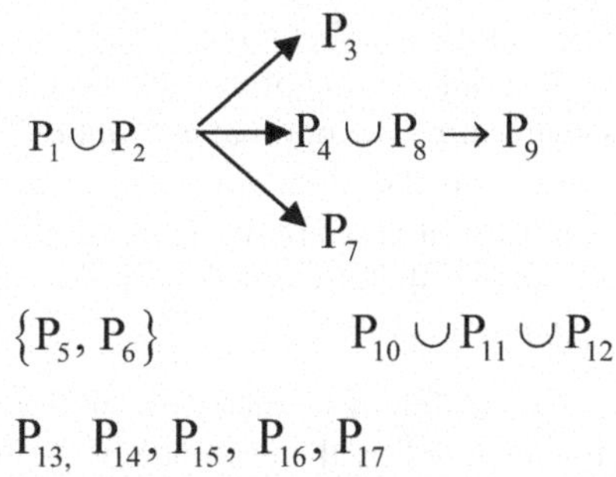

Fig. 1 Logical network of connections

battery power limit for robot 5", P_5 = "Set 25% battery power limit for robot 5". Any battery power limit should be set for robot 5.

The description of any situation is a logical network [2] containing a set of sentences, some of which are united by logical connections, while others are independent of the rest. An example of a logical network is shown in Fig. 1.

In this logical network, sentences P_{13}, P_{14}, P_{15}, P_{16}, P_{17} are logically independent from other sentences, in other words, they do not form complex sentences with other sentences. All other sentences form 3 connected subnets (3 complex sentences). The first subnet includes sentences P_1, P_2, P_3, P_4, P_5, P_6, P_7, P_8, P_9. The second subnet includes sentences P_5 and P_6. The third subnet includes sentences P_{10}, P_{11}, P_{12}.

3.2 The Description of "Combinatorics" Module

The structure of "Decision" tool system represents a set of software tools designed for the use of optimization models of simulation modelling [7].

"Decision" tool system includes two modules:

- "Equilibrium" module is designed to solve equilibrium problems, which is based on the evolutionary simulation methodology of decision-making under conditions of uncertainty;
- "Combinatorics" module is designed for solving combinatorial problems, allowing to conduct research on various options for activities under resource and logical conditions that should be selected in such a way as to achieve the specified goals in the best way. In this case, an "event" means an integral part of an investment, technical, legal or other project that requires resource expenditures, brings results and is connected with other events by logical connections. The module is based on a modified dynamic programming algorithm for solving Boolean problems (DPASBP).

The application of "Combinatorics" module allows not only to solve combinatorial problems, but also to study them comprehensively. Thus, for example, when drawing up a plan for the distribution of robots by territorial areas, one can:

- establish whether it is expedient to increase and whether it is permissible to reduce the total volume of energy resources;
- determine the optimal amount of financial expenditures in terms of profitability;
- perform the simulation of various planning options for the survey of various territorial areas by robots under different changes in conditions (changes in the indices of individual robots or areas, changes in logical conditions, etc.) in a dialogue mode;

Any problem in the class of Boolean problems, which we are considering, is characterized by the presence of the following items:

- firstly, the initial set of elements (of any nature);
- secondly, a set of indices characterizing the elements and allowing addition (the additive ones);
- thirdly, the restrictions on the allowable aggregate value of some indices (the resource restrictions);
- fourthly, the logical connections (LC) between elements;
- fifthly, a goal, expressed in the aggregate value of one of the indices.
- a goal, in the form of a requirement to achieve the minimum or the maximum value of one of the indices.

"Combinatorics" module allows to solve various problems using dialogue procedures for the interaction between an analyst and a computer. Dialogue procedures allow to load initial data, set targets (maximize or minimize) and limit systems, set various options for logical links, if required, to automatically issue intermediate distribution options, as well as the best option. The module includes Help block, which contains "Training" block.

3.3 Technology for Solving Problems Using "Combinatorics" Module

Assignment problem (4)–(10) with logical connections (LC) being taken into account can be solved in a dialogue, within the framework of "Combinatorics" module of "Decision" tool system, which allows you to overcome the specified difficulties in a modified task.

Let us consider some techniques for solving problem (4)–(9) using the Combinatorics module. Let us start by solving problem (1)–(3), and afterwards we will sequentially consider the ways of easing conditions (1)–(3) and introducing additional conditions. Let us renumber areas $i = \overline{1, I}$ and robots $j = \overline{1, J}$ using index $k = \overline{1, I \times J}$ as it is shown in Table 3. Each value of k corresponds to an "event", which consists in the fact that a certain robot surveys a certain specific area (the value of k uniquely determines the values of i and j).

The assignment problem now appears as the problem of compiling a set of events that minimizes total expenditures. When numbering, the total number of events can

Table 3 The options of events for area i to be surveyed by robot j

j/i	1	2	...	I
1	k = 1	k = 2	...	k = I
2	k = J + 1	k = J + 2	...	k = J + I
...	...	...	...	...
J	k = J*(I − 1)+1	k = J*(I − 1)+2	...	k = J*I

be significantly reduced, given that in reality, not every robot can survey any area, since it has a certain specialization and qualification.

Let us assume that robots can survey the areas shown in Table 4. The (+) sign means that the robot-performer can be assigned to the corresponding area; and the (−) sign means that the robot-performer cannot be assigned to the corresponding area.

Let us introduce a list of events to "Combinatorics" module in accordance with the data provided in Table 4. For each event, we indicate the expenditures of a survey in a given area. In addition, let us introduce "Additional index" to be considered and to be attached equal importance for all events (Table 5). Let us choose "Expenditures" as a limited resource and set a deliberately overestimated limit for them, for example, 50. Let us choose the additional index as the target one (Table 6), which determines the number of robots participating in the survey of areas.

Let us form logical connections (LC) in accordance with Table 4. LC [2, 4] expresses the requirement that area 1 should not be surveyed twice (events 2 and 4 are alternatives). LC [5, 6, 9], [3, 10, 12], [7, 11] and [8, 13] have a similar meaning in relation to areas 2, 3, 4 and 5, respectively. LC [2, 3] expresses the requirement that robot 1 be involved only in the 1st zone. LC [4–8], [9, 10, 11] and [12, 13] have a similar meaning in relation to the 2nd, 3rd, 4th and 5th robots, respectively.

Now, in order to obtain the optimal distribution of areas between robots, the maximization calculation is performed in a dialogue mode (Table 6).

Since we have not introduced requirements that all areas are to be surveyed (this could be done by entering the following LC: {2, 4}, {5, 6, 9}, {3, 10, 12}, {7, 11}, {8, 13}), the system reports a complete list of optimal assignments for different levels of total expenditures. Thus, should we have set a limit for total expenditures of 6 units, we could only perform events 2 and 5, and this would mean that the 1st robot

Table 4 The options of events for surveying the areas

I j	Area 1	Area 2	Area 3	Area 4	Area 5
Robot 1	+	−	+	−	−
Robot 2	+	+	−	−	−
Robot 3	−	+	−	+	+
Robot 4	−	+	+	+	−
Robot 5	−	−	+	−	+

Table 5 The list of events and the preset values of indices

1	Events	Expenditures	Additional indices
2	Robot 1, area 1	1.00	1.00
3	Robot 1, area 3	6.00	1.00
4	Robot 2, area 1	2.00	1.00
5	Robot 2, area 2	5.00	1.00
6	Robot 3, area 2	10.00	1.00
7	Robot 3, area 4	7.00	1.00
8	Robot 3, area 5	8.00	1.00
9	Robot 4, area 2	11.00	1.00
10	Robot 4, area 3	9.00	1.00
11	Robot 4, area 4	11.00	1.00
12	Robot 5, area 3	10.00	1.00
13	Robot 5, area 5	9.00	1.00

Table 6 The optimal options for assignments at different levels of total expenditures

1	Expenditures	Additional indices	TI/LR	Events				
2	1.00	1.00	1.00	2				
3	6.00	2.00	0.33	2	6			
4	13.00	3.00	0.23	2	6	7		
5	22.00	4.00	0.18	2	6	7	10	
6	31.00	5.00	0.16	2	6	7	10	13

surveys the 1st area, and the 2nd robot surveys the 2nd area, as it can be seen from Table 5.

If we want all areas to be surveyed, we should use the assignment from the last line of Table 6. The optimal assignments are events 2, 5, 7, 10 and 13: thus, the 1st robot surveys the 1st area; the 2nd robot surveys the 2nd area; the 3rd robot surveys the 4th area; the 4th robot surveys the 3rd area; the 5th robot surveys the 5th area. In other words, we have found the same option that was previously found in example 1 (Table 1). This is not surprising, since this problem has a unique optimal solution. During searching for the optimum, we have already seen that it is possible to solve the problem upon the absence of the requirement to survey all areas. This demonstrates the possibility of solving the problem not only under condition (2), but also under condition (5).

Let us suppose now that some robots have sufficient power to survey not one but several areas. This means replacing condition (1) with condition (4). For example, let robot 1 be able to survey both the 1st and 3rd areas simultaneously. To solve the problem under these changed conditions, we perform the dialog procedure:

Parameters → *Logical Connections* → *Delete LC* [2, 3] → *Calculation* → *Maximum*

The dialog procedure involves executing the sequence of the given commands in "Combinatorics" module of "Decision" tool system. The result is shown in Table 7.

The assignment presented in the last line of Table 7 requires less total expenditures than the assignment presented in the last line of Table 6.

Condition (6) can be exemplified by the requirement that area 2 be surveyed after area 3, or not surveyed at all. This requirement is expressed using the set of the following LC: (3) → (5, 6), (10) → (5, 6), (12) → (5, 6). Let us perform the dialog procedure:

Parameters → *Logical connections* → *Add LC:* (3) → (5, 6) → *Add LC:* (10) → (5, 6) → *Add LC:* (12) → (5, 6) → *Calculation* → *Maximum.*

The list of LC is shown in Table8, and the result of the optimization calculation is shown in Table 9.

Let us perform the dialog procedure:

Table 7 The optimal assignment, upon which all areas are surveyed, and two of the ones are surveyed by the 1st robot (represented in the last line)

1	Expenditures	Additional indices	TI/LR	Events				
2	1.00	1.00	1.00	2				
3	6.00	2.00	0.33	2	5			
4	12.00	3.00	0.25	2	3	5		
5	19.00	4.00	0.21	2	3	5	7	
6	28.00	5.00	0.18	2	3	5	7	13

Table 8 The list of LC, the last 3 of which express the requirement that area 2 be surveyed after area 3

1	Events	Logical network
2	Robot 1, area 1	[2, 4]
3	Robot 1, area 3	[5, 6, 9]
4	Robot 2, area 1	[3, 10, 12]
5	Robot 2, area 2	[7, 11]
6	Robot 3, area 2	[8, 13]
7	Robot 3, area 4	[4, 5]
8	Robot 3, area 5	[6–8]
9	Robot 4, area 2	[9, 10, 11]
10	Robot 4, area 3	[12, 13]
11	Robot 4, area 4	(3) → (5, 6)
12	Robot 5, area 3	(10) → (5, 6)
13	Robot 5, area 5	(12) → (5, 6)

Table 9 Assignment options with the requirement to perform work 3 after work 4 being taken into account

1	Expenditures	Additional indices	TI/LR	Events				
2	1.00	1.00	1.00	2				
3	7.00	2.00	0.29	2	3			
4	12.00	3.00	0.25	2	3	5		
5	19.00	4.00	0.21	2	3	5	7	
6	28.00	5.00	0.18	2	3	5	7	13

Table 10 Assignment options in a text entry. Area 3 is presented in every assignment, where area 2 is presented

1	Expenditures	Additional indices	TI/LR	Events				
2	1.00	1.00	1.00	Robot 1, area 1				
3	7.00	2.00	0.29	Robot 1, area 1	Robot 1, area 3			
4	12.00	3.00	0.25	Robot 1, area 1	Robot 1, area 3	Robot 2, area 2		
5	19.00	4.00	0.21	Robot 1, area 1	Robot 1, area 3	Robot 2, area 2	Robot 3, area 4	
6	28.00	5.00	0.18	Robot 1, area 1	Robot 1, area 3	Robot 2, area 2	Robot 3, area 4	Robot 5, area 5

Parameters → Options → Text.

We can see from Table 10 that there are assignment options without area 2 (the first 2 lines), as well as the options with area 2 (the last 3 lines) in each of which area 4 is presented simultaneously.

4 The Problems of Optimizing the Time of All Areas Being Surveyed by a Team of Robots

Compose the optimal assignment under the following conditions: there are 3 robots and 3 areas. Table 11 provides for the time expenditures matrix. The "-" sign means that the robot cannot (for some reason or another) perform the survey in the corresponding area. Table 12 shows the power of the battery of robots and the powers required for surveying the corresponding areas.

In order to determine the optimal distribution of robots by territorial areas, we draw up a list of events (Table 13).

Table 11 Expenditures matrix

Robot	Area 1	Area 2	Area 3
Robot 1	13	–	5
Robot 2	–	9	7
Robot 3	15	8	–

Table 12 The power of the robot and the required power for the survey of the area

Robot No.	Battery power	Area No.	Power required to survey the area
P1	135	31	121
P2	211	32	67
P3	147	33	115

Table 13 The events for the implementation of works

Events	Contents	Expenditures	Power
2	Robot 1 surveys area 1	13	121
3	Robot 1 surveys area 3	5	115
4	Robot 2 surveys area 2	9	67
5	Robot 2 surveys area 3	7	115
6	Robot 3 surveys area 1	15	121
7	Robot 3 surveys area 2	8	67

LC expressing the requirement that no zone be surveyed twice: [2–7]. Let us load "Combinatorics" module and enter the initial data. Upon that, we will introduce a column for each robot in which we indicate the power required for surveying a particular area so that we could determine the power of each robot, which is the required to survey the areas under the accepted assignment, according to the results of the calculation (Table 14).

Table 14 The list of events and values of indices selected as LR and TI

Events		Survey duration	Additional index	Power for P1	Power for P2	Power for P3
1	**2**	**3**	**4**	**5**	**6**	**7**
2	P1, 31	13	1	121.00	1.00	1.00
3	P1, 33	5	1	115.00	1.00	1.00
4	P2, 32	9	1	1.00	67.00	1.00
5	P2, 33	7	1	1.00	115.00	1.00
6	P3, 31	15	1	1.00	1.00	121.00
7	P3, 32	8	1	1.00	1.00	67.00

Table 14 also shows that, firstly, the “Survey duration” is selected as the limited resource (LR), and “Additional index” is selected as the target index (TI). Since empty fields are not allowed, units, i.e. negligible (compared to the required powers) values, are entered into the corresponding positions of columns “Power for P1”, “Power for P2” and “Power for P3”.

We set the value of the limited resource somewhat larger than the sum of the values specified in column 3 of Table 14 (equal to 57), namely 59, and perform an optimization calculation. These actions are included in the following dialog procedure:

Parameters → Limit of Limited Resource → 59 → Calculation → Maximum → Parameters → Options → Text

The calculation results are presented in Table 15. The optimal assignment is specified in the last line.

Table 15 shows that under such logical conditions, the 1st robot will be engaged twice and the impossible power of 237 will be required (robot 1, as can be seen from Table 6, has a power of 135).

In order to take into account the limitations on the power of the robot, we will select the “Power for P1” index as LR and indicate its actual power (i.e. 135) as the limit. For this purpose, let us perform the following dialog procedure:

Switch to the “Data” Page → place the Cursor on “D1” → Parameters → Limited Resource → Limit of Limited Resource → 135.

Let us set the power limits for other performers in a similar way (Table 16).

We perform the optimization calculation using the following dialog procedure:

Calculation → Maximum

The result is presented in Table 17.

The survey of areas No. 2 and No. 3 by robot No. 2, and the survey of area No. 1 by robot No. 3 is optimal.

Table 15 The result of the optimization calculation to the maximum. The optimal assignment is specified in the last line

1	Survey duration	Additional indices	Power for P1	Power for P2	Power for P3	TI/LR	Events		
2	5.00	1.00	115.00	1.00	1.00	0.20	P1, 33		
3	13.00	2.00	116.00	2.00	68.00	0.15	P1, 33	P2, 33	
4	26.00	3.00	237.00	3.00	68.00	0.15	P1, 31	P1, 33	P3, 32

Table 16 Introduced power limits for each performer

Events		Survey duration	Additional index	Power for P1	Power for P2	Power for P3
1	2	3	4	5	6	7
2	P1, 31	13	1	**121.00**	1.00	1.00
3	P1, 33	5	1	**115.00**	1.00	1.00
4	**P2, 32**	9	1	1.00	**67.00**	1.00
5	**P2, 33**	7	1	1.00	**115.00**	1.00
6	**P3, 31**	15	1	1.00	1.00	**121.00**
7	P3, 32	8	1	1.00	1.00	**67.00**

Table 17 Optimal assignment

1	Optimal assignment	Optimal assignment	Power for P1	Power for P2	Power for P3	Events		
2	31	3.00	3.00	183.00	123.00	P2, 32	P2, 33	P3, 31

5 Conclusions

The problems of organizing the team work of the teams of robots, when their behavior is entirely subordinated to the achievement of a common team goal related to the survey of emergency areas, can be considered the closest to MAS theory. Upon that, the goal itself is decomposed (in space and time) into many interconnected territorial areas, the survey of which is assigned to various robots of the team. It is implicit that the distribution of areas between robots, as well as the coordination of their behavior, should be performed by the means of the team itself, without any external interference.

The proposed approach to controlling a team of robots in emergencies is presented as a task of assigning functionally-oriented robots to survey territorial zones. The classical variant of the assignment and the procedure for reducing it to a new modified optimization problem are considered. Upon that, the nature of the tasks performed by robots, their software and hardware capabilities, as well as the specifics of the environment under study are taken into account. It is shown that an increase in efficiency in solving real problems can be achieved by taking into account important logical conditions that are often encountered in practice. A way of describing possible logical relationships between robots and territorial areas is given. A technique for solving a discrete distribution problem is proposed, in which the analyst can choose any option for distributing robots over territorial areas, relying only on his own intuition, experience, etc., in a dialogue. The examples of solving some problems are given, taking into account situations that are often encountered in practice.

References

1. Vasilyev, I.A., Polovko, S.A., Smirnova, E.Yu.: The organization of group control of mobile robots for special robotics tasks. Sci. Techn. J. Inform. Telecommun. Control. SPbSPU 1', 119–123 (2013)
2. Ross, G., Konyavsky, V.: The methods for emergency robot self-organization control. In: Advances in Digital Science, pp. 5–25 (2022). https://doi.org/10.33847/978-5-6048575-0-2_1
3. Beloglazov, D.A., Solovyov, V.V., Titov, A.E., Finaev, V.I., Shapovalov, I.O.: The method of distribution of goals in groups of intelligent mobile robots. Bull. TulSU. Techn. Sci. **11**, 3 (2016)
4. Topolsky, N.G., Vilisov, V.Ya.: Algorithm for optimal scheduling of operations of eliminating the consequences of fires and emergencies by a heterogeneous group of autonomous mobile robots. Eng. Bull. Don **4** (2021)
5. Cherchman, U., Akof, R., Arnof, L.: Introduction to Operations Research//Translated from English. "Science" Ed., Chief Editorial Board of Physical and Mathematical Literature, M., 488p. (1967)
6. Ross, G.V.: The Modeling of Production and Socio-Economic Systems Using the Apparatus of Combinatorial Mathematics. M.: Mir (2001)
7. Liechtenstein, V.E., Ross, G.V.: Information Technologies in Business. Workshop: Application of Decision system in Solving Applied Economic Problems. Finance and Statistics, 560p. (2009)
8. Ross, G., Antipova, T., Konyavsky, V.: Methodology for innovative projects' financing in IT business. In: Antipova, T. (eds.) Digital Science. DSIC 2021. Lecture Notes in Networks and Systems, vol 381, pp. 257–268. Springer, Cham (2022). https://doi.org/10.1007/978-3-030-93677-8_22.

A Model for Identifying Exchange Transactions with Signs of Manipulation

Vladimir Oniskiv, Vladislav Dolgirev, Ludmila Oniskiv, and Sergei Fedoseev

Abstract The work is related to the creation of an effective algorithm that allows you to identify stock transactions with a sign of manipulation. The fundamental reasons for the research are the growing volume of the securities market and the emergence of digital technologies. The chapter presents a general statement of the problem. The criterion used is based on the concept of information entropy. It is calculated as a function of the various time series patterns identified during the trading process. A sign of manipulation is the probability that a sequence of length m will be part of a longer pattern. To compare patterns, the finite deterministic automata method procedure was used. The emergence of patterns in stock trading is usually associated with attempts to manipulate prices in order to change the latter in a favorable direction. The proposed algorithm is based solely on real trading transactions and, of course, does not cover all possible types of manipulation. However, it does not depend on the type of traded instrument. The chapter presents the results of calculations demonstrating the effectiveness of the algorithm.

Keywords Information entropy · Time series · Patterns · Stock trading · Criterion of alleged manipulation · Algorithm of finite automata

1 Introduction

Stock trading is one of the essential elements of a market economy. It brings together buyers and sellers for the purpose of conducting trade transactions and must ensure the formation of fair pricing solely on the basis of the balance of supply and demand

V. Oniskiv (✉) · L. Oniskiv · S. Fedoseev
Perm National Research Polytechnic University, Perm, Russia
e-mail: oniskivf@gmail.com

L. Oniskiv
e-mail: oniskivf@gmail.com

V. Dolgirev
Perm State National Research University, Perm, Russia

T. Antipova (ed.), *Digital Technology Platforms and Deployment*, Information Systems Engineering and Management 36, https://doi.org/10.1007/978-3-031-86547-3_21

[1]. Exchange activities are among the licensed types of activities. The current trading mechanisms allow for the balancing of supply and demand volumes and the formation of a market price for a certain product on this basis. There are different types of exchange platforms [1]. They differ in the composition of participants, the type of exchange commodity traded, access to the trading procedure and the purpose of the activity (commercial / non-commercial). A classic example of exchange activity is a stock exchange, the activity of which is associated exclusively with trading in securities. Securities include shares of enterprises, government and other bonds, bills of exchange, savings certificates and some other documents. The rules for trading in securities are a key element of exchange trading and contain procedures for concluding and registering transactions, as well as measures to prevent manipulative transactions and the use of insider information. The above-mentioned control functions are assigned to the exchange by the state. It is important to add that the exchange, as a legal entity, cannot be a party to any of the purchase and sale transactions, that it cannot make transactions on its own behalf and at the expense of the exchange itself. This circumstance, among other things, ensures its independence and objectivity of trading transactions. The exchange has no right to determine any limits on the prices and volumes of exchange goods, as well as on the amount of the broker's remuneration for his intermediary services. At the same time, the rapid development of the Internet and computer technologies has led to new opportunities in exchange trading. A trading platform installed on a computer allows you to significantly speed up the process of exchanging commercial information and eliminates the need to be present in the trading floor. The first electronic exchange in 1971 was the American NASDAQ. Its appearance made it possible to significantly increase the volume of trading operations and the liquidity of securities. At the same time, transaction costs were reduced and the speed of order execution increased, which opened up new opportunities for brokers and dealers. The emergence of digital financial assets also made a significant contribution to exchange trading [2]. The capitalization of this sector is currently estimated at 1.6 trillion dollars. In addition to the usual trade purchase and sale transactions, there are transactions of a different nature, such as: futures transactions, forward transactions, options transactions [3]. These are the so-called derivative instruments that appeared in the 80 s of the last century and have received quite extensive practice by now. However, neither the unconditional innovativeness of financial instruments, nor the high technology of the trading procedure, do not rid this market of existing popular manipulation schemes. Unfortunately, established manipulations took place on almost all stock markets. The assessment of the damage to investors is quite comparable with the losses brought by crises on the largest stock and commodity exchanges. The main reason for this state of affairs, in addition to legal aspects, is certainly associated with the lack of prompt identification of transactions with suspicion of manipulation. Thus, the task of prompt computer analysis of transactions for manipulation is certainly important and relevant.

Manipulation is usually understood as artificial interference in the price formation process. If manipulation is successful, securities are sold at an inflated price, bringing speculators unjustified profits and causing losses to ordinary investors and other stock market participants. It should be noted that there is no single legal or economic definition of this term today. This chapter examines one of the types of manipulation, namely manipulation based exclusively on real trading. Trading manipulation is one of the most difficult to detect, since any of the transactions in the full tuple is not associated with any violation of the trading rules. At the same time, a "price precedent" is created, forming a tendency to increase the price and volume of subsequent transactions. Moreover, as current practice shows, the more low-liquid shares there are, the greater the likelihood that they will be manipulated, and manipulation, in turn, will increase the volatility of the share price.

The approach proposed in this study is quite general, does not depend on the type of the traded instrument and allows us to study problems related to other areas of knowledge and technology [4]. The current approach is based on the hypothesis that the emergence of manipulations in trading operations leads to the emergence of a certain regularity in the market and to a decrease in the inherent entropy [5]. In other words, the emergence of manipulations reduces the degree of chaos in the system. Probably, the first initiators of the concept of using the concept of entropy to assess the degree of market manipulation were the authors of the work [6]. They also introduced the concept of sample entropy, for the calculations of which a number of intraday transactions were used. Having estimated the values of sample entropy over seven months, they made a forecast of possible manipulations on those days when the indicator was not high. Note that undetected manipulations probably still remain in the sample and thus "contaminate" it somewhat.

It should be noted that the assertion that a particular transaction is manipulative is an evaluative legal judgment and requires research related to the analysis of the motivational aspects and intentions of the trader. The presence of a specific intent to increase or decrease the price of a financial instrument, as well as the degree of guilt of the participants, must be obvious to the court. In this process, mathematical modeling should only accurately identify the necessary signs of such suspicious transactions and detect them. Of course, the model only works with a posteriori information and in this sense, the concept of the model working in "real time" is quite conditional.

2 The General Formulation of the Problem of Identifying Manipulative Transactions

As noted above, entropy in a certain sense reflects the degree of orderliness of the system. It is known that entropy increases with an increase in the degree of chaos in the distribution of information resources and decreases with their ordering [7]. Of course, the measure can be defined in a variety of ways and therefore one should not expect unambiguity.

Within the framework of the general paradigm of identifying manipulations using repeating patterns [8, 9], we introduce in the event space an event Y- which corresponds to the appearance in the considered time series M patterns of length m each.

We also introduce an event X—the appearance in the same time series M′ patterns of length (m + 1). It is easy to see that the product of events XY is equal to X, since the set X can only be narrower than Y, that is, $X \subseteq Y$. It seems obvious that the longer the pattern and the more often it occurs in the time series, the greater the expectation that trading manipulation is taking place. In other words, the sign of manipulation is the probability that a sequence of length m will be part of a longer pattern of length (m + 1).

The probability of such an event can be defined as the conditional probability of event X, provided that event Y has already occurred, i.e.: P(X|Y) = P(XY)/P(Y). Given that $X \subseteq Y$, it follows: P(XY) = P(X). Thus, the expression for conditional probability can be simplified:

$$P(X|Y) = P(X)/P(Y)$$

In more detail, we can write:

$$P(Y) = \lim_{N\to\infty}\left(\sum_{i=1}^{N-m+1} n_i^m(r)/(N-m+1)\right),$$

$$P(X) = \lim_{N\to\infty}\left(\sum_{i=1}^{N-m} n_i^{m+1}(r)/(N-m)\right),$$

Each individual outcome of the event X|Y is assigned its own entropy value: H(X|Y) = −ln (P(X|Y)). As a result, the entropy can be written:

$$H(m, r) = -\ln(P(X)/P(Y)) =$$

$$= \lim_{N\to\infty}\left(-\ln\left(\frac{\left(\sum_{i=1}^{N-m} n_i^{m+1}(r)\right) * (N-m+1)/}{\left(\sum_{i=1}^{N-m+1} n_i^m(r) * (N-m)\right)}\right)\right).$$

Here: r is the pattern similarity parameter ($\rho(u_m(i), u_m(j)) \le r$), N is the length of the time series, $n_i^m(r)$ is the number of patterns of length m with the similarity coefficient r and the beginning at point i, $u_m(j)$ is a subsequence of length m that begins with the *j*th member of the sequence. The choice of the base of the logarithm in the ratio for entropy is not of fundamental importance, since the changes are reduced to multiplying the function by some constant coefficient $\log_a b$.

From the analysis of the above ratio, it is easy to conclude that as soon as, P(X) → P(Y), then immediately H (m, r) → 0. Thus, the entropy decreases and the probability of manipulation increases accordingly. On the other hand, individual entropy increases if the probability of the event (X|Y) is small and thus the probability of the presence of manipulation decreases. A similar estimate was previously proposed in [8] as a criterion for the analysis of time series in the field of physiology. It was defined by the authors as selective entropy.

The use of the above estimate is possible not only for the real initial values of the time series, but also for incremental ratios ($x_{i+1} - x_i$), which is quite often used in the analysis. You can also build additional series of the form: (($x_{i+1} - x_i$)/x_i) – a series of returns, or logarithmic ratios log(x_{i+1}/x_i), etc. In the analysis of a series for manipulation, this is certainly very important, since the general pattern usually does not depend on the value at the starting point and can be implemented for different initial price values, while maintaining a typical pattern. In addition, when analyzing sufficiently long time series, it can be expected that a number of absolute prices contain a certain trend in a given trading session, which may make it difficult to identify patterns.

The most expensive part of the computational algorithm is the procedure for searching (comparing) vector patterns. In fact, this is a comparison of each n-dimensional vector with each other n-dimensional vector for "proximity". The time of this procedure increases non-linearly with an increase in the number of members of the time series. Of course, direct calculation, although not algorithmically complex, is certainly computationally very expensive, since the number of patterns is small, and calculations must be done for each pair of vectors. Given this feature, it makes sense to use a special algorithm with higher computational efficiency and without leading to a significant loss of accuracy. We are talking about using algorithms of deterministic finite automata (DFA) [10].

Formally, a deterministic finite automaton is usually represented as a tuple of 6 elements: a finite set of input signals X (this set is usually called the input alphabet); a finite set of output signals Y (by analogy, the output alphabet); a finite set of internal states of the automaton Z; the initial state q_0. The tuple is closed by the transition function δ(.) and the output function γ(.). In the case under consideration, the input alphabet is formed by a pair: "coincidence of components" and "non-coincidence of components". The internal state is characterized by the parameters: "the first values matched", "the second values matched", ..., "the *n*th values" of the components matched. The detected patterns of size m should appear at the output of the finite automaton. Obviously, the search for a pattern of size m requires knowledge of the coincidence of (m − 1) components as a necessary condition. Thus, the finite automaton should have some memory (more than one state). The term "coincidence of components" has a formal content corresponding to the fulfillment of the inequality:

$$x_i - y_i|/|x_i| \leq \varepsilon, \forall i = 1,\ m.$$

It is easy to show that component by component proximity of two vectors guarantees their proximity in the Euclidean metric. The converse is also true. A noticeable analogy with the k-nearest neighbors (KNN) method is certainly present in this algorithm [11]. Of course, this is not only an analogy, but also some ideological intersection. In both cases, "similar" vectors of different dimensions are formally sought, which form manipulation patterns (as in the present approach), or are compactly located in a multidimensional space, that is, form the so-called "set of nearest neighbors". This remark should not be interpreted as a statement about the identity of the approaches. Most likely, we can talk about their consistency.

3 Test Verification of the Software Product

The main problem of the algorithm being created is the procedure for searching for patterns of length m in a time series. The test algorithm uses a generated time series of prices of exchange purchase and sale transactions, which is exclusively probabilistic in nature. The parameters of the series are determined by the values of the mathematical expectation M_x and the value of the variance σ_x.

To perform calculations of the entropy parameters H (r, m), a time series of 45,000 values was generated. It was supposed to reflect the results of trading for 90 days with 500 transactions daily. The parameters of the random process were determined as follows: the mathematical expectation was assumed to be 5, and the variance parameter was chosen to be 2. It is very important to note that the resulting time series is devoid of any manipulativeness and reflects an exclusively random process, without any premeditation. In the present study, the similarity parameters of the pattern patterns had the following range of values: 0.1, 0.15, 0.20. The patterns themselves with sizes (length, number of components) from 2 to 4 in an amount from 6 to 9 randomly entered into trading transactions on the 42nd, 48th and 71st days. The authors' results of the calculations are shown in the graphs below (Figs. 1, 2 and 3).

Attention is drawn to a fairly large number of matching pairs of transactions (patterns of two values), m = 1, m = 2. In all likelihood, this reflects the innate entropy of the time series and the random nature of its values. The entropy values for manipulative transactions on the specified days (t = 42, 48, 71) have clearly expressed minimum values for similarity parameters $\varepsilon = 0.1$ and $\varepsilon = 0.15$ and for

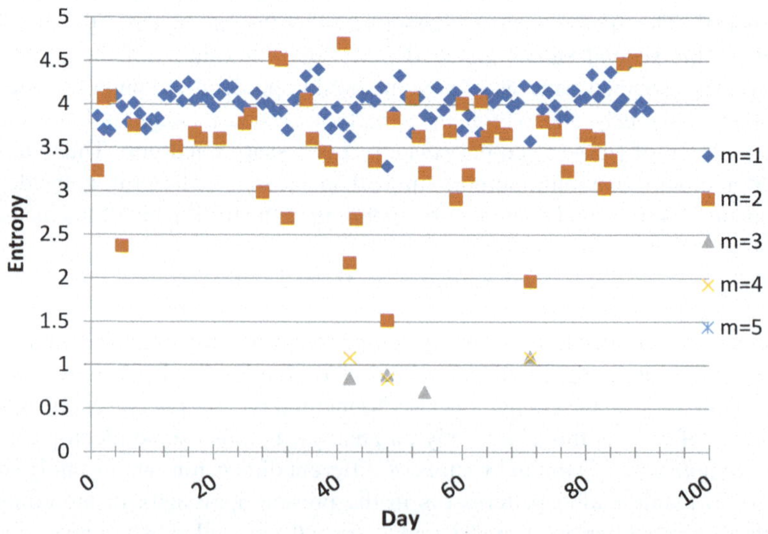

Fig. 1 The dependence of entropy on the day (values of H (m, r) for m = 1, 2, 3, 4, 5; $\varepsilon = 0.1$)

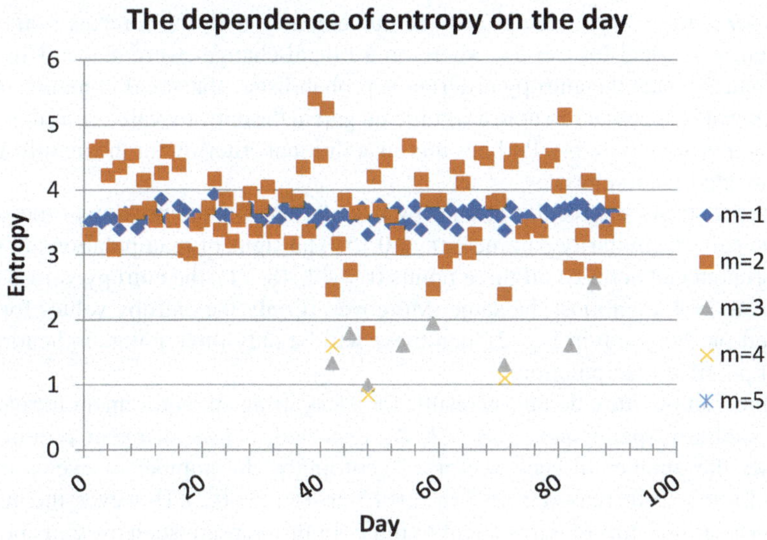

Fig. 2 Values of H (m, r) for m = 1, 2, 3, 4, 5; ε = 0.15

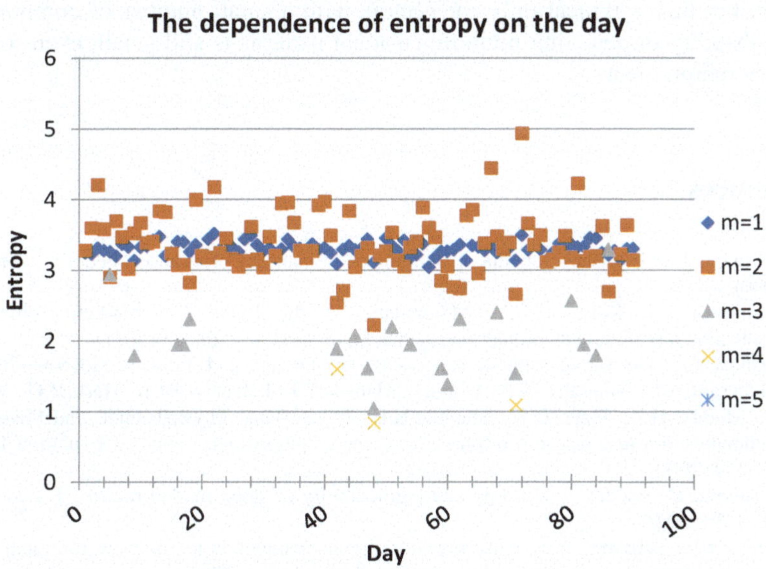

Fig. 3 Values of H (m, r) for m = 1, 2, 3, 4, 5; ε = 0.2

pattern sizes m = 2, m = 3, m = 4. Simultaneously with the forecast points, the minimum is marked for t = 54, where no artificial changes were made. This once again indicates that the entropy criterion is probabilistic, statistical in nature. At the same time, it is necessary to note a significant gap in the entropy values for m = 2 and m = 3 at this point (t = 54). Perhaps it is an additional criterion for the manifestation of unintended randomness.

Figure 2 shows the results of entropy calculations for the previous time series, with the pattern similarity parameter $\varepsilon = 0.15$. The signs of manipulation are much more pronounced here. At all three points (t = 42, 48, 71), the entropy values for m = 3 and m = 4 are almost the same. Moreover, if only the entropy values for m = 4 are left on the graph in Fig. 2, then these will be only three points indicating the days of possible manipulation.

Figure 3 shows the calculation results for a less stringent requirement for patterns in the similarity parameter, $\varepsilon = 0.2$. As expected, a less stringent requirement increases the number of such vectors. Accordingly, the number of events of this quality for m = 3 increased from 7 ($\varepsilon = 0.15$) to 18 ($\varepsilon = 0.2$). However, the number of patterns m = 4 still remains equal to three. In this variant, such a sign is perhaps the only sure indicator of the presence of signs of manipulation. Thus, reducing the accuracy requirements for the similarity parameters of patterns increases their number, but this is typical only for vectors with a small number of components. The probability of detecting multidimensional patterns is still small, even for low accuracy requirements.

References

1. Wolfela, C.J. (eds.): Encyclopedia of Banking and Finance. Corporation "Fedorov", Moscow (2000)
2. Khisamova, Z.I., Komova, E.Y.: Manipulation of the digital asset market: problems of establishing a criminal law protection regime. State Service. **2**, 68–73 (2021)
3. Galanov, V.A.: Derivative of the derivatives market. Finance and statistics, Moscow (2008)
4. Goldberger, A.L., Amaral, L.A.N., Glass, L., Hausdoff, J.M., Ivanov, P.Ch., Mark, R.G., Mietus, J.E., Moody, G.B., Peng, C.K., Stanley, H.E.: PhysioBank, PhysioToolkit, and PhysioNet: components of a new research resource for complex physiologic signals. Circulation **10**(23), 215–220 (2000)
5. Maasoumi, E., Racine, J.: Entropy and predictability of stock market returns. J. Econ. **107**, 291–312 (2002)
6. Reddy, Y.V., Sebastin, A.: Measuring stock price manipulation using entropy analysis: a conceptual framework. ICFAI J. Appl. Financ. **12**(5), 48–61 (2006)
7. Electronic resource. http://profbeckman.narod.ru/InformLekc.htm. Accessed 23 June 2024
8. Richman, J.S., Murman, J.R.: Analysis of physiological time series using approximate entropy and selective entropy. Am. J. Physiol. Physiol. Heart Blood Circul. **278**(6), 2039–2049 (2000)
9. Chen, J.: Generalized: entropy theory of information and market patterns. Corp. Financ. Rev. **9**(3), 21–32 (2004)
10. Hopcroft, J., Motwani, R., Ulman, J.: Introduction to the theory of automata, languages and calculations. Publishing house "Dialectics", Moscow (2019)
11. Hastie, T., Tibshirani, R., Friedman, J.: The elements of statistical learning, 2nd edn. Springer, New York (2017)

Effect of Gamma Irradiation on Mechanical Properties of Metals and Alloys: Mechanisms and Applications

Vladimir Oniskiv, Valerii Stolbov, Vladimir Makarov, and Mikhail Pesin

Abstract The mechanisms of gamma irradiation effect on the structure and properties of metals and alloys are considered. The main mechanisms of irradiation effect that affect the wear resistance of the material are identified. The results of testing samples for electrical resistance, hardness and wear resistance before and after radiation modification are presented. The possibility of increasing the wear resistance of tools made of hard alloy materials using gamma irradiation during turning by cutting is shown.

Keywords Gamma irradiation · Metals and alloys · Mechanisms of action · Physical and mechanical properties · Wear of cutters · Turning

1 Introduction

Gamma radiation is one of the types of electromagnetic radiation with the highest energy and the shortest wavelength in the electromagnetic spectrum [1]. It occurs during the discharge of nuclear energy levels, namely, during the transition of atomic nuclei from a state with higher energy to a state with lower energy. The energy range of the emitted quanta is quite wide: from 1 keV to 10 MeV, and the corresponding wavelengths vary from 10^{-9} to 10^{-13} m. The electric charge and rest mass of gamma quanta are zero, they do not interact with electric and magnetic fields. This explains the outstanding penetrating power of gamma radiation (up to 2.5 km in air). Its effective interaction cross-section is much smaller than the cross-section of charged pchapters. Of course, this type of irradiation has no analogues for the process of modifying thick-walled products or metals and alloys.

Radiation effects on various materials are the subject of numerous current studies [2–6]. The main objective of such experiments is related to radiation modification of

V. Oniskiv · V. Stolbov (✉) · V. Makarov · M. Pesin
Perm National Research Polytechnic University, Perm, Russia
e-mail: valeriy.stolbov@gmail.com

T. Antipova (ed.), *Digital Technology Platforms and Deployment*, Information Systems Engineering and Management 36, https://doi.org/10.1007/978-3-031-86547-3_22

the properties of the irradiated material. This is a very promising and dynamically developing industry with annual growth rates of 7–10%. At the same time, there are still many questions related to the explanation of the mechanisms of gamma-quanta effects on the microstructure and physical and mechanical properties of metals and alloys. Of interest are also questions about possible applications of the gamma-irradiation procedure to the management of the operational characteristics of various metal products. In this regard, the topic of the chapter is certainly important and relevant.

2 Mechanisms of Gamma Quanta Impact on Material

It is common to distinguish the following main mechanisms of gamma quanta interaction with material: the photoelectric effect, the Compton effect, the creation of electron–positron pairs, and the nuclear photoelectric effect.

The photoelectric effect dominates over other types of gamma-quantum absorption when the photon energy is no more than 500 keV. In this case, it is realized only on the bound electrons of the inner shells of atoms: M, L, K, … At gamma-quantum energy values greater than the noted level, the probability of the photoelectric effect reaction is low. The result of the photoelectric effect is positive ionization of the atom. In this case, a free electron knocked out of its orbit can create new electron–hole pairs [7].

The Compton effect, in turn, reflects the interaction of incident quanta with the electrons of the outer electron shell of the atom [7]. The binding energy of these electrons with the atom is minimal (a "weakly bound" electron) and can often be neglected. In this case, the electron also leaves the atom and an electron–hole pair is formed. In addition, the Compton photon (scattered quantum) also leaves the atom, of course, with an energy lower than the energy of the incident photon. In this case, the difference in the energies of the primary and scattered photons is equal to the kinetic energy of the electron, which Compton called the "recoil electron". It is these electrons with an energy of less than 100 eV that transmit the bulk of the radiation energy (up to 90%). The probability of the Compton effect being realized has a maximum for quanta with an energy of about 1 MeV. Note that in light substances for quanta with an energy of 150 keV–5 MeV, in most cases the Compton scattering mechanism is realized. Of course, the change in the energy of gamma quanta in the noted process is random in nature, but in any case, this leads to a weakening of their flow by any given number of times.

At a threshold value of gamma quantum energy of more than 1.02 MeV and its movement near the nucleus of an atom, a reaction can occur where the photon is converted into an electron–positron pair. This phenomenon is usually called the effect of electron–positron pair formation [7].

If the energy exceeds 10 MeV, the gamma quantum is capable of knocking nucleons out of the nucleus of an atom. This interaction mechanism is called the nuclear photoelectric effect.

It is important to add that the processes of interaction of gamma quanta with electron shells are predominant, since there are Z (charge number) times more electrons than nuclei.

It should be noted that the average energy of a gamma quantum is considered to be 1.25 MeV, and the average energy transferred during irradiation of the medium per unit length of the pchapter track is 0.2 eV/nm. If we take into account that the average energy of interatomic bonds in a substance is 2–5 eV, then it is reasonable to assume both the possibility of breaking existing bonds and the probability of the emergence of new bonds during the process of radiation exposure to gamma irradiation. In fact, the combination of these two processes determines the possible modification of the properties of the material.

It should be noted that the main industrial sources of gamma radiation are ^{60}CO and ^{137}Cs. The half-life of Cobalt is 5.25 years, while that of Cesium is significantly longer—30.17 years. The nucleus of radioactive ^{60}CO contains 27 protons and 33 neutrons and is a decay product of ^{60}Fe. In turn, the decay product of ^{60}CO is stable ^{60}Ni.

The interaction of gamma radiation with metals and alloys includes two competing processes: on the one hand, there is an accumulation of radiation defects, on the other hand, their annihilation and ordering of the crystal structure takes place. Let us consider some features of the noted processes in more detail.

First of all, it should be noted that during the irradiation process, gamma quanta initially excite the electron system of the atom and only after that various mechanisms of their displacement begin to turn on and work [1]. Excitation of the electron shell occurs several feto-fractions of a second after the start of irradiation, while the duration of existence of the resulting radicals or defects can be calculated in months. If we consider that a steel sheet 28 mm thick is capable of weakening the photon flow by about half, then the conclusion necessarily follows that half of the flow energy remains in the material or metal in one form or another. In other words, the processes of relaxation of the internal state occur for quite a long time.

As noted above, the photon flux generates electron activity, and the energy of the electrons incident on the atom is often quite sufficient to displace it. This is probably the most obvious mechanism of interaction between an electron and an atom. It leads to the rupture of interatomic bonds and displacement of atoms, as a result of which a pair is formed: a vacancy and an interstitial atom. Such a pair of defects is usually called a "Frenkel pair" [1, 2]. The displacement of an ionized atom with the formation of a pair is also possible as a result of electrical repulsion from a similarly charged and closely located impurity atom or a previously ionized neighboring atom-ion [1]. Point defects formed by radiation exposure are no different from similar defects that already existed in the material. They also interact with other imperfections in the metal structure, in particular with edge dislocations. Note that during irradiation, both vacancies and interstitial atoms are generated in equal quantities. However, in general, the concentration of interstitial atoms is significantly lower compared to the concentration of vacancies, since the energy required for the formation of an interstitial atom is significantly greater than the energy of a vacancy.

Of course, displacement of atoms is also possible in photonuclear reactions. In this case, the energy of the primary displaced atom can reach 10–80 eV, and it is sufficient to form a cascade of collisions along the trajectory of motion and the formation of so-called “radiation clusters” [8, 9]. The size of the cluster is 10^{-6}–10^{-5} cm, and the concentration of Frenkel pairs in the cluster can reach values of 10^{21}–10^{22} cm^{-3}. The formation of a new nanophase during irradiation is usually called radiation-induced phase transformation.

As mentioned above, vacancies and interstitial atoms that arise during radiation exposure can mutually annihilate. Point defects in the form of vacancies will move to the compression region, and interstitial atoms—to the tension region. Thus, the nanostructure is ordered, and the internal stresses in the crystal are reduced [2, 10, 11]. A characteristic feature of the ordering of the metal structure is a decrease in its specific electrical resistance [9]. In addition, nonequilibrium vacancies can diffusely migrate to sinks, practically without annihilation with interstitial atoms, the concentration of which is noticeably lower. As a rule, grain boundaries are sinks for almost all defects and zones of maximum concentration of stress fields. Note that one of the causes of the initial interphase micro-stresses is the difference in the temperature expansion coefficients of the binder and filler material.

In addition to the drop in electrical resistance, when irradiated with relatively small doses, the energy stored in the crystal is released [12]. Thus, in addition to the natural radiation heating that accompanies the irradiation process, the reduction in the number of imperfections, annihilation and reconstruction of defects leads to additional heat generation processes. For example, a Fresnel pair releases energy at a level of 3–10 eV during its annihilation. Moreover, it is important to note that additional heat generation occurs only in the first irradiation cycle and is not observed in any subsequent ones. Experimental studies have shown that the difference in sample temperatures between the first and subsequent irradiations under identical conditions is in the range of 2–4 degrees °C [10, 12].

The above-mentioned mechanisms of imperfection drainage, together with annihilation processes, lead to a decrease in the number of defects and a restructuring of the crystalline bodies into a more harmonious, equilibrium state, significantly increasing intergranular adhesion. The latter circumstance plays a dominant role, for example, in the processes of reducing the wear of cutting tools. On the other hand, with an increase in the concentration of single defects, as a rule, the metal is strengthened. Hardness parameters can increase five times or more, but at the same time, plasticity is noticeably reduced. In some cases, when significant doses of radiation are reached, intergranular cracking occurs in alloys and embrittlement of the material occurs [2]. Long periods of constant radiation exposure, lead to the manifestation of the so-called “radiation swelling”, which can significantly change the dimensions of products, cause their warping and cracking. In particular, this is typical for austenitic steels and is associated with the accumulation of radiogenic gases, for example, helium. As a rule, it accumulates in microbubbles. Its formation is the result of not only neutron irradiation, but is also quite acceptable with gamma irradiation. The carbon present in the alloy has an alpha cluster nucleus. Intense gamma irradiation can lead to the decay of $^{12}C_6$ with the formation of 3 alpha pchapters, which, having captured electrons, actually turn into a helium atom [13]. Note that the binding energy of alpha pchapters in the carbon nucleus does not exceed 2 MeV. Simultaneously with the accumulation of helium, there is also an accumulation of hydrogen. This is especially characteristic of welded seam areas. In a number of technologies, in order to sensitize the nuclear system of the metal, hydrogen atoms are deliberately introduced into the metal environment using electrolysis for subsequent gamma irradiation [13, 14]. In this case, the excited hydrogen atmosphere actively interacts with the nuclear system of the solid, chain processes of defect annihilation occur. This leads the internal structure to self-organization and a more balanced state.

It should be added that irradiation also changes the physical and chemical properties of the surface of solids [11]. First of all, this refers to properties such as adsorption, oxidation and catalytic reactions, and corrosion resistance.

In this regard, the question of the environment in which the radiation exposure process occurs seems to be very important [8, 15]. It is obvious that ions and radicals arise both on the surface of metals and in the environment surrounding the irradiated material. Moreover, the result of the interaction of atoms of two contacting environments certainly depends on the parameters of the external environment (pressure, temperature, etc.) [16].

Taking into account all of the above, it must be recognized that the complete physical picture of the phenomena observed during radiation modification is still far from being fully understood and requires additional research.

3 Results and Discussion

The effect of gamma-ray irradiation on the properties of replaceable multi-faceted plates CNMG 120,408-46 manufactured by Sandvik MKTS (Russia) made of hard alloy type WC8, intended for turning by cutting, was investigated (Fig. 1).

For the purposes of the study, they were subjected to radiation modification by irradiation using ^{60}CO sources with the following doses: 5, 10 and 15 Mrad. For samples with a radiation dose of 10 Mrad, the exposure was carried out in an open atmosphere and in a nitrogen atmosphere.

First of all, the electrical resistance of the cutting plates was directly measured before and after irradiation. The measurement accuracy was 0.01 Om. The data obtained in the experiment are given in Table 1.

The obtained results correspond to the results presented in the work [12]. They confirm the drop in electrical resistance in the range of radiation doses of 0–5 Mrad, in accordance with the ordering of the structure and the reduction in the number of defects in the material. Further radiation exposure leads to an increase in resistance and, accordingly, to an increase in the number of imperfections.

The material hardness index is a very important parameter for assessing the wear resistance of cutters. In this regard, the hardness of tools above the specified grade was investigated. In accordance with GOST 9013-59, the hardness of the studied alloy

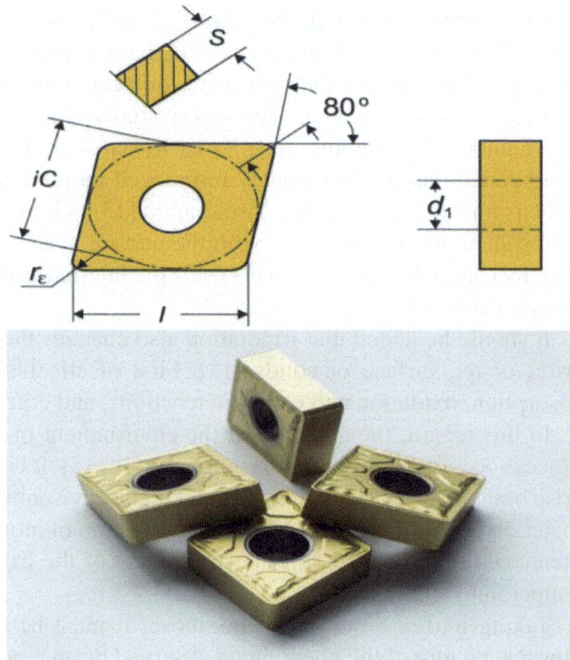

Fig. 1 Carbide plate CNMG 120,408-HS LF6081. *Source* Authors' elaboration

Table 1 Effect of dose on electrical resistance

Radiation doses, Mrad	0.0	5.0	10.0 (air)	10.0 (nitrogen)	15.0
Electrical resistance, Om	0.2	0.10	0.11	0.12	0.15

WC8 was determined according to Rockwell, scale—HRC. The averaged results are given in Table 2.

It is easy to notice that in the studied range of radiation doses, there is no significant pattern that would allow us to draw a conclusion about the correlation of the parameters under consideration. Most likely, we are talking about the statistical scatter of experimental data relative to the average value of 41.34 HRC. In general, these results correlate with the data of work [17].

Next, we will consider the effect of radiation modification on the operational characteristics of cutter wear resistance. Figure 2 shows the results of tests of modified cutters for wear during cutting. It shows the dependence of cutter edge wear on

Table 2 The material hardness index

Radiation doses, Mrad	0.0	5.0	10.0 (air)	10.0 (nitrogen)	15.0
Hardness, HRC	41.7	41.9	40.3	41.3	41.5

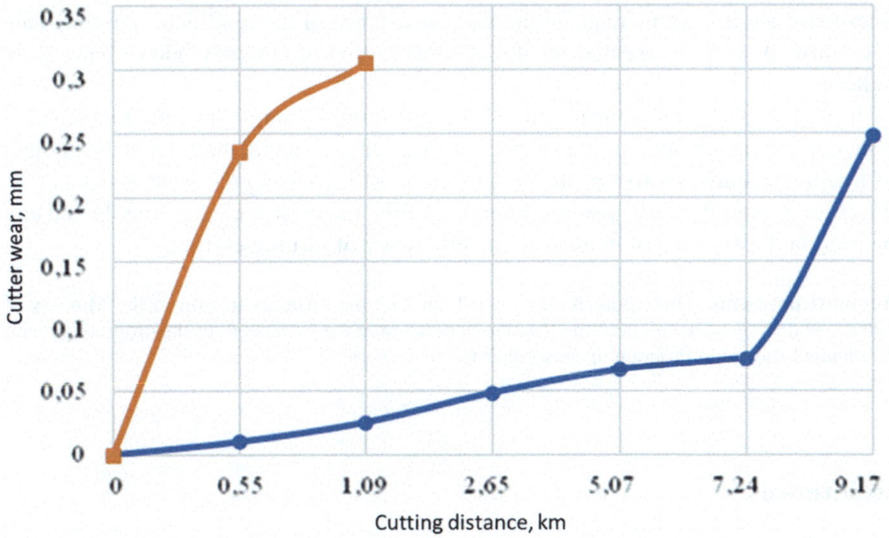

Fig. 2 Wear of the cutter depending on the cutting path (Red color—before radiation modification, blue color—after)

the cutting path. It is evident from the figure that the modified cutter is subject to significantly less wear compared to the original.

At the same time, as tests showed (Table 2), the hardness of the cutter material remained practically unchanged. Thus, the increase in the wear resistance of the cutter after radiation modification is associated not with a change in the hardness of the material, but with other mechanisms of the effect of gamma quanta on the material at the structural level.

In accordance with the above descriptions of the main mechanisms of gamma irradiation impact on the material, the authors believe that the increase in wear resistance is caused by the mechanism of imperfection drain in combination with their annihilation processes. This leads to a decrease in the number of defects and the reorganization of the structure of crystalline bodies into a new ordered state, significantly increasing intergranular adhesion and reducing the level of accumulated micro-stresses. This statement is indirectly confirmed by the results of testing samples for electrical resistance before and after modification.

4 Conclusion

Possible mechanisms of gamma irradiation impact on metals and alloys are considered. It is shown that among a large variety of mechanisms, the most interesting from the point of view of applications are the mechanisms of intergranular adhesion in the crystalline structure of the material, leading to a decrease in the level of accumulated micro-stresses and an increase in the wear resistance of the modified material. This is confirmed by the presented results of wear testing of cutters made of hard-alloy material.

It should be noted that the required dose of gamma radiation, leading to a positive effect of increasing operational characteristics, for each metal material is the subject of a separate study. However, the general trend of increasing the wear resistance of the cutting tool after radiation modification of the material is clearly visible and can be used in the practice of managing the efficiency of turning cutting.

Acknowledgements This research was carried out with the financial support of the Ministry of Science and Higher Education of the Russian Federation in the framework of the program Federal Academic Leadership Program «Priority-2030».

References

1. Electronic resource. http://profbekman.narod.ru
2. Vas, G. S.: Fundamentals of Radiation Materials Science. Metals and Alloys. World of Materials and Technologies, 991p. M: Technosfera (2014)
3. Obodovsky, I.M.: Radiation technologies. Application in laboratory research, materials science and nanotechnology, 296p. ID Intellect (2015)
4. Maruuchi, K., Cheng, S.: Radiation Processing of Polymer Materials and its Industrial Applications, 401p. Jong Wiley and Sons Inc., Hoboken, New Jersey (2012)
5. Woods, R., Pikaev, A.: Applied Radiation Chemistry: Radiation Processing, 346p. Jong Wiley and Sons Inc., New York (1994)
6. Naikwadi, A.T., Sharma, B.K., Bhatt, K.D., Mahanwar, P.A.: Gamma radiation processed polymeric materials for high performance applications: a review. Front. Chem. **10**, 1–15p. (2022)
7. Feynman, R., Leighton, R., Sands, M.: Feynman lectures on physics. Quantum Mech. **9**(2), 256p. MIR (1965)
8. Timoshenko, Yu.A., Klopotov, A.A., Ivanov, Yu.F.: Changes in the structural and phase state of the WC8 alloy under the influence of a gamma-quantum flux. News Univ. Ferrous Metall. **4**, 40–43p. (2001)
9. Chernov, I.P., Mamontov, A.P., Botaki, A.A.: Anomaly effects of low doses of ionizing radiation on metals and alloys. At. Energy. **57**(1), 56–58p. (1984)
10. Chernov, I.P., Timoshenkov, Yu.A., Mamontov, A.P., Korotchenko, V.A., Lapsker, I.A., Semukhin, B.S.: Changes in the structure of the WC alloy when exposed to low doses of gamma irradiation. At. Energy. **57**(1), 58–69p. (1984)
11. Oskolkova, T.N., Glezer, A.M.: Current state of the scientific problem of surface hardening of tungsten carbide hard alloys (review). Bull. Univ. Ferrous Metall. **60**(12), 980–991p. (2017)
12. Chernov, I.P., Mamontov, A.P.: Ordering of the crystal structure by ionizing radiation (the effect of low doses of ionizing radiation). Bull. Tomsk Polytech. Univ. **303**(1), 74–80p. (2000)
13. Oniskiv, V., Stolbov, V., Ibragimova, E.: Effect of gamma irradiation on morphology and local element composition of basalt-based composite material. J. Digit. Sci. **5**(2), 8–18 (2023). https://doi.org/10.33847/2686-8296.5.2_2
14. Mamontov, A.P., Ryabchikov, S.Ya., Chernov, I.P.: Effect of hydrogen on the ordering of the structure of inorganic materials under irradiation // Coll. materials. In: 10th International Conference on Radiation Physics and Chemistry of Inorganic Materials (RFH-10). - Tomsk, 329–331p. (1999)

15. Vorobyov, V.L., Bykov, P.V., Gilmutdinov, F.Z., Bayakin, V.Ya.: Effect of N+ ion implantation on the surface morphology, chemical composition and atomic structure of the surface layers of stainless steel and titanium alloy. In: 14th International Ural Seminar Radiation Physics of Metals and Alloys. Abstracts of reports. 25–29 April 2022, 61–62p. (2022)
16. Didyk, A.Yu., Vishnevsky, R., Wilczynska-Kitowska, T., Myshinsky, G.V., Semin, V.A.: Synthesis of chemical elements under palladium gamma-ray irradiation in a condensed gas environment. In: RENSIT **11**(2), 143–157p. (2019)
17. Ermolaev, V.A., Mamontov, A.P., Ryabchikov, S.Ya.: Features of the structure and texture of hard alloys before and after gamma irradiation. Dep. – M.: Central Research Institute of Economic and Informational Non-Ferrous Metallurgy, - No. 1771–84. BU VINITI "Deposited Scientific Works" **10**(156), 28p. (1984)

Optical Wireless Battery Management System for Enhanced Performance and Safety

Simona Riurean and Cosmin Rus

Abstract The Battery Management System (BMS) is critical for the efficient, safe, and extended operation of rechargeable batteries, especially in electric vehicles (EVs), renewable energy systems, and advanced electronics. The BMS system monitors key parameters such as current, voltage, temperature, and State of Charge (SoC), ensuring that batteries operate within optimal and safe limits. It provides critical protection against overcharging, over discharging, and overheating, thereby safeguarding battery integrity and extending lifespan. Emerging trends in BMS technology emphasize the integration of advanced algorithms, IoT connectivity, enhanced safety features, and the shift towards wireless BMS (wBMS). Wireless BMS reduces wiring complexity, enhances reliability, and offers greater flexibility in battery packaging. Optical Wireless BMS (OW-BMS), on the other hand, is a light-based communication, an advanced technology that offers immunity to electromagnetic interference (EMI), high data rates, and improved security. OW-BMS significantly reduces system weight and complexity while maintaining robust and secure data transmission, and real-time monitoring capabilities. As research progresses, OW-BMS is envisioned to revolutionize battery management, offering a glimpse into the future of more efficient, secure, and scalable energy storage solutions.

Keywords Battery management system · Electric Vehicles (EVs) · Wireless Battery Management System (wBMS) · Optical wBMS (OW-BMS) · State of Charge (SoC) · Electromagnetic interference

1 Introduction

A BMS is a critical component in any application that relies on rechargeable batteries, especially those incorporating multiple cells, such as electric vehicles (EVs), renewable power stations, uninterruptible power supplies (UPS), and other advanced electronical systems [1]. As an intricate electronic system, a BMS is designed to monitor

S. Riurean (✉) · C. Rus
University of Petroşani, Petroşani, Romania
e-mail: sriurean@yahoo.com

T. Antipova (ed.), *Digital Technology Platforms and Deployment*, Information Systems Engineering and Management 36, https://doi.org/10.1007/978-3-031-86547-3_23

and manage the performance, safety, and longevity of rechargeable batteries. It ensures that batteries operate within their designated safe operating conditions, optimizing their performance to suit the specific requirements of each application. Consequently, the BMS is indispensable in battery-powered applications, providing essential protection, reliability, and durability to the batteries. The BMS plays a vital role in the operational maintenance and safety of a battery pack.

The system continuously monitors parameters (voltage, current, temperature, and State of Charge (SoC)) to ensure that the battery pack operates at its optimal level [2]. This cautious monitoring helps maintain the "health" of the battery pack throughout its lifetime, preventing issues that could lead to reduced performance or potential failures. The integration of a BMS not only enhances the efficiency and safety of battery packs but also extends their lifespan, making it a cornerstone technology in modern energy storage solutions [3].

Emerging trends in BMS are advanced algorithms and wBMS. Advanced algorithms aim to improve SoC and State of Health (SoH) estimation [4] using machine learning and AI to enhance accuracy, integration with IoT remote monitoring and management, improved safety features for enhanced protection mechanisms to prevent thermal runaway and other failure modes. The wBMS goal is to replace wires between the BMS and battery cells with wireless communication.

This chapter presents the components, key characteristics, and applications of the BMSs, wireless technologies applied in wBMS and a number of case studies of industrial applications. It also explores the possibility of using optical wireless communication novel technologies for an efficient BMS monitoring system in industrial areas with hazard environments as mining industry is.

2 BMS Components, Functions, and Applications

2.1 Components of a BMS

The components of a BMS work together to monitor, protect, and manage the performance of rechargeable battery packs.

By integrating these components, a BMS ensures that batteries operate safely, efficiently, and with optimal longevity, which is essential for applications ranging from EVs to renewable energy systems [5].

A BMS is composed by several components, each responsible for different functions that contribute to the overall management and protection of the battery pack. These components are the Battery Monitoring Unit (BMU), Control Unit (CU), balancing circuitry, protection circuitry, communication interface, power management, User Interface (UI) and display, Thermal Management System (TMS), housing and connectors, safety and compliance features, integration and configuration tools.

The BMU consists of voltage sensors that measure the voltage of each cell or group of cells to ensure they are within safe operating limits, temperature sensors that monitor the temperature of cells or modules to prevent overheating and thermal runaway and current sensors that measure and detect overcurrent conditions and manage charging/discharging [6].

CU, represented by microcontroller/microprocessor which processes data from the sensors, executes algorithms for battery management, and communicates with external systems, and by the firmware/software which is the embedded code that runs on the microcontroller, implementing battery management algorithms, state estimation, and communication protocols.

The balancing circuitry consists of the passive balancers that use resistors to dissipate excess energy from higher voltage cells, equalizing the charge across cells, and active balancers use inductors, capacitors, or transformers to transfer energy from higher voltage cells to lower voltage ones more efficiently than passive balancers.

The protection circuitry consists of the overvoltage protection that disconnects or stops charging if cell voltages exceed safe limits, the undervoltage protection that disconnects or stops discharging if cell voltages drop below safe limits, the overcurrent protection that detects and interrupts excessive current flow to prevent damage or safety hazards, the short circuit protection that is important to detect short circuits and disconnect the battery pack to prevent fires or explosions, and the temperature protection which shuts down or limits operations if temperatures exceed safe thresholds.

Communication interface consists of the communication bus (e.g., Controller Area Network—CAN, System Management Bus—SMBus, MODBUS) [7] enables data exchange between the BMS and other systems like vehicle CUs, power management systems, or user interfaces [2], and data logger records operational data for diagnostics, performance analysis, and historical tracking.

The power management contains the DC-DC converters regulate power for the BMS itself and possibly other auxiliary systems, and power supply ensures that the BMS has a stable and reliable power source to operate effectively.

UI and display (optional) provide real-time information on battery status, such as SoC, SoH, and any fault conditions and control buttons/touch interface allow users to interact with the BMS, adjust settings, or retrieve data.

TMS can be optional and has a cooling fan or liquid cooling system to manage the temperature of the battery pack, ensure that cells remain within optimal operating temperatures, and heating elements to prevent cells from getting too cold, which can affect performance and longevity.

Housing protects the BMS components from environmental factors such as dust, moisture, and physical damage and connectors and wiring provide secure and reliable connections between the BMS, battery cells, and external systems.

Safety is provided by fuses and circuit breakers to protect against electrical faults and compliance features and the isolation monitoring ensures electrical isolation between the battery pack and the vehicle chassis or other systems, preventing short circuits or electrical hazards.

Integration and configuration tools consist of configuration interface often via software tools, allows for setting parameters specific to the battery pack and application and diagnostics and maintenance tools provide the ability to diagnose issues, update firmware, and perform maintenance tasks [1].

2.2 Components of a BMS

The BMS electronic system manages a rechargeable battery (cell or battery pack), such as by monitoring the battery's state, calculating secondary data, reporting that data, protecting the battery, controlling its environment, balancing the battery and authenticating it. The key functions of a BMS (Fig. 1) are monitoring, protection, balancing, state calculation, and communication [8].

The BMS continuously monitors the battery's voltage, current, temperature, and other parameters to ensure safe operation. Therefore, BMS measures the voltage of individual cells and the overall pack, tracks the charge and discharge current to ensure it stays within safe limits, and monitors the temperature of the cells to prevent overheating [9].

The protection role is to safeguard the battery by preventing overcharging, over discharging, and excessive temperature. If any limits are exceeded, the BMS takes corrective actions. Therefore, protection is related to the overcharge protection over-discharge protection overcurrent protection thermal protection. Overcharge protection prevents cells from being charged beyond their maximum voltage, over-discharge protection stops cells from being discharged below their minimum voltage, overcurrent protection limits the current to prevent damage during charge and discharge and thermal protection shuts down or reduces power to prevent overheating.

In multi-cell battery packs, the BMS balances individual cell voltages to maintain uniform charge levels and extend battery life, therefore, balancing maximizes the pack's performance and lifespan.

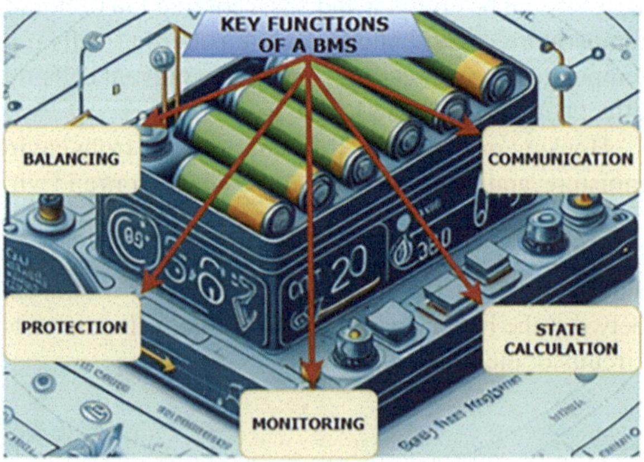

Fig. 1 The key functions of a BMS. *Source* Authors' elaboration

The state calculation applies to SoC that estimates the remaining charge in the battery, SoH that assesses the overall health and capacity of the battery, and state of power (SoP) that evaluates the battery's ability to deliver power, based on measurements and algorithms [10].

The last key function of a BMS, but not the least important, communication, is relevant for data reporting that provides real-time data to external devices or systems (e.g., electric vehicle CUs), and User Interface that displays battery status to the user, often via LEDs, screens, or connected devices, thus BMS communicates with external devices (such as inverters or chargers) in order to provide data for monitoring, diagnostics, and control.

2.3 Applications of BMSs

BMSs are critical for managing the large battery packs used in electrical vehicles (EVs), ensuring safety and extending battery life. With the growing importance of safety in EVs, compliance with the ISO 26262 functional safety standard has become a critical requirement for passenger cars, road vehicles, and other EVs. Given the BMS's pivotal role in ensuring the safety of these applications, certification at the ASIL C level or higher is emerging as a market necessity for BMS systems. This stringent safety standard underscores the critical role of the BMS in EV safety, and its compliance is essential to ensure reliability and minimize risks associated with malfunctioning or compromised batteries. ISO 26262 is a comprehensive standard that governs the functional safety of electrical/electronic systems in cars, covering their entire life cycle [11].

These standard demands that all requirements be proven through proper documentation, review, verification, and validation. It also requires that potential failures be analyzed, risks specified, and quantified. The standard divides failures into four categories based on the potential impact of errors/hazards (i) ASIL A: Minor injuries, (ii) ASIL B: Serious injuries, (iii) ASIL C: Life-threatening injuries, (iV) ASIL D: Fatal injuries [10].

Renewable energy storage used in solar and wind energy systems to manage energy storage in batteries, consumer electronics found in smartphones, laptops, and other portable devices to manage their rechargeable batteries, uninterruptible power supplies (UPS) used to ensure reliable backup power by managing the batteries in UPS systems and various industrial applications with battery-powered systems for monitoring and protection are few examples of MBSs applications [11].

3 Wireless Technologies for BMS (wBMS) and Case Studies

The technology surrounding BMS systems has undergone a steady evolution, with a particularly noteworthy shift being the shift from traditional wired BMS to wireless BMS (wBMS) over the past five years [12]. In a traditional wired BMS, data from each cell in the battery pack is transmitted via cable to a monitor, and subsequently, the monitor data is relayed to the microcontroller via wired communication paths [4]. Common communication protocols used in wired BMS systems include Controller Area Network (CAN), Inter-Integrated Circuit (I2C), and Serial Peripheral Interface (SPI).

As battery packs grow in capacity and voltage, the number of cells also increases, leading to a corresponding rise in the number of wires and connectors required in wired BMS systems. Although such systems are effective, they introduce added complexity, mass, and price, and reduce flexibility, particularly in scenarios where space optimization and scalability are critical. According to [6], connectors account for a significant portion (8%) of the electrical component cost in an EV propulsion system [2]. Moreover, wired systems are more prone to physical failures such as wire breaks and poor connector contact, especially in environments where vibration is prevalent. The implementation of wBMS has numerous advantages over wired BMS systems, offering greater flexibility in battery packaging, simplifying wiring, reducing package weight and volume, and lowering service and maintenance costs [4].

Its simplified connections and absence of communication wiring harnesses also facilitate quicker removal of battery modules for repurposing and recycling. Moreover, wBMS integration allows for real-time monitoring and management of battery packs across various devices and platforms, contributing to efficient operation, predictive maintenance, and seamless integration with IoT-based applications [4].

Bluetooth Low Energy (BLE) is designed for low-power, short-range connectivity, operating in the 2.4 GHz ISM band. It is widely used in consumer electronics; however, it has limited range and potential interference with other 2.4 GHz devices. In BMS applications, BLE can transmit battery status, cell voltages, temperatures, and other critical data to a central CU or a mobile device [2].

Zigbee is a specification for a suite of high-level communication protocols using low-power digital radios, designed for personal area networks with small, low-power digital radios. Zigbee can form a mesh network to connect multiple battery cells or modules, facilitating robust and scalable communication [2].

In BMS applications, Wi-Fi can be used for high-data-rate tasks such as firmware updates, diagnostic data transmission, and real-time monitoring. The advantages of Wi-Fi include high data rate, widespread availability, and robust security features. However, it has high power consumption and potential interference in crowded RF environments [2].

LoRa is designed for long-range communication with low power consumption, enabling long-range communication between battery modules, low power consumption, and good penetration through obstacles. However, LoRa has a low data rate and potential latency issues [2].

LTE and 5G offer can be used for remote monitoring, diagnostics, and management of battery systems, especially in stationary applications. While cellular networks offer wide-area coverage, high data rates, and low latency, they have high power consumption and potential costs associated with data plans [2].

Tesla's introduction of a wireless BMS for its next-generation electric vehicles in 2020 marked a significant advancement in the field. This innovative system uses proprietary wireless communication protocols to connect battery modules, offering several key benefits over traditional wired systems. Real-time data provided by the wireless BMS is transmitted to the vehicle's central control system, allowing for dynamic adjustments to improve performance and safety [13].

Visteon, a global automotive electronics supplier, transformed its wired BMS solution into a first-to-market smart wireless system in 2020. This system communicates wirelessly with key EV components, eliminating the need for traditional low-voltage wiring harnesses. The wireless smartBMS by Visteon enhances battery cell measurement accuracy, maximizes energy use per cell, and reduces overall vehicle weight, offering scalability across different vehicle brands and segments [14]. General Motors (GM) uses Visteon's wireless BMS technology in all its planned EV models powered by Ultium batteries [15].

Analog Devices has developed a wireless BMS platform that utilizes a combination of BLE and proprietary technologies to ensure robust communication within battery systems [16].

Other notable implementations of wireless BMS technologies include those by companies such as NXP Semiconductors and Texas Instruments. This solution supports various battery chemistries and configurations, making it versatile for different types of EVs [17].

The integration of wireless BMS with IoT platforms can provide real-time data analytics, predictive maintenance, and enhanced control over battery systems [18].

4 Optical Wireless Battery Management Systems (OW-BMS)

An OW-BMS represents a cutting-edge approach in battery management, utilizing light-based communication [19–21] to monitor and control battery systems efficiently and reliably. The core idea revolves around replacing traditional electrical wiring with optical links that use light, such as infrared (IR) or visible light, to transmit data between battery cells and the central CU.

This model enhances data transmission reliability by being immune to EMI, which is a common issue in environments with high electrical noise. In this conceptual

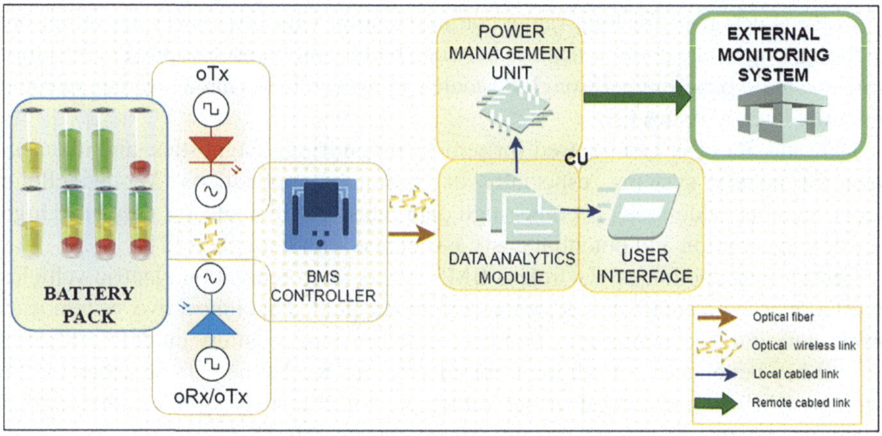

Fig. 2 Diagram of an OW-BMS. *Source* Authors' elaboration

model (Fig. 2), each battery cell set is equipped with a small wireless optical transmitter (oTx) that sends data to the wireless optical receiver (oRx) that acts as an optical (wireless or cabled) transmitter, as well.

These components use light to send and receive data about the cell's voltage, temperature, and SoC to a central optical hub. The BMS controller, positioned within the battery pack, collects data from all cells and communicates with the vehicle's main CU via optical fibers or wireless optical link.

The use of light for data transmission allows for higher data rates, enabling more detailed and frequent monitoring of each cell. This results in better overall battery management, ensuring optimal performance and extending the battery's lifespan.

The diagram connections between battery pack, oTx, oRx, BMS controller, CU (data analytics module, power management unit, user interface), and the external monitoring system.

The OW-BMS enhances security, as light-based communication is less susceptible to interception compared to traditional RF methods [22]. This ensures that sensitive data remains protected from potential hacking attempts.

The prevalent of the direct line-of-sight requirement for optical communication means that the system design must account for the placement and orientation of optical components to avoid physical obstructions.

The graphic (Fig. 3) shows the data flow from collection to real-time adjustments being a time graph of data transmission and processing in OW-BMS. Data are collected by battery cells (DC), the optical transmission (oTx) and optical reception from oRx is active while the data are processed by the CU and the adjustments are made in real-time (R-TA).

Advanced algorithms in the central CU process the real-time data received from the battery cells, making dynamic adjustments to optimize charging, discharging, and thermal management. This ensures that each cell operates within safe limits, preventing issues such as overcharging, overheating, and cell imbalance. As research

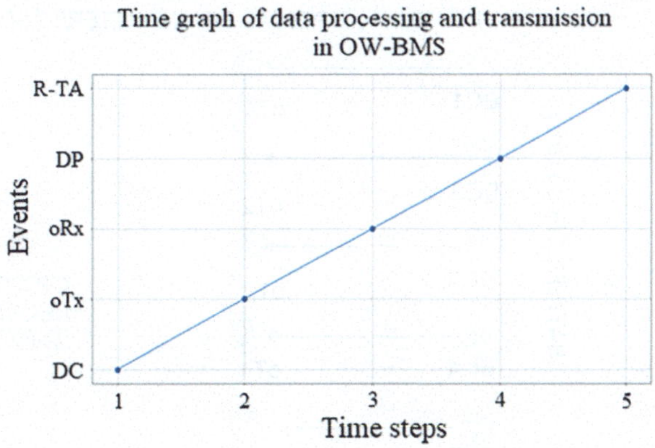

Fig. 3 Time graph of data transmission and processing in OW-BMS. *Source* Authors' elaboration

and development in optical communication technologies progress, the feasibility and efficiency of optical OW-BMS are expected to improve, making it a viable and attractive option for next-generation battery management solutions. This conceptual model illustrates the potential of light-based communication to revolutionize battery management systems, offering a glimpse into the future of more efficient, secure, and reliable energy storage solutions [23].

5 OW-BMS in Mining Industrial Areas

In mining industrial environments where the risk of accidents caused by short circuits or other electrical faults is significant, the implementation of a robust and reliable OW-BMS is essential. Traditional RF wireless BMS solutions, while effective in many contexts, can pose serious safety risks in these areas. The use of GSM, LoRa, Zigbee, and other RF wireless technologies can be problematic due to potential interference and the susceptibility to environmental factors prevalent in mining industrial areas. In this regard, OW-BMS (Fig. 4) offer a revolutionary approach, leveraging light-based communication to enhance safety and reliability in hazardous environments. In Fig. 4, the oTx convert the electrical signal (received from the BC) in optical signal that is sent wirelessly to the oRx.

According to [24], nearly 50% of mining vehicle sales will be EVs (haul trucks, dump trucks, wheel loaders, underground loaders, underground trucks, mining light vehicles, other mining vehicles and so on) in 2044, valued at over US$23 billion.

In underground mining, trucks and LHDs (load-haul-dump vehicles) can be electrified in two primary ways.

The first method involves sourcing power from an external supply, such as a trailing cable or an overhead catenary line, a technique that has been used for many years, particularly with cable-tethered LHDs.

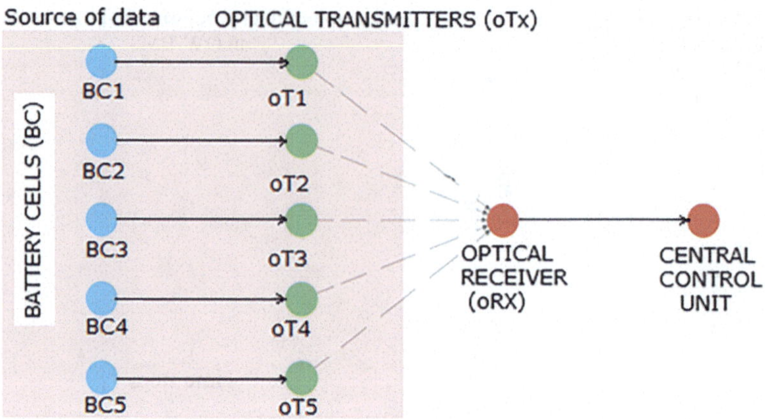

Fig. 4 Conceptual representation of the OW-BMS in mining environment. *Source* Authors' elaboration

The second method relies on onboard battery packs to supply power, a technology that has gained increasing popularity in recent years. Currently, the largest combined populations of electric LHDs and trucks are found in Canada (101), followed by China (80), Russia (52), and Sweden (34).

Several factors are driving the deployment of electric LHDs and trucks, including significant reductions in underground mine ventilation costs, more efficient energy consumption, and lower operating costs.

Additionally, the benefits of reduced maintenance requirements, decreased noise pollution, and minimal CO2 and heat emissions further support the adoption of these electric vehicles in underground mining [25].

OW-BMS systems eliminate the need for electrical cabled connections, thus eliminating a potential source of ignition of a potential explosion. Light-based communication does not produce sparks, making it an intrinsically safe technology for hazardous environments. This feature alone makes OW-BMS a superior choice for industries such as petrochemicals, mining and grain processing, where explosion hazards are a constant threat [26].

In addition, OW-BMS can maintain high data integrity and security in environments where traditional wireless communication might fail. In industrial settings, wireless communication systems such as GSM, LoRa and Zigbee can suffer from interference due to the presence of heavy machinery often equipped with electromagnets, metal structures (Faraday cage effect) and other sources of radio frequency (RF) noise. Such interference can degrade the performance of wireless systems, resulting in data loss or corruption, which in turn can compromise the safety and efficiency of the battery management system. Optical communication, however, is not affected by RF noise or other types of magnetic or other disturbances, ensuring that data transmission remains accurate and secure. This reliability is crucial for real-time monitoring and control of battery systems, enabling precise management of charging, discharging and thermal regulation (monitoring the temperature of each cell).

As mentioned before, the proposed configuration (Fig. 2) ensures that data on the voltage, temperature and state of charge of each cell is continuously monitored and transmitted without the risk of electrical interference or sparking [27].

The high data transfer rates achievable through optical communication enable detailed and real-time monitoring of each battery cell. Advanced algorithms in the central CU can analyze this data in real time, enabling dynamic adjustments to optimize battery performance and safety during both working and charging periods. This capability is particularly important in industrial applications where precise control over battery systems is required to prevent breakdowns and ensure the continuous operation of various technical and technological equipment or assemblies.

Furthermore, the use of OW-BMS can lead to significant reductions in system weight and complexity. Traditional wired systems require extensive cables and connectors, which add to the overall weight and complexity of the battery. In contrast, optical systems are much lighter and easier to install, resulting in easier maintenance and lower costs over the life of the system. This reduction in weight and complexity also increases overall system reliability, as there are fewer components that can fail or require replacement.

6 Conclusions

The advancement of BMS has been crucial in ensuring the efficient, safe, and extended operation of rechargeable batteries, particularly in EVs. Among the most notable innovations in this field is the RF wireless BMS and the emergence of OW-BMS. This cutting-edge technology leverages light-based communication to enhance the performance, reliability, and safety of battery systems. OW-BMS represents a significant step from traditional wired and RF wireless BMS technologies since using light, such as IR or visible light for data transmission, overcomes many of the limitations associated with EMI (a common issue in high-electrical-noise environments). The immunity to EMI ensures more reliable and accurate data transmission, which is critical for maintaining optimal battery performance. The high data rates achievable with optical communication enable more detailed and frequent monitoring of each battery cell. This allows for better state estimation and management, including SoC, SoH, and SoP. Advanced algorithms in the central CU can process this real-time data to make dynamic adjustments, optimizing charging, discharging, and thermal management, ensuring that each cell operates within safe limits, preventing issues such as overcharging, overheating, and cell imbalance.

OW-BMS can be a revolutionary step forward in the evolution of wireless BMS. By offering superior performance, reliability, and security through light-based communication, OW-BMS play a critical role in the future of industrial mining EV. This enhances the operational capabilities of battery systems and also supports the broader goals of efficiency, safety, and sustainability in various equipment that operate in industrial mining areas.

References

1. Li, A., Ponchant, M., Sturm, J., Jossen, A.: Reduced-order electro-thermal battery model ready for software-in-the-loop and hardware-in-the-loop BMS evaluation for an electric vehicle. World Electr. Veh. J. **11**, 75 (2020). https://doi.org/10.3390/wevj11040075
2. Cao, Z., Gao, W., Fu, Y., Mi, C.: Wireless battery management systems: innovations, challenges, and future perspectives. Energies **17**, 3277 (2024). https://doi.org/10.3390/en17133277
3. Samanta, A., Williamson, S.S.: A survey of wireless battery management system: topology, emerging trends, and challenges. Electronics **10**, 2193 (2021). https://doi.org/10.3390/electronics10182193
4. Li, M., Zhang, Y., Hu, Z., Zhang, Y., Zhang, J.: A battery SOC estimation method based on AFFRLS-EKF. Sensors **21**, 5698 (2021). https://doi.org/10.3390/s21175698
5. Liu, J.: Understand the BMS components and functions. https://www.mokoenergy.com/understand-the-bms-components-and-functions/. Accessed 14 Feb 2024
6. Xing, Y., Ma, E.W.M., Tsui, K.L., Pecht, M.: Battery management systems in electric and hybrid vehicles. Energies **4**, 1840–1857 (2011). https://doi.org/10.3390/en4111840
7. Fan, R.: SMBus quick start guide. https://www.nxp.com/docs/en/application-note/AN4471.pdf
8. https://wiki.dfrobot.com/Modbus_vs_CAN_bus
9. Gabbar, H.A., Othman, A.M., Abdussami, M.R.: Review of battery management systems (BMS) development and industrial standards. Technologies **9**(2), 28 (2021)
10. Sanino, E.: Introduction to battery management systems. https://www.allaboutcircuits.com/technical-chapters/. Accessed 08 Feb 2021
11. John, P.C.; Kumar, B.P.N.: Functional safety design and ISO26262 compliance for BMS in EV and HEV. In: Siano, P., Williamson, S., Beevi, S. (eds.) Intelligent Solutions for Smart Grids and Smart Cities, pp. 329–339. Springer Nature, Singapore (2023)
12. Tehnical paper, ISO 26262 challenges for Battery Management System. https://www.e-motec.net/iso-26262-certified-bms
13. https://www.lifepo4-battery.com/News/tesla-battery.html
14. Shell, C., Henderson, J., Verra, H., Dyer, J.: Implementation of a wireless battery management system (WBMS). In: 2015 IEEE International Instrumentation and Measurement Technology Conference (I2MTC) Proceedings, May 2015, pp. 1954–1959. IEEE (2015). https://www.gartner.com/doc/3100227/. Accessed 31 Mar 2019
15. Afzal, Y.: Wireless BMS technology for electric vehicles. https://www.wheelsatev.com/2020/11/wireless-bms-technology-for-electric.html
16. https://www.allaboutcircuits.com
17. Chamari, L., Petrova, E., Pauwels, P.: A web-based approach to BMS, BIM and IoT integration: a case study, May 2022 (2022). https://doi.org/10.34641/clima.2022.228
18. Prüfling, S., Steffenhagen, T., Kazanci, O., et al.: Elektronikdesign und funktionale Sicherheit bei drahtlosen Batteriemanagementsystemen. ATZ Automobiltech Z **126**, 42–47 (2024). https://doi.org/10.1007/s35148-024-2084-z
19. Leba, M., Riurean, S., Lonica, A.: LiFi—The path to a new way of communication. In: 2017 12th Iberian Conference on Information Systems and Technologies (CISTI), June 2017, pp. 1–6. IEEE ((2017)
20. Riurean, S., Antipova, T., Rocha, Á., Leba, M., Ionica, A.: VLC, OCC, IR and LiFi reliable optical wireless technologies to be embedded in medical facilities and medical devices. J. Med. Syst. **43**(10), 308 (2019)
21. Riurean, S.M., Nagy, A.A., Leba, M., Ionica, A.C.: A small step in VLC systems–a big step in Li-Fi implementation. IOP Conf. Ser.: Mater. Sci. Eng. **294**(1), 012047). IOP Publishing (2018)

22. Je Lee, S., Lee, Y.H., Park, J., Park, N.: Analyzing RF interference on wireless BMS in high-congestion environments. In: 2023 IEEE Green Energy and Smart Systems Conference (IGESSC), Long Beach, CA, USA, pp. 1–6 (2023). https://doi.org/10.1109/IGESSC59090.2023.10321762
23. Thomas, R., Heiries, V., Franchi, R., Ranieri, M.: An IR-based wireless optical communication system for new generation battery pack. In: Advanced Battery Power Conference, April 2023, Aix-la-chapelle (Aachen), Germany. ⟨cea-04149493⟩ (2023). https://cea.hal.science/cea-04149493
24. Jaswani, P., Jeffs, J., Electric vehicles in mining 2024–2044: technologies, players, and forecasts. https://www.idtechex.com/en/research-report/electric-vehicles-in-mining-2024-2044-technologies-players-and-forecasts/994
25. https://www.globaldata.com/store/report/development-of-electric-vehicles-in-surface-and-underground-mining-market-analysis/
26. Stoicuta, O., Riurean, S., Burian, S., Leba, M., Ionica, A.: Application of optical communication for an enhanced health and safety system in underground mine. Sensors **23**, 692 (2023). https://doi.org/10.3390/s2302069
27. Komsiyska, L., Buchberger, T., Diehl, S., Ehrensberger, M., Hanzl, C., Hartmann, C., Hölzle, M., Kleiner, J., Lewerenz, M., Liebhart, B., et al.: Critical review of intelligent battery systems: challenges, implementation, and potential for electric vehicles. Energies **14**, 5989 (2021). https://doi.org/10.3390/en14185989

The manufacturer's authorised representative in the EU is Springer Nature Customer Service Centre GmbH, Europaplatz 3, 69115 Heidelberg, Germany. If you have any concerns regarding our products, please contact ProductSafety@springernature.com

Printed and bound by CPI Group (UK) Ltd, Croydon, CR0 4YY

30/09/2026

02225569-0012